Healing from the Depths

Marine Metabolites in Disease Management

nova
science publishers

www.novapublishers.com

Marine Science and Technology

More information about this series can be found at
https://novapublishers.com/product-category/series/marine-science-and-technology/

Biochemistry Research Trends

More information about this series can be found at
https://novapublishers.com/product-category/series/biochemistry-research-trends/

Pranav Kumar Prabhakar
Tarun Kumar Upadhyay
Editors

Healing from the Depths

Marine Metabolites in Disease Management

nova
science publishers
www.novapublishers.com

DOI: https://doi.org/10.52305/QYZO1641

Additional color graphics may be available in the e-book version of this book.

Library of Congress Cataloging-in-Publication Data

ISBN: 979-8-89530-832-5 (Hardcover)
ISBN: 979-8-89530-983-4 (eBook)

Published by Nova Science Publishers, Inc. † New York

Contents

Preface

The world's oceans, covering more than 70% of the Earth's surface, represent an immense reservoir of biodiversity and biochemical diversity. Marine ecosystems have long fascinated scientists for their potential to provide novel compounds with significant therapeutic value. In recent decades, marine-derived metabolites have emerged as promising agents in the prevention and management of various human diseases, from cancer and neurological disorders to inflammation and infectious diseases.

Healing from the Depths: Marine Metabolites in Disease Management brings together contemporary research exploring the potential of marine-derived compounds in modern medicine. This edited volume highlights not only the diversity of marine metabolites but also their underlying mechanisms of action, therapeutic applications, and future potential in clinical settings. It reflects the collective effort of interdisciplinary collaborations spanning marine biology, biotechnology, pharmacology, and clinical sciences. The chapters in this book delve into areas such as cancer therapy, neurological health, cardiovascular disease management, and antimicrobial strategies, all centered around the unique contributions of marine bioactives. Moreover, the book also addresses sustainable approaches to marine drug development, ensuring that the therapeutic potential of the oceans is harnessed responsibly.

We express our sincere gratitude to all contributing authors for their valuable insights, to the reviewers for their critical feedback, and to Nova Science Publishers for their support in publishing this volume. It is our belief that this book will serve as an important resource for researchers, academicians, clinicians, and students engaged in marine biotechnology and drug discovery.

As we continue to explore the vast, largely untapped potential of the marine world, we hope this work inspires further research and innovation aimed at addressing some of the most pressing healthcare challenges of our time.

Dr. Pranav Kumar Prabhakar
Dr. Tarun Kumar Upadhyay
Editors

Chapter 1

Marine Metabolites as Anti-Infective Agents

Ghanshyam Parmar[1,*]
Jay Mukesh Chudasama[1]
Ashish Shah[1]
Chintan Aundhia[1]
and Tarun Kumar Upadhyay[2]

[1]Department of Pharmacy, Sumandeep Vidyapeeth Deemed to be University, Piparia, Waghodia, Vadodara, Gujarat, India
[2]Department of Biotechnology, Parul Institute of Applied Sciences and Research and Development Cell, Parul University, Vadodara, Gujarat, India

Abstract

Marine-sourced organisms are vital sources of bioactive compounds, especially anti-infective agents. These compounds, like alkaloids, terpenoids, peptides, and polyketides, have unique structures and work against various germs in different ways. These are palau'amine, sourced from marine sponges, which shows mitochondrial dysfunction; cyclopamine, sourced from *Myxococcus xanthus* which suppresses hedgehog signaling; and salinosporamide-A, from *Salinispora tropica* which affects the proteasome. Marine biotechnology, in association with genomics and metagenomics, synthetic biology, and advanced extraction methods, plays a pivotal role in discovering and deriving these compounds. However, the process of marine drug discovery is confronted with various challenges, including technical difficulties, practical issues related to sustainable sources, legal complications, and ethical considerations. Nevertheless, constant issues such as competition with microbial endophytes make possible the clinical application of marine-derived anti-infective agents, which is very promising since some of the compounds are under trial and bear the potential of novel drugs. It is projected that the future trends in this field will involve the development of new interdisciplinary strategies aimed at escalating the discovery and use of marine metabolites sustainably. Based on current advancements in research, it is predicted that marine-derived compounds will significantly contribute towards solving the crisis of antibiotic resistance and enhancing people's well-being. This abstract summarizes the broad prospect of marine metabolites in the synthesis of innovative

---

* Corresponding Author's Email: ghanstaurus22@gmail.com

In: Healing from the Depths Marine Metabolites in Disease Management
Editors: Pranav Kumar Prabhakar and Tarun Kumar Upadhyay
ISBN: 979-8-89530-832-5

drugs and stresses the necessity of long-term preservation of marine organisms and active research in this direction.

Keywords: marine biotechnology, marine metabolites, drug discovery, anti-infective agents

1. Introduction

The sea, ocean, and other estuaries of the earth account for about 70 percent of the earth's surface, and it's within this aquatic world that almost 80% of life on earth exists. In turn, this abundance has led to the development of a broad spectrum of secondary metabolites, which have high bioactivity and include anti-infective compounds. Because of the urgent necessity of developing new drugs as antibiotic resistance and the emergence of new infectious diseases, the study of marine metabolites as anti-infective agents has gained great momentum in the last few decades. Spongouridine, the first identified bioactive marine chemical, was isolated from a sponge back in the 1950s, although the contemporary approaches to matrix metalloproteinases (MMPs) started emerging in the middle of the twentieth century (Landrigan et al., 2020). Due to this discovery, it grew emphasis that the ocean can also be a source of new chemical substances. Many distinctive structures and bioactive compounds of marine origin have been discovered and characterized due to advancements in marine biotechnology, improved techniques of diving and sampling, and the availability of better analytical techniques. Marine organisms, including sponges, algae, corals, tunicates, mollusks, and water microbes, are other producers of a wide array of secondary metabolites (Karthikeyan, Joseph, and Nair 2022a; Maslin et al., 2021). These molecules' structures are often complex and dissimilar to those of other terrestrial organisms; this is because aquatic habitats hold few environmental pressures and countermeasures. Peptides, terpenoids, sterols, polyketides, and alkaloids are examples of marine metabolites. The uniqueness of their structures makes them have multiple bioactivities. The constituents of metabolism taking place in the ocean hold a variety of antimicrobial as well as antiviral characteristics. Examples of antibiotics that disrupt the bacterial membrane or inhibit bacterial cell wall synthesis, protein synthesis, or DNA replication are cephalosporin, derived from marine fungi, and the cyclic peptide teixobactin. For instance, the marine fungus-derived equisetin would be toxic to the fungal organisms by making a cell membrane porous or preventing the construction of ergosterol. For instance, homarine, a marine alkaloid, has displayed activity against herpes simplex virus and is but one of the marine compounds that can inhibit viral replication through interaction with viral enzymes, viral entry into host cells, or viral assembly (Gross and König 2006; Núñez-Pons et al., 2020). For example, the marine alkaloid manzamine A shows that substances from the ocean can stop the life cycles of parasites that cause malaria and leishmaniasis.

Despite such enormous potential, there are many challenges for the complete realization of marine metabolites as therapeutic compounds. Among them are the effects on the marine organisms, collecting the organisms and later chemical synthesis, and the challenge of getting an adequate number of bioactive compounds from the marine organisms. As to these problems, advances in marine biotechnology, which encompasses genetic engineering, aquaculture, and fermentation technology, protrude the accessibility of marine-derived substances and contribute to the swell in production. In this regard, to further enhance the process of drug discovery to treat the anti-infectious ailments, the research work carried out on the marine natural products will need to incorporate other strategies such as synthetic biology, genomics, and metabolomics (Karan et al., 2020; Hamann 2007). Thus, as academic institutions, pharmaceutical businesses, and conservation groups get the best out of marine biodiversity while at the same time conserving the marine ecosystem, there is a need to collaborate. In conclusion, anti-infective drugs are in the vanguard of innovative drugs, and marine metabolites are a new horizon. They have many opportunities to develop new treatments for usage in combating infectious diseases because of the differences in the mode of action and structures. That is why now, essentially searching for a solution to the global problem of antibiotic resistance and, accordingly, improving the overall population's health, sustainable development of marine resources must continue (Lindequist 2016; National Research 1999).

2. Overview of Marine Metabolites

Marine animals include sponges, algae, corals, tunicates, mollusks, microbes, and many other various kinds of animals that are endemic to the extensive and continually transforming marine environment. To produce a huge spectrum of secondary metabolites, which have complex chemical formulas and tremendously effective biological effects or a set of effects, these creatures evolved complex biosynthetic routes. This structural oddity is considered to arise from the perils of the ocean's environment that create this structural oddity, which often has abnormal functional groupings that involve out-of-the-ordinary parts or components. Some of the groups of marine metabolites that have been shown to be useful in the search for new pharmaceutically active molecules and particularly antimicrobial agents include polyketides, alkaloids, peptides, terpenoids, and sterols (König et al., 2006; Gerwick and Moore 2012). These substances fight infections by stopping bacteria from building their cell walls, breaking down fungal cell walls, preventing viruses from making copies of themselves, and disrupting the life cycle of parasites. Plinabulin, which is an anti-cancer agent being developed; bryostatin-1, used for both cancer and Alzheimer's disease; ziconotide, an efficient analgesic; largazole, a drug with anticancer potential; and salinosporamide-A, an anticancer molecule for multiple myeloma testing (Figure 1). However, there are challenges such as issues with synthesis, solutions to environmental impact, and sustainable resources that hamper development despite advancements in this field with a lot of potential. Research and development of these molecules, as well as their efficient production, is now becoming feasible through marine biotechnology such as genomics, metabolomics, and synthetic biology (Reygaert 2018; Guilhelmelli et al., 2013). Unlocking the full potential of marine cell-based products to affect society while at the same time protecting the environment is a complex process that needs the efficiencies of universities, drug manufacturing companies, and nonprofit organizations. The continued discovery of new chemicals offering promising therapeutic applications provides hope for overall improvement of people's well-being from such sea metabolites, drawing on the efficient practices and continuous investigations, besides the fight against a number of world health issues, including antibiotic resistance (Sigwart et al., 2021).

3. Importance of Anti-Infective Agents

Thus, anti-infective agents include antibiotics, antivirals, antifungals, and antiparasitic medications for combating infectious diseases. They are vital in the management of bacterial, viral, fungal, and parasitic infections in care delivery settings since they reduce mortality and morbidity and prevent crises from transforming into epidemic or pandemic proportions. From simple, often treatable bacterial infections to life-threatening ones like sepsis and bacterial meningitis, those antibiotics play a critical role in managing and eradicating a spectrum of diseases. They are highly relevant in the function and outcome of surgeries and medical procedures due to their preventive effect against postoperative infections, and they are relevant in the treatment of diseases like HIV/AIDS, influenza, and COVID-19. Positive results were obtained when anti-infectives were used for the treatment and care of those patients with compromised immunity or chronic ailments, resulting in better quality of life and control of infections (Sanchez Armengol, Harmanci, and Laffleur 2021; Huang et al., 2024). For this reason, they aid farmers in keeping their animals healthy and food safe to ensure that food chains within and between producers run effectively. The appearance of antibiotic-resistant bacteria is a threat to global health; therefore, the search for innovative approaches to manage infections belongs to the goals of anti-infective new drug design. Costs that can also be alleviated through these medicines include the impact of healthcare costs due to infections, lost productivity, and chronic disability in the affected areas over time. Hence, in low- and middle-income countries, where infectious diseases are the major cause of mortality, it is imperative to have good anti-infectives to reduce morbidity and mortality. In conclusion, anti-infective agents play a crucial role in modern world medicine, serving as essential tools for treating infections, enhancing various therapeutic procedures, stabilizing the economy of a country or region, and managing chronic diseases. New infections are emerging, and there is also the issue of antibiotic resistance; therefore, research and development must continue so that treatment is available to people (Manyi-Loh et al., 2018; Almansour et al., 2023).

4. Marine Environment as a Source of Bioactive Compounds

Currently the world's estuaries, oceans, and seas can be viewed as a wide and still very much unexplored vessel for the bounty of maritime space. In this environment, such organisms as sponges, corals, algae, mollusks, and microbes are just some of the species found in this environment. All of them adapted to the special and severe circumstances that marine environments may offer. Consequently, to live and reproduce within their habitats, these various organisms have developed numerous bioactive molecules, or secondary metabolites (Ghosh et al., 2022). The high interest in bioactive chemicals is due to their diverse chemical structures and extremely potent bioactivity documented in marine areas. Since marine environments are different in the effect of various ecological forces like salinity, temperature, and pressure, very often these compounds show some structural features that are quite unique to animals of the ocean. Some of the many bioactivities shown by these compounds include polyketides, alkaloids, peptides, terpenoids, and sterols. Specific activities of these substances are anti-inflammatory, anti-cancer, antiviral, and antibacterial (Ameen, AlNadhari, and Al-Homaidan 2021).

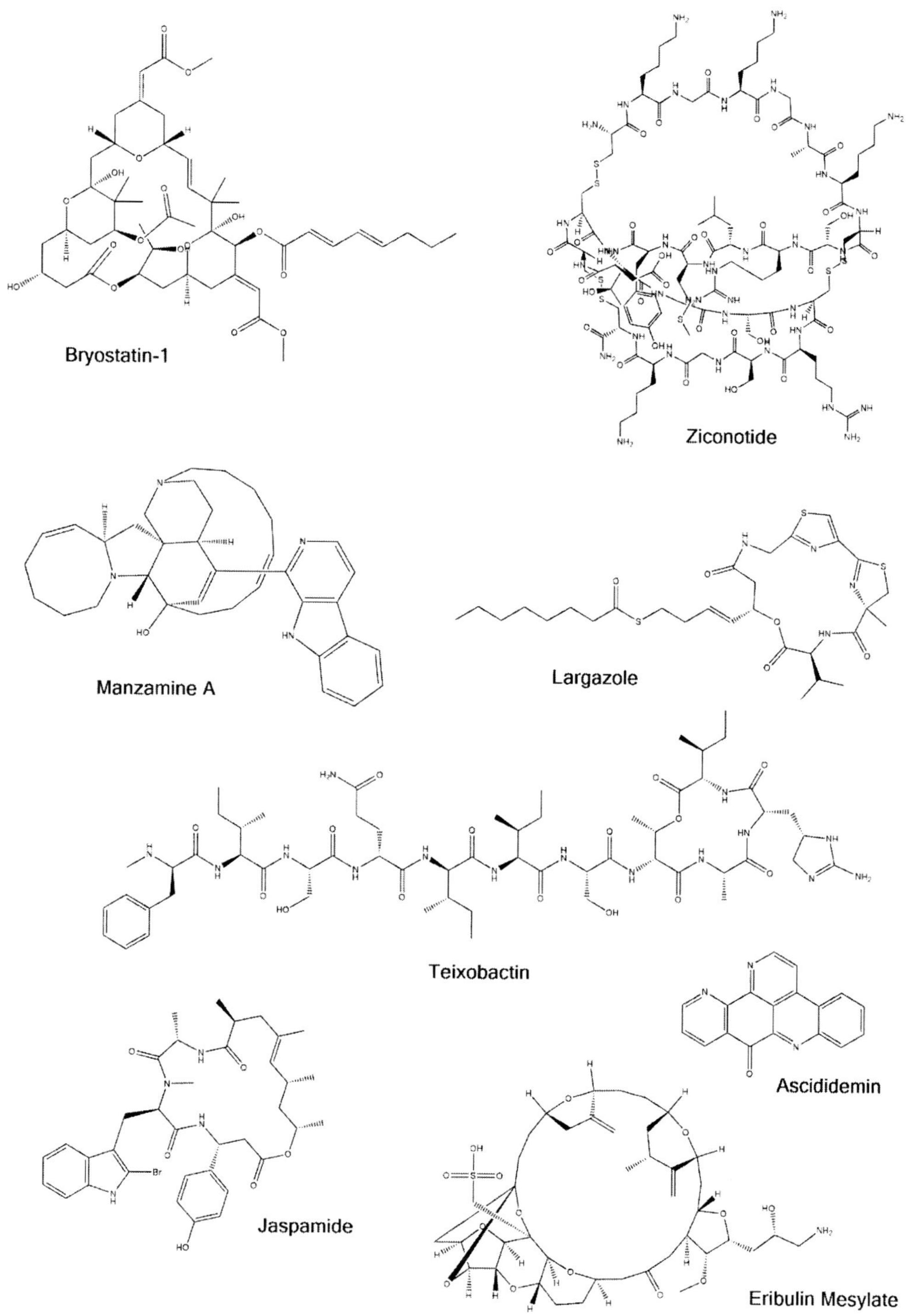

Figure 1. Marine metabolites as anti-infective agents.

Researchers have documented several new bioactive chemicals with pharmacological properties from the marine realm. For instance, the protein extracted from the marine bryozoan is bryostatin-1, which holds promise in cancer and Alzheimer's disease treatment. They include ziconotide, which is an analgesic obtained from the cone snail venom and given to patients with extreme chronic pain. The antibiotic of marine actinomycete, salinosporamide A, which in cancer treatment has bright anticancer prospects in different kinds of cancers, principally multiple myeloma Figure 1. These outcomes emphasize the problem of the prospects of appearing in the focus of interest as new sources of medications, marine species. The complex chemical structures required for their synthesis, issues related to the environment from the point of view of the extraction technique, and the challenge of obtaining sufficiently enormous quantities are some of the challenges that complicate the study of bioactive compounds of marine origin. Priorities for future growth in marine biotechnology include enhancing the processes of identification and synthesis of new chemical structures and increasing the efficiency of their use in the long term (Sruthi et al., 2020; Tian et al., 2018).

5. Diversity of Marine Ecosystems

Namely, marine ecosystems cover the coastal zone, the open sea, ocean depths, and abyss, so this list of habitats defines this type of ecosystem as one of the most diverse and productive environments on our planet. These factors include temperature changes, salinity changes, light changes, and pressure changes, as well as the differences in aquatic ecosystems biotic interactions that are present in the environment. The coral reef, mangrove vegetation, and seagrass bed are some of the most productive ecosystems in the coastal zone. Coral reefs, or the sea rainforests, support fish, crustaceans, plants, and various other forms of life. Besides sheltering the shores and absorbing CO_2, mangroves and seagrasses offer the opportunity for the development of many sea species (Cochrane et al., 2016). The open ocean is the largest biome in the pelagic zone, covering regions such as the epipelagic, mesopelagic, and bathypelagic zones, right from the chunkiest plank and float to the darkest of the deep sea. Each zone's unique issues have led to the adaptation of various societies. Stressors in the deep-sea environment include low temperatures and high pressure, and many peculiar forms of life exist in these habitats, including bioluminescent forms under communities of gigantic hydrothermal vents. Also, estuaries are very productive because of the opportunity to find, for instance, fish, birds, and invertebrates due to their geographical location, which is the connection of freshwater sources and the ocean. They are equally vital in the life cycle of several species, given that they function as transitional areas. In general, the variety of marine ecosystems can be a sign of an increased and constant evolution of marine organisms. Therefore, understanding the significance and equal importance of the protection and sustained utilization of this unbelievably rich biodiverse region cannot be overemphasized, as it is a very vital key to the maintenance of the balance of nature and provision of essential resources and services to mankind (Parmar et al., 2023).

6. Unique Characteristics of Marine Organisms

Due to the variegated and often severe environment in which they exist, sea creatures have a vast number of peculiar characteristics. One characteristic is the consideration of the pressures that influence the organization. For instance, organisms inhabiting the deep water have specific enzyme systems and fluid membranes to adapt to this unthinkably high pressure. Other useful adaptations include bioluminescence, where through chemical reactions deep-sea fish and jellyfish emit light, which is used in signals, privacy, and in fending off animals that are prey to them. The extremophiles are animals, and some species are capable of preferring the extreme priority of temperature as well as salinity, and they live in various places such as hydrothermal vents as well as the Arctic Sea. Moreover, it is necessary to point out several peculiarities of reproduction that have been seen, for example, the fact that some marine species are able to reproduce asexually or undergo complex ontogeny. For instance, squid employ their siphon to create a jet by which they swim in water, and fish use their poisonous spines for protection, and innumerable organisms in the sea have structures amazingly designed or adapted for the above functions. Apart from increasing the rate of differentiation and the complexity of living organisms in the sea, these adaptations make it possible for marine species to live in almost any type of environment (Rampelotto 2013; Dalmaso, Ferreira, and Vermelho 2015).

7. Types of Marine Metabolites

The variety of chemical frameworks and highly desirable biological functions can be developed due to the abundance of metabolite-producing marine species. The anticancer drug halichondrin-B and the antibiotic erythromycin are both made from compounds in the polyketide family, which is defined by many structures. Some examples of nitrogen base substances include manzamine-A, which is an antibacterial and antiparasitic chemical with alkaloid characteristics, and sarcotoxin, an insecticidal molecule derived from marine sponges. Peptides are the heads of the battlefield against cancer and microorganisms. Bactericidal drugs that target gram-negative bacteria, protein phosphatase inhibitors, and harmful agents like calyculin-A all belong to this group. Polymyxin B was made by microorganisms in the water bodies. The isoprene moieties that make up terpenoids are common in marine settings. A natural substance called Sarcodictyin has anticancer properties, while another called largazole is an inhibitor of histone deacetylase. These sterols influence membrane signaling and structure. The marine sterol squalamine showed antimicrobial and antiviral effects. Beyond the data's primary use in illuminating relationships between marine life and its surroundings, the information about marine metabolites' medicinal properties—including the quantity of chemical compounds and the various ways in which they exert their effects—points to opportunities for future study and practical application (Parmar et al., 2023).

Among the notable chemicals found in marine alkaloids is palau'amine, a molecule generated from the marine sponge that owns potent cytotoxic and antiviral characteristics. This alkaloid stops cancer cells from growing and viruses from multiplying by activating natural cell death processes and disrupting how mitochondria work, both of which are crucial for cell death. A second significant marine alkaloid, cyclopamine, is produced by the marine bacterium *Myxococcus xanthus*. In some cancers, cyclopamine stops tumors from growing by attaching to a protein in the hedgehog signaling system called smoothened, which keeps it from being activated. For cancer cells to proliferate, this pathway is essential (Mioso et al., 2017; Bouyahya et al., 2024).

Marine terpenoids, such as cembranoid, have anti-inflammatory, immunomodulatory, and anti-cancer effects; they are derived from marine soft coral *Sinularia* spp. Cembranoids reduce inflammation and slow down the growth of cancer cells by controlling the NF-κB signaling pathway. Fijiolide, a part of marine sponges *Ircinia* spp., shows promising antiviral and anticancer effects. As with other terpenoids, it interferes with viral assembly and alters microtubule dynamics, which in turn inhibits cell proliferation and death in cancer cells. It may possibly have antiviral properties (Nurrachma et al., 2021; Putra and Murniasih 2016).

Nereistoxin, an insecticide with strong neurotoxic effects, is a marine peptide and protein derived from the marine polychaete worm *Nereis diversicolor*. Nereistoxin is useful for pest management because it blocks ion channels in insects, which interferes with neurotransmitter function and causes paralysis and death. Didemnin B, derived from the marine tunicate *Didemnum granulosum*, is another instance. It attaches to the ribosomal subunit, stopping protein production; it interferes with protein translation; and it causes cancer cells to die while reducing the spread of viruses; these effects show strong anticancer and antiviral abilities (Kem et al., 2022; Cusick and Sayler 2013).

One marine polyketide that has strong anticancer action is salinosporamide A, which is produced by the marine bacteria *Salinispora tropica*. Proteasome is an essential part of the ubiquitin-proteasome cascade that breaks down proteins. When it is inhibited, pro-apoptotic molecules build up and cancer cells die. *Jaspis johnstoni*, a type of marine sponge, produces jaspamide, a compound that fights viruses and cancer by stopping the spread of viruses, slowing down cell division and death in cancer cells, and affecting how microtubules work (Gulder and Moore 2010).

Nesiotoxin A, a newly discovered secondary metabolite, has strong cytotoxic and antibacterial characteristics; it originates from the marine sponge *Nesiotissa* sp. A potent anticancer drug, E7389, derived from the marine sponge *Halichondria okadai*, has been proved to be effective against a range of malignancies. The E7389 offers a new way to treat cancer by interfering with microtubule dynamics, which causes cell cycle arrest and cell death. The nesiotoxin-A, while showing antibacterial efficacy, disrupts microbial cell wall formation and promotes apoptosis by activating intrinsic apoptotic pathways and altering mitochondrial function. (Selim, Abdelhamid, and Mohamed 2021; Jamal and Satheesh 2022).

8. Mechanisms of Action

Bioactive compounds with anti-infective qualities are abundant in the marine environment. These compounds are known for their intricate chemical structures, distinct biosynthesis pathways, and specialized ways of acting against diverse types of infections.

This marine sponge is a source of alkaloids, one of which is palau'amine. With a complex polycyclic framework that includes many nitrogen rings, this molecule proves significant cytotoxic and antiviral activities. A complex series of enzyme cyclization characterizes its polyketide mechanism of production. In cancer cells, palau'amine affects energy metabolism and cell death by interfering with mitochondrial activity. Its polyketide mechanism of production is characterized by a complex series of enzyme-catalyzed cyclizations. The marine bacterium *Myxococcus xanthus* is the source of another alkaloid called cyclopamine. This compound has an unusual steroidal structure with four rings. Specialized enzymes change steroid precursors during their production. The primary mechanism by which cyclopamine suppresses tumor-promoting hedgehog signaling is its limited direct antibacterial action; nevertheless, it may have indirect effects on bacterial pathogenesis (Figure 1) (Elissawy et al., 2021; Towle et al., 2001).

Cembranoid is one example of a terpenoid with a bicyclic sesquiterpene structure; it is present in the marine soft coral *Sinularia* spp. Cycles involving farnesyl pyrophosphate engage in the production of cembranoids. These substances influence the NF-κB pathway, which is important for inflammation and cancer development, and they have properties that can fight cancer, reduce inflammation, and adjust immune responses. They might also help against viruses and fungi through their effects on cell signaling and processes. Through their effects on cellular signaling and processes, they may also have antiviral and antifungal properties. The intricate bicyclic structure of fijolide is derived from marine sponges, specifically *Ircinia* sp. Terpene precursors undergo oxidation and cyclization throughout their production. As an antiviral agent, fijiolide interferes with viral assembly; it also inhibits cell division and induces death in cancer cells by disrupting microtubule dynamics. (Rodrigues, Miguel, and Mnif 2019).

Peptides and proteins include nereistoxin, extracted from the marine polychaete worm *Nereis diversicolor*. Nereistoxin, a peptide with several amino acid residues, is synthesized via peptide synthesis pathways specific to marine organisms. It functions as an insecticide by blocking ion channels, leading to paralysis and death in insects. Although its antimicrobial activity is less pronounced, it plays a crucial role in pest control. Didemnin-B, which comes from the sea creature *Didemnum granulosum*, is a special type of cyclic peptide made through a process that doesn't involve ribosomes. Didemnin-B inhibits protein synthesis by binding to the ribosomal subunit, disrupting translation, and inducing apoptosis in cancer cells, while also exhibiting antiviral properties by inhibiting viral replication (Figure 1) (Sable, Parajuli, and Jois 2017).

Polyketides like salinosporamide-A, made by the marine bacterium Salinispora tropica, have a complicated structure that includes a β-lactone ring. Its biosynthesis involves a polyketide synthase pathway. Salinosporamide-A stops the proteasome, which is an important part of the ubiquitin-proteasome pathway, causing a build-up of factors that promote cell death in cancer cells. Though its primary application is in cancer therapy, it may also have indirect antibacterial effects through proteasome inhibition. Jaspamide, isolated from the marine sponge *Jaspis johnstoni*, has a polyketide backbone with several functional groups. Its biosynthesis involves complex polyketide synthase pathways. Jaspamide disrupts microtubule dynamics, impairs cell division, and induces apoptosis in cancer cells. It also demonstrates antiviral activity by interfering with viral replication, although its primary use is in oncology (Figure 1) (Eustáquio et al., 2009; Gao, Deng, and Jiang 2022).

Other secondary metabolites include nesiotoxin-A, obtained from the marine sponge *Nesiotissa sp.*, which has a polycyclic structure and is synthesized via a polyketide pathway. Nesiotoxin A causes cancer cells to die by messing up how their mitochondria work and triggering their natural cell death processes, and it also fights bacteria by blocking the building of their cell walls. The Halichondrin-B Derivative (E7389), which comes from the marine sponge *Halichondria okadai*, has a complicated ring-like structure that is changed from halichondrin B. The E7389 disrupts microtubule dynamics, leading to cell cycle arrest and apoptosis in cancer cells, with potential antiviral effects due to its interference with cellular machinery used by viruses (Figure 1) (Lee, Cho, and Tran 2021).

These marine metabolites show the variety of chemical structures, how they are made, and how they work, which all help their strong ability to fight infections, highlighting the significance of marine biodiversity in creating new treatments, as seen in Table 1.

Table 1. The main marine compounds mechanisms with their antimicrobial and antiviral activity

Metabolite	Source	Antibacterial Mechanism	Antifungal Mechanism	Antiviral Mechanism	Antiparasitic Mechanism
Palau'amine	Marine sponge *Palau'amine*	Disrupts mitochondrial function (indirect)	Limited direct action.	May affect viral replication via mitochondrial interference	Limited direct action
Cyclopamine	Marine bacterium *Myxococcus xanthus*	Inhibits hedgehog signaling (indirect)	Limited direct action	Inhibits hedgehog signaling (indirect)	Limited direct action
Polymyxin B	Marine bacteria	Disrupts bacterial cell membrane	Limited direct action	Limited direct action	Limited direct action
Salinosporamide A	Marine bacterium *Salinispora tropica*	Proteasome inhibition (indirect)	Limited direct action	Limited direct action	Limited direct action
Fijiolide	Marine sponge *Ircinia spp.*	Limited direct action	Disrupts microtubule dynamics	Interference with viral assembly	Limited direct action
Didemnin B	Marine tunicates *Didemnum granulosum*	Limited direct action	Inhibits protein synthesis	Inhibits viral protein synthesis	Limited direct action
Halichondrin B Derivative (E7389)	Marine sponge *Halichondria okadai*	Limited direct action	Limited direct action	Disrupts microtubules (potential indirect effect)	Limited direct action
Manzamine A	Marine sponge *Halichondria okadai*	Limited direct action	Limited direct action	Limited direct action	Disrupts cellular processes and membrane integrity
Sarcotoxin	Marine sponges	Limited direct action	Limited direct action	Limited direct action	Disrupts ion channels, causing paralysis
Jaspamide	Marine sponge *Jaspis johnstoni*	Limited direct action	Limited direct action	Limited direct action	Inhibits microtubule dynamics

9. Challenges in Marine Drug Discovery

The potential of marine organisms as a source of drugs is very bright but is accompanied by challenges that are immense. Owing to the nature of marine environments as often inhospitable and physically challenging to access, collection of samples that could give a good sign of the marine species is a herculean task that often requires sophisticated equipment and techniques. This creates several practical and technical challenges. One potentially important reason for this is that the oceans cover a large area of the earth's surface, and a vast range of environments is characterized by sea-related ecosystems: substantial investigation and the appearance of new technology are needed for the collection and analysis of sea samples. Also, because marine metabolites have complicated molecular structures, studying and recreating them can be difficult, needing advanced tools and a lot of knowledge about pharmacology and chemistry (Malve 2016; Karthikeyan, Joseph, and Nair 2022b).

As for various similar fields, including the research of marine organisms as potential sources of drugs, the problem of reproduction and the minimization of negative impacts on the environment are extremely high priority. Such as elements of consideration: Some degree of ecological imbalance and the extermination of critical forms of sea life may occur due to the exploitation of oceanic goods and services. Efficiently collecting marine species without harming ecosystems requires conserving these species. For these issues, it is possible to apply modern methods of aquaculture and biotechnology to produce marine metabolites in a suitable manner. However, we still need to improve and advance these approaches to a level where they can be considered economically feasible. The third challenge that researchers face while hunting marine drugs is the ethical and regulatory issues. When using chemicals synthesized from marine sources in clinical research, there are several technical and legal factors that must be met, and they vary from one country or jurisdiction to another. Drugs should be kept safe and effective, and their effect on the natural environment must be controlled to extreme measures. Often acquiring the rights and consents from the locals and other stakeholders is an ethical imperative, especially when taking advantage of the marine activities leading to exploitation of the marine resources. In order to deal with these ethical and regulatory issues, regularity and cooperation between the researchers, regulatory governmental bodies, and conservation institutions are required (Kayode-Afolayan, Ahuekwe, and Nwinyi 2022; Perhar and Arhonditsis 2014).

10. Advances in Marine Biotechnology and Drug Development

10.1. Methods Based on Genomic and Metagenomic Data

Metagenomics and genomic data have been of immense benefit in pharmaceutical research in the marine biotechnology field. Genomic approaches involve obtaining the sequence data of the DNA from organisms living in the ocean, such as the large initiates (corals, sponges, etc.) and small initiates (fungi and bacteria). Due to this, scientists have been able to find the genes that code for bioactive chemicals. Future improvements in high-throughput sequencing and its use in studying genomes have opened up new opportunities to explore metabolism and identify groups of genes involved in creating special substances in the ocean. Genetic studies, for instance, have led to the development of new antibiotics and anticancer medications by illuminating the genes responsible for their production. Taken to the next level, metagenomics treats the genetic content of various marine microorganisms all at once. This technology opens the door to addressing genetic variation and the functional ability of an entire microbial community without necessarily isolating the species of interest. Executives have named previously unknown biosynthetic routes and new classes of bioactive compounds in studying metagenomics data. For instance, metagenomics approaches have recently identified new groups of genes that may have pharmacological application, for instance, clusters for non-ribosomal peptides and polyketides (Kodzius and Gojobori 2015; Alma'abadi, Gojobori, and Mineta 2015).

11. Bioprospecting and Synthetic Biology

Within this context, synthetic biology is a versatile approach in marine biotechnology due to the ability to engineer marine microbes to produce valuable chemicals. Scientists can synthesize compounds with lengthy synthesis procedures or scarcity in the natural world through improving and mimicking the biochemical pathways in a host organism. For instance, salinosporamide A and any other marine-sourced anticancer drugs have been biosynthesized in bacterial ones using synthetic biology platforms. This has made it possible to have voluminous production and close structural analysis.

Bioprospecting methods hunt marine organisms for potential novel activity chemicals. Progress in molecular biology has enhanced this procedure in recent years; it has extended the new efficiencies of the hitherto unexplored marine environment as the reservoir for medicinal discoveries. Thus, modern bioprospecting applies coupled environmental sampling, high-throughput screening, and bioinformatics to seek new chemicals that may have medicinal value. This approach has found interesting marine metabolites, which include new antibiotics and antiviral drugs. New Technologies for Marine Metabolite Extraction and Purification: In the last two decades, the extraction and purification technology for marine metabolites has shown tremendous progress. Traditional methods often resulted in low yields and complicated composites, which posed significant challenges. High-throughput techniques have also enabled the extraction of bioactive compounds from marine sources. The recent innovations, such as the ASE and the SPME, have enhanced production and productivity. Separation of pure metabolites for further analysis is possible due to modern chromatography techniques like UHPLC and HPLC, by which complex mixtures can be accurately separated. To enhance the efficiency and scalability of extraction operations, new solvent systems and extraction methods, such as pressurized liquid extraction (PLE) and supercritical fluid extraction (SFE), have been developed (Yan et al., 2023; Seghal Kiran et al., 2018).

Due to advances in extraction and techniques in purification, these metabolites have become more available, and these compounds are more accessible to integrate them in drug development. New natural compounds and plant-based treatments from the ocean are being discovered more often because of improvements in molecular biology, especially genetic engineering, and extraction methods that are set to change marine biotechnology (Najmi et al., 2022).

12. Clinical Applications and Future Prospects

12.1. Clinical Trial Status of Anti-Infective Agents Derived from Marine Sources

In one way or another, the therapeutic potential of the anti-infective medicines from marine sources is being acknowledged as more of such drugs are formulated and evaluated for clinical application. Several marine metabolites are already at various phases of clinical trial. Concerning other cancers such as liposarcoma and breast cancer, for instance, Halichondrin B derivatives, particularly E7389 (Halaven), have promising effects. Another compound that has been used to cure some types of lymphoma and leukemia originates from marine sponges, and this is cytarabine. It is attainable to note that these medicines' potentials of safety and effectiveness as treatments for severe infections and malignancies have been cross-checked by the clinical trial adoptions. Some of the challenges likely to be encountered frequently while going from the preclinical to clinical phase include how to come up with an optimal dosage, how to avoid the side effects, and how to ensure that synthesis of drugs got from marine sources is constant. Marine-derived metabolites show high potential in the synthesis of novel drugs, perhaps because of their diverse structures and activities. Sea creatures synthesize numerous compounds that are bioactive and exhibit reactions not commonly found in medicines. This diversity offers the chance to search for new groups of medication. That is especially so with chronic diseases, especially diseases for which antibiotics have proved to be ineffective, new diseases viral in nature, and different classes of cancer. With the help of contemporary trends in synthetic biology and omics, these microbes are produced to synthesize such chemicals, which help in the formation of new medications originating from marine resources. With this method, not only are the metabolites that contain these valuable substances for medical use made more available, but their medical properties can be enhanced while their side effects are decreased through changes to their structures (Gupta et al., 2013).

Some current trends and directions in the research that are going to be useful in the future of creating drugs for marine use are the following. AI and ML in drug development is a new avenue of research that could shorten the search for new marine metabolites and predict how well they might work as a medication. One more young discipline that could lead to better outcomes and reduced side effects in patients is the concept of personalized medicine, which might use, for example, the molecules derived from marine organisms, tailored to patients' genetic background. Increased use of pharmaceutical products derived from marine organisms means that proper sourcing and management of marine resources are increasingly significant. If people wish fish sources of bioactive substances to be preserved for future generations, researchers need to continue studying efficient fishing practices and designing marine fish farms. Copyright 2007: A snapshot of the direction of future research would include penetrating more uncharted regions of the oceans, like deep-sea and extreme environments, in the hope of discovering new metabolites with peculiar medicinal properties. Improving pharmaceuticals derived from marine sources in terms of manufacturing and effectiveness will continue to be defined highly by advances in chemistry and genetic engineering. Thus, researchers, pharmaceutical companies, and conservation organizations must collaborate to fully exploit marine biodiversity and use marine resources sustainably (Paul et al., 2021).

Conclusion

Last, but not least, there is a wide array of chemicals that can be considered chemicals with bioactivities that possess potent anti-infectious attributes inside the marine environment. Apart from primary metabolites that induced suspicion due to similarities with human metabolism, research on marine metabolites has unveiled secondary metabolites having different chemical compositions and begetting anti-infection procedures against different diseases. Some of these chemicals are alkaloids, terpenoids, peptides, polyketides, and others. Salinosporamide-A, palau'amine, cyclopamine, and cembranoid are some of the marine-derived compounds that have offered varying potentials of fighting cancer, bacterial infections, and viral disorders. In the aspect of anti-infectives, the metabolites from marine sources could turn the tide. The latter gives a chance to develop new therapeutic agents that use molecular structures and the mechanisms of action to avoid the existing medication's flaws. Besides, substances extracted from the bottomless sea are also bright candidates for novel treatments for new and resistant diseases; in addition, thanks to the advance of synthetic biology and biotechnology, these natural products offer brand-new architecture and methods to create drugs. But there are prospects that can be considered for the enhancement of MDD in the future. Research and technical advancements in marine metabolites of new species or new generations and synthesis, as well as optimization, is anticipated to be achieved soon, providing a chance to invent efficient medicines for various diseases. Sustainable use and conservation of marine resources require cooperation and sustainability in the methods of getting these resources. Marine metabolites' significance in adding to the chances of treatment and improving the health of the public will increase progressively as the niche evolves.

Disclaimer

None.

References

Alma'abadi, A. D., T. Gojobori, and K. Mineta. 2015. "Marine Metagenome as A Resource for Novel Enzymes." *Genomics Proteomics Bioinformatics* 13 (5):290-5. doi: 10.1016/j.gpb.2015.10.001.

Almansour, A. M., M. A. Alhadlaq, K. O. Alzahrani, L. E. Mukhtar, A. L. Alharbi, and S. M. Alajel. 2023. "The Silent Threat: Antimicrobial-Resistant Pathogens in Food-Producing Animals and Their Impact on Public Health." *Microorganisms* 11 (9). doi: 10.3390/microorganisms11092127.

Ameen, Fuad, Saleh AlNadhari, and Ali A. Al-Homaidan. 2021. "Marine microorganisms as an untapped source of bioactive compounds." *Saudi Journal of Biological Sciences* 28 (1):224-231. doi: https://doi.org/10.1016/j.sjbs.2020.09.052.

Bouyahya, Abdelhakim, Saad Bakrim, Imane Chamkhi, Douae Taha, Nasreddine El Omari, Naoual El Mneyiy, Naoufal El Hachlafi, Mohamed El-Shazly, Asaad Khalid, Ashraf N. Abdalla, Khang Wen Goh, Long Chiau Ming, Bey Hing Goh, and Tarik Aanniz. 2024. "Bioactive substances of cyanobacteria and microalgae: Sources, metabolism, and anticancer mechanism insights." *Biomedicine & Pharmacotherapy* 170:115989. doi: https://doi.org/10.1016/j.biopha.2023.115989.

Cochrane, Sabine K. J., Jesper H. Andersen, Torsten Berg, Hugues Blanchet, Angel Borja, Jacob Carstensen, Michael Elliott, Herman Hummel, Nathalie Niquil, and Paul E. Renaud. 2016. "What is marine biodiversity? Towards common concepts and their implications for assessing biodiversity status." *Frontiers in Marine Science* 3:248.

Cusick, Kathleen D., and Gary S. Sayler. 2013. "An overview on the marine neurotoxin, saxitoxin: genetics, molecular targets, methods of detection and ecological functions." *Marine Drugs* 11 (4):991-1018.

Dalmaso, G. Z., D. Ferreira, and A. B. Vermelho. 2015. "Marine extremophiles: a source of hydrolases for biotechnological applications." *Mar Drugs* 13 (4):1925-65. doi: 10.3390/md13041925.

Elissawy, A. M., E. Soleiman Dehkordi, N. Mehdinezhad, M. L. Ashour, and P. Mohammadi Pour. 2021. "Cytotoxic Alkaloids Derived from Marine Sponges: A Comprehensive Review." *Biomolecules* 11 (2). doi: 10.3390/biom11020258.

Eustáquio, A. S., R. P. McGlinchey, Y. Liu, C. Hazzard, L. L. Beer, G. Florova, M. M. Alhamadsheh, A. Lechner, A. J. Kale, Y. Kobayashi, K. A. Reynolds, and B. S. Moore. 2009. "Biosynthesis of the salinosporamide A polyketide synthase substrate chloroethylmalonyl-coenzyme A from S-adenosyl-L-methionine." *Proc Natl Acad Sci USA* 106 (30):12295-300. doi: 10.1073/pnas.0901237106.

Gao, Qi, Sizhe Deng, and Tianyu Jiang. 2022. "Recent developments in the identification and biosynthesis of antitumor drugs derived from microorganisms." *Engineering Microbiology* 2 (4):100047. doi: https://doi.org/10.1016/j.engmic.2022.100047.

Gerwick, William H, and Bradley S Moore. 2012. "Lessons from the Past and Charting the Future of Marine Natural Products Drug Discovery and Chemical Biology." *Chemistry & Biology* 19 (1):85-98. doi: https://doi.org/10.1016/j.chembiol.2011.12.014.

Ghosh, Susmita, Tanmay Sarkar, Siddhartha Pati, Zulhisyam Abdul Kari, Hisham Atan Edinur, and Runu Chakraborty. 2022. "Novel bioactive compounds from marine sources as a tool for functional food development." *Frontiers in Marine Science* 9:832957.

Gross, Harald, and Gabriele M. König. 2006. "Terpenoids from marine organisms: unique structures and their pharmacological potential." *Phytochemistry Reviews* 5:115-141.

Guilhelmelli, Fernanda, Nathália Vilela, Patrícia Albuquerque, Lorena da S. Derengowski, Ildinete Silva-Pereira, and Cynthia M. Kyaw. 2013. "Antibiotic development challenges: the various mechanisms of action of antimicrobial peptides and of bacterial resistance." *Frontiers in microbiology* 4:353.

Gulder, T. A., and B. S. Moore. 2010. "Salinosporamide natural products: Potent 20 S proteasome inhibitors as promising cancer chemotherapeutics." *Angew Chem Int Ed Engl* 49 (49):9346-67. doi: 10.1002/anie.201000728.

Gupta, Ajai Prakash, Pankaj Pandotra, Rajni Sharma, Manoj Kushwaha, and Suphla Gupta. 2013. "Chapter 8 - Marine Resource: A Promising Future for Anticancer Drugs." In *Studies in Natural Products Chemistry*, edited by Rahman Atta ur, 229-325. Elsevier.

Hamann, M. T. 2007. "The manzamines as an example of the unique structural classes available for the discovery and optimization of infectious disease controls based on marine natural products." *Curr Pharm Des* 13 (6):653-60. doi: 10.2174/138161207780162818.

Huang, Yujing, Xiaohan Guo, Yi Wu, Xingyu Chen, Lixiang Feng, Na Xie, and Guobo Shen. 2024. "Nanotechnology's frontier in combatting infectious and inflammatory diseases: prevention and treatment." *Signal Transduction and Targeted Therapy* 9 (1):34. doi: 10.1038/s41392-024-01745-z.

Jamal, Mamdoh T., and Sathianeson Satheesh. 2022. "Antibiofilm activity of secondary metabolites of sponge-associated bacterium Alcanivorax sp. from the Red Sea." *Frontiers in Marine Science* 9:980418.

Karan, Dev, Seema Dubey, Lucia Pirisi, Alexis Nagel, Ivett Pina, Yeun-Mun Choo, and Mark T. Hamann. 2020. "The marine natural product manzamine A inhibits cervical cancer by targeting the SIX1 protein." *Journal of natural products* 83 (2):286-295.

Karthikeyan, A., A. Joseph, and B. G. Nair. 2022a. "Promising bioactive compounds from the marine environment and their potential effects on various diseases." *J Genet Eng Biotechnol* 20 (1):14. doi: 10.1186/s43141-021-00290-4.

Karthikeyan, Akash, Abey Joseph, and Baiju G. Nair. 2022b. "Promising bioactive compounds from the marine environment and their potential effects on various diseases." *Journal of Genetic Engineering and Biotechnology* 20 (1):14. doi: 10.1186/s43141-021-00290-4.

Kayode-Afolayan, Shola D., Eze F. Ahuekwe, and Obinna C. Nwinyi. 2022. "Impacts of pharmaceutical effluents on aquatic ecosystems." *Scientific African* 17:e01288. doi: https://doi.org/10.1016/j.sciaf.2022.e01288.

Kem, W. R., K. Andrud, G. Bruno, H. Xing, F. Soti, T. T. Talley, and P. Taylor. 2022. "Interactions of Nereistoxin and Its Analogs with Vertebrate Nicotinic Acetylcholine Receptors and Molluscan ACh Binding Proteins." *Mar Drugs* 20 (1). doi: 10.3390/md20010049.

Kodzius, Rimantas, and Takashi Gojobori. 2015. "Marine metagenomics as a source for bioprospecting." *Marine Genomics* 24:21-30. doi: https://doi.org/10.1016/j.margen.2015.07.001.

König, G. M., S. Kehraus, S. F. Seibert, A. Abdel-Lateff, and D. Müller. 2006. "Natural products from marine organisms and their associated microbes." *Chembiochem* 7 (2):229-38. doi: 10.1002/cbic.200500087.

Landrigan, P. J., J. J. Stegeman, L. E. Fleming, D. Allemand, D. M. Anderson, L. C. Backer, F. Brucker-Davis, N. Chevalier, L. Corra, D. Czerucka, M. D. Bottein, B. Demeneix, M. Depledge, D. D. Deheyn, C. J. Dorman, P. Fénichel, S. Fisher, F. Gaill, F. Galgani, W. H. Gaze, L. Giuliano, P. Grandjean, M. E. Hahn, A. Hamdoun, P. Hess, B. Judson, A. Laborde, J. McGlade, J. Mu, A. Mustapha, M. Neira, R. T. Noble, M. L. Pedrotti, C. Reddy, J. Rocklöv, U. M. Scharler, H. Shanmugam, G. Taghian, Jajm van de Water, L. Vezzulli, P. Weihe, A. Zeka, H. Raps, and P. Rampal. 2020. "Human Health and Ocean Pollution." *Ann Glob Health* 86 (1):151. doi: 10.5334/aogh.2831.

Lee, Yeon-Ju, Yeonwoo Cho, and Huynh Nguyen Khanh Tran. 2021. "Secondary metabolites from the marine sponges of the genus Petrosia: A literature review of 43 years of research." *Marine drugs* 19 (3):122.

Lindequist, U. 2016. "Marine-Derived Pharmaceuticals - Challenges and Opportunities." *Biomol Ther (Seoul)* 24 (6):561-571. doi: 10.4062/biomolther.2016.181.

Malve, H. 2016. "Exploring the ocean for new drug developments: Marine pharmacology." *J Pharm Bioallied Sci* 8 (2):83-91. doi: 10.4103/0975-7406.171700.

Manyi-Loh, C., S. Mamphweli, E. Meyer, and A. Okoh. 2018. "Antibiotic Use in Agriculture and Its Consequential Resistance in Environmental Sources: Potential Public Health Implications." *Molecules* 23 (4). doi: 10.3390/molecules23040795.

Maslin, Mathilde, Nabila Gaertner-Mazouni, Cécile Debitus, Nicole Joy, and Raimana Ho. 2021. "Marine sponge aquaculture towards drug development: An ongoing history of technical, ecological, chemical considerations and challenges." *Aquaculture Reports* 21:100813. doi: https://doi.org/10.1016/j.aqrep.2021.100813.

Mioso, R., F. J. Marante, R. S. Bezerra, F. V. Borges, B. V. Santos, and I. H. Laguna. 2017. "Cytotoxic Compounds Derived from Marine Sponges. A Review (2010-2012)." *Molecules* 22 (2). doi: 10.3390/molecules22020208.

Najmi, A., S. A. Javed, M. Al Bratty, and H. A. Alhazmi. 2022. "Modern Approaches in the Discovery and Development of Plant-Based Natural Products and Their Analogues as Potential Therapeutic Agents." *Molecules* 27 (2). doi: 10.3390/molecules27020349.

National Research, Council. 1999. "Marine-Derived Pharmaceuticals and Related Bioactive Agents." *From Monsoons to Microbes: Understanding the Ocean's Role in Human Health.*

Núñez-Pons, Laura, Andrew Shilling, Cinzia Verde, Bill J. Baker, and Daniela Giordano. 2020. "Marine terpenoids from polar latitudes and their potential applications in biotechnology." *Marine drugs* 18 (8):401.

Nurrachma, Marsya Yonna, Deamon Sakaraga, Ahmad Yogi Nugraha, Siti Irma Rahmawati, Asep Bayu, Linda Sukmarini, Akhirta Atikana, Anggia Prasetyoputri, Fauzia Izzati, and Mega Ferdina Warsito. 2021. "Cembranoids of soft corals: Recent updates and their biological activities." *Natural Products and Bioprospecting* 11 (3):243-306.

Parmar, Ghanshyam, Jay M. Chudasama, Vaidehi Shah, Ashish P. Shah, and Ashish Patel. 2023. "A Mini-Review on Marine Bioactive Compounds as a Source of Antiviral Drug Lead." *Current Traditional Medicine* 9 (5):116-119.

Paul, D., G. Sanap, S. Shenoy, D. Kalyane, K. Kalia, and R. K. Tekade. 2021. "Artificial intelligence in drug discovery and development." *Drug Discov Today* 26 (1):80-93. doi: 10.1016/j.drudis.2020.10.010.

Perhar, G., and G. B. Arhonditsis. 2014. "Aquatic ecosystem dynamics following petroleum hydrocarbon perturbations: A review of the current state of knowledge." *Journal of Great Lakes Research* 40:56-72. doi: https://doi.org/10.1016/j.jglr.2014.05.013.

Putra, Masteria Yunovilsa, and Tutik Murniasih. 2016. "Marine soft corals as source of lead compounds for anti-inflammatories." *J. Coast. Life Med* 4:73-77.

Rampelotto, P. H. 2013. "Extremophiles and extreme environments." *Life (Basel)* 3 (3):482-5. doi: 10.3390/life3030482.

Reygaert, W. C. 2018. "An overview of the antimicrobial resistance mechanisms of bacteria." *AIMS Microbiol* 4 (3):482-501. doi: 10.3934/microbiol.2018.3.482.

Rodrigues, I. G., M. G. Miguel, and W. Mnif. 2019. "A Brief Review on New Naturally Occurring Cembranoid Diterpene Derivatives from the Soft Corals of the Genera Sarcophyton, Sinularia, and Lobophytum Since 2016." *Molecules* 24 (4). doi: 10.3390/molecules24040781.

Sable, R., P. Parajuli, and S. Jois. 2017. "Peptides, Peptidomimetics, and Polypeptides from Marine Sources: A Wealth of Natural Sources for Pharmaceutical Applications." *Mar Drugs* 15 (4). doi: 10.3390/md15040124.

Sanchez Armengol, Eva, Melisa Harmanci, and Flavia Laffleur. 2021. "Current strategies to determine antifungal and antimicrobial activity of natural compounds." *Microbiological Research* 252:126867. doi: https://doi.org/10.1016/j.micres.2021.126867.

Seghal Kiran, G., P. Ramasamy, S. Sekar, M. Ramu, S. Hassan, A. S. Ninawe, and J. Selvin. 2018. "Synthetic biology approaches: Towards sustainable exploitation of marine bioactive molecules." *Int J Biol Macromol* 112:1278-1288. doi: 10.1016/j.ijbiomac.2018.01.149.

Selim, M. S. M., S. A. Abdelhamid, and S. S. Mohamed. 2021. "Secondary metabolites and biodiversity of actinomycetes." *J Genet Eng Biotechnol* 19 (1):72. doi: 10.1186/s43141-021-00156-9.

Sigwart, Julia D., Robert Blasiak, Marcel Jaspars, Jean-Baptiste Jouffray, and Deniz Tasdemir. 2021. "Unlocking the potential of marine biodiscovery." *Natural Product Reports* 38 (7):1235-1242.

Sruthi, V., N. S. Grace, N. Monica, and V. M. Kumar. 2020. "Marine pharmacology: an ocean to explore novel drugs." *Int. J. Basic Clin. Pharmacol* 9.

Tian, X. R., H. F. Tang, X. L. Tian, J. J. Hu, L. L. Huang, and K. R. Gustafson. 2018. "Review of bioactive secondary metabolites from marine bryozoans in the progress of new drugs discovery." *Future Med Chem* 10 (12):1497-1514. doi: 10.4155/fmc-2018-0012.

Towle, Murray J., Kathleen A. Salvato, Jacqueline Budrow, Bruce F. Wels, Galina Kuznetsov, Kimberley K. Aalfs, Susan Welsh, Wanjun Zheng, Boris M. Seletsky, and Monica H. Palme. 2001. "In vitro and in vivo anticancer activities of synthetic macrocyclic ketone analogues of halichondrin B." *Cancer research* 61 (3):1013-1021.

Yan, Xu, Xu Liu, Cuihuan Zhao, and Guo-Qiang Chen. 2023. "Applications of synthetic biology in medical and pharmaceutical fields." *Signal Transduction and Targeted Therapy* 8 (1):199. doi: 10.1038/s41392-023-01440-5.

Chapter 2

Marine Metabolites and Their Emerging Role in Antimicrobial Resistance and Infection Control

Manish Dwivedi[1,2,*]
Divyansh Verma[1]
Palak Mishra[1]
Somali Sanyal[1]
and Radhey Shyam Kaushal[3]
[1]Amity Institute of Biotechnology, Amity University Uttar Pradesh, Lucknow Campus, Gomtinagar Extension, Lucknow, India
[2]Research Cell, Amity University Uttar Pradesh, Lucknow Campus, Lucknow, India
[3]Department of Life Sciences, Parul Institute of Applied Sciences and Biophysics and Structural Biology Laboratory, Research & Development Cell, Parul University, Gujarat, India

Abstract

Infectious diseases remain a significant global health challenge, necessitating the continuous exploration of novel therapeutic strategies. Marine ecosystems, teeming with diverse life forms, have emerged as a prolific source of bioactive compounds with remarkable pharmacological properties. This review represents the application and significance of marine metabolites in combating infectious diseases, offering insights into their immense potential as a therapeutic arsenal. Marine organisms, including bacteria, fungi, algae, and invertebrates, produce an array of secondary metabolites as part of their defense mechanisms and ecological interactions. These metabolites exhibit a diverse range of bioactivities, including antimicrobial, antiviral, antifungal, and immunomodulatory effects. Their structural complexity and unique chemical scaffolds render them promising candidates for drug discovery and development. Many marine-derived compounds have demonstrated potent antimicrobial activity against various pathogens, including bacteria, viruses, and fungi. The significance of marine metabolites in infectious disease therapeutics extends beyond their antimicrobial and immunomodulatory effects. Various other marine-derived compounds exhibit synergistic interactions with conventional antimicrobial agents, enhancing their efficacy and overcoming resistance mechanisms. Moreover, the marine environment offers a vast and largely unexplored reservoir of bioactive compounds,

* Corresponding Author's Email: mdwivedi@lko.amity.edu

In: Healing from the Depths Marine Metabolites in Disease Management
Editors: Pranav Kumar Prabhakar and Tarun Kumar Upadhyay
ISBN: 979-8-89530-832-5

presenting ample opportunities to discover novel drugs to combat emerging infectious threats. Their multifaceted mechanisms of action, including antimicrobial, antiviral, and immunomodulatory effects, underscore their significance as a promising avenue for drug discovery and development. Continued exploration of marine biodiversity and innovative drug discovery approaches hold the key to unlocking the full potential of marine metabolites in infectious disease management.

Keywords: marine metabolite, infections, therapy, drug, antimicrobial

1. Introduction

With its wide range of ecosystems, the marine environment offers an undiscovered source of special bioactive substances. These marine metabolites contain a broad spectrum of biological functions and hold great promise for treating infectious disorders. They are derived from a variety of marine animals, including sponges, algae, bacteria, and fungi (Mayer A. M.-S., 2017). In recent decades, there has been a surge in the investigation of marine natural products due to the growing demand for novel therapeutic agents due to the rise in antimicrobial resistance (Blunt, 2018).

To survive in their harsh and competitive surroundings, marine species have evolved sophisticated chemical defence mechanisms. The generation of secondary metabolites with strong antibacterial, antiviral, antifungal, and antiparasitic qualities is the result of these natural defence mechanisms (Hu G.-P. Y.-G.-H.-J.-C.-P., 2020). For example, substances like ecteinascidins, which come from the tunicate *Ecteinascidia turbinata*, and *bryostatins*, which come from the marine bryozoan *Bugula neritina*, have demonstrated encouraging properties against infectious illnesses and cancer (Pettit R. K., 2011); (Newman D. J., 2016).

The variety of structures and unique ways that marine metabolites function present a good opportunity for therapeutic development. Marine organisms, in contrast to terrestrial sources, frequently produce metabolites with distinctive chemical structures that can be used as models to create novel antibiotics and other treatments (Molinski T. F., 2009). For instance, it has been discovered that polyketides and peptides originating from marine sources can impede the development of bacteria resistant to several drugs, providing possible remedies for the worldwide health emergency caused by antibiotic resistance (Wright, 2019).

Technological developments like synthetic biology and metagenomics have made it easier to find and produce marine natural goods. The sustainable synthesis of marine metabolites is made possible by these technologies, which allow for the discovery and modification of the biosynthetic gene clusters responsible for the creation of bioactive chemicals (Medema, 2015). In addition, the combination of bioinformatics and high-throughput screening technologies has expedited the process of identifying potential medication candidates (Watkins, 2010).

Using marine metabolites as a weapon against infectious diseases seems like a viable tactic. These compounds' distinctive chemical variety and strong bioactivities offer a useful resource for the creation of novel medicinal medicines. Research on marine natural products has the potential to address some of the most important issues facing contemporary medicine with continued study and innovation.

The vast diversity of chemicals found in the ocean presents an almost unexplored chemical diversity, making it an unrivalled prospect for drug development. Innovative methods for chemical extraction and synthesis combined with recent developments in marine biotechnology have made it easier to isolate and characterise a wide range of marine metabolites that may have therapeutic value (Hu G. K., 2021). Promising activity against HIV has been demonstrated by the marine-derived compound bryostatin, which was initially isolated from the marine bryozoan *Bugula neritina*. This highlights the potential of marine metabolites in addressing emerging infectious diseases (Gutiérrez-Rodríguez, 2019).

Over 70% of the Earth's surface is made up of marine settings, which are home to a remarkably wide variety of species. These creatures create a wide variety of distinct metabolites as part of their survival strategies in hostile and competitive environments. These organisms range in size from microbes to larger marine animals. These marine metabolites have attracted a lot of attention lately because of possible medical uses, especially in the fight against infectious illnesses.

Because of their possible medical uses, marine metabolites—bioactive substances made by algae, sponges, marine microbes, and other marine organisms—have attracted a lot of attention from scientists. An increasing amount of study conducted in the last few decades has revealed the potential of these chemicals originating from marine sources as a powerful defence against infectious diseases, which continue to be a major danger to world health. The introduction of new diseases and the resurgence of old ones, along with the rising incidence of antibiotic-resistant bacteria, highlight the pressing need for novel treatments. Secondary metabolites produced by marine species, including sponges, algae, and marine bacteria, display a diverse array of biological activity. For example, numerous chemicals derived from marine sources have demonstrated strong antimicrobial, antifungal, antiviral, and anticancer effects. The diverse chemical structures of these substances are frequently different from terrestrial ones.

In summary, using marine metabolites as a weapon against infectious diseases seems like a good idea. These compounds' distinctive chemical variety and strong bioactivities offer a useful resource for the creation of novel medicinal medicines. Research on marine natural products has the potential to address some of the most important issues facing contemporary medicine with continued study and innovation.

2. Natural Products in the Marine System

Marine natural products encompass a wide range of chemical compounds that are obtained from marine organisms, such as bacteria, fungi, algae, corals, and sponges. These products have shown great promise for use in biotechnological and medicinal applications. Because of the particular environmental factors found in marine habitats—such as high pressure, fluctuating temperatures, and little light—these organisms have evolved unique metabolic pathways that have produced novel and structurally different chemicals (Hu G. K., 2021). These metabolites are promising targets for therapeutic development due to their diverse biological actions, which include antibacterial, antifungal, antiviral, anticancer, and anti-inflammatory effects (Blunt, 2018). For instance, ovarian cancer and soft tissue sarcoma can be treated using the substance trabectedin, which is produced from the sea squirt *Ecteinascidia turbinata* (Newman D. J., 2016). Furthermore, manzamine A, a molecule produced from marine sponges, has demonstrated strong antimalarial action, underscoring the medicinal promise of marine natural products in treating infectious disorders (Gutiérrez-Rodríguez, 2019). The creation of novel marine-derived medications is being facilitated by advancements in chemical synthesis and marine biotechnology, notwithstanding the difficulties in isolating, characterising, and procuring these substances sustainably (Leal M. C., 2016).

2.1. Importance of Marine Metabolites in Drug Discovery

The marine metabolites have substantial implications in the drug development process, considering their distinct chemical compositions and strong biological properties that present encouraging answers to various medical problems. Because they live in a variety of harsh and varied settings, marine species have evolved to produce a vast range of bioactive chemicals that are not present in terrestrial organisms. These substances, referred to as marine metabolites, have demonstrated incredible promise for the creation of cutting-edge medicines. For instance, ovarian cancer and soft tissue sarcoma can be treated with the marine-derived medication trabectedin, which is generated from the sea squirt *Ecteinascidia turbinata* (Gomes, 2015). In a similar vein, the identification of marine peptides possessing antibacterial qualities has created new opportunities for tackling antibiotic resistance, an increasingly pressing global health issue (Blunt, 2018). The identification and characterisation of these metabolites have been greatly accelerated by developments in marine biotechnology and high-throughput screening techniques, opening the door for their incorporation into the pharmaceutical pipeline (Hu G. K., 2021). Moreover, compared to conventional medications, distinct mechanisms of action displayed by marine metabolites frequently result in increased efficacy and decreased resistance development, highlighting their crucial significance in contemporary drug research initiatives (Martins A. V., 2014). Therefore, the discovery and application of marine metabolites constitute an important and cutting-edge field in the search for novel, efficient therapies for a wide range of illnesses.

2.2. Historical Context and Discovery

The exploration of marine metabolites and their potential anti-infective effects has a lengthy and fascinating past spanning multiple decades. Realising that marine species produce distinct chemical compounds not found in terrestrial organisms despite surviving in a variety of frequently harsh conditions sparked the early interest in marine natural products. The first marine natural product, spongothymidine, was identified from a Caribbean sponge in the early 1950s by chemist Werner Bergmann, whose groundbreaking work signalled the beginning of the systematic study of these metabolites in the mid-20th century (Bergmann W. &., 1951). This discovery was revolutionary since it paved the way for the creation of significant antiviral medications and provided the groundwork for further marine bioprospecting initiatives.

Thanks to developments in scuba diving and underwater exploration equipment, scientists were able to gather and study marine species on a larger scale during the 1960s and 1970s. Many bioactive substances with antibacterial qualities were found at this time, such as cephalosporin C from marine fungi, which served as the inspiration for the creation of the popular cephalosporin class of antibiotics (Fleming A. B., 1970). Research on marine natural products increased dramatically in the 1980s and 1990s as a result of the development of advanced chemical extraction and analytical methods. Numerous marine-derived substances with strong anti-infective effects were found by researchers. For example, pseudopterosins from the soft coral *Pseudopterogorgia elisabethae* were found to have strong analgesic and anti-inflammatory effects (Look, 1986).

The emphasis on marine microorganisms, especially bacteria and fungi, as abundant sources of new antibiotics and antifungal drugs has increased over the late 20th and early 21st centuries. For instance, the identification of marinomycins from a marine *actinomycete* demonstrated the potential of chemicals taken from the sea to battle diseases that are resistant to several drugs (Kwon, 2006).

2.3. Mechanism of Action

The many modes of action exhibited by marine metabolites are responsible for their strong anti-infective characteristics. One of the main ways of action is to cause lysis and death of microorganisms by breaking their cell membranes. For example, bacterial cell membrane lipid bilayers interact with antimicrobial peptides derived from marine species, such as tachyplesins and defensins, increasing permeability and inducing cell death (Noga, 2011). The suppression of vital bacteria enzymes and metabolic pathways is another important strategy. For instance, phospholipase A2 is inhibited by the marine-derived chemical manoalide, obtained from the sponge *Luffariella variabilis*. This interferes with bacterial lipid metabolism and inflammatory reactions (Mayer A. M., 2010). Furthermore, marine metabolites can prevent infections from synthesising proteins and nucleic acids. It has been demonstrated that the nucleoside analogue cytarabine, which is generated from the marine sponge *Cryptotethya crypta*, inhibits DNA synthesis, making it useful against some viruses and cancer cells (Newman D. J., 2020).

Some marine metabolites also have indirect antimicrobial effects by strengthening the host's immunological response, which goes beyond these pathways. For instance, it has been discovered that the substance fucoidan, which is taken from brown seaweed, increases immune cell activity and improves the synthesis of antibodies, supporting the body's defence against illnesses (Fitton, 2011). Additionally, the prevention of viral entrance and replication is how marine polysaccharides, such as carrageenans from red algae, have their broad-spectrum antiviral effect (Berri, 2016). These many processes highlight the adaptability and promise of marine metabolites in the creation of novel anti-infective treatments.

3. Types of Marine Metabolites with Anti-Infective Activity

3.1. Terpenoids

One of the most varied families of natural compounds, terpenoids, have drawn a lot of notice due to their strong anti-infective properties. The compounds in question, predominantly sourced from marine organisms, fungus, and plants, are renowned for their diverse biological activities and intricate structures. These include antiviral, antifungal, antibacterial, and antiparasitic qualities (Thoppil, 2011). Terpenoids can disrupt several essential processes in pathogens, including cell membrane integrity, protein synthesis, and enzyme activity, which is largely responsible for their anti-infective capability. For example, the quick and powerful antimalarial effects of the sesquiterpenoid artemisinin, which was discovered from the plant

Artemisia annua, have revolutionised the treatment of malaria (Meshnick, 1996). Similar to this, Andrographis paniculata's diterpenoid andrographolide shows promise as a therapeutic candidate since it has broad-spectrum antibacterial activity against a range of bacteria and viruses (Islam, 2018). Terpenoids originating from marine sources have demonstrated exceptional potential in combating diseases that are resistant to drugs. One important lead molecule for the creation of novel anti-infective drugs is manoalide, a sesterterpenoid that was discovered from the marine sponge *Luffariella variabilis* (Figure 1). Manoalide has shown strong antibacterial and anti-inflammatory properties (Gunasekera, 1980). These results highlight the important role terpenoids play in the continuous hunt for cutting-edge, efficient cures for infectious diseases.

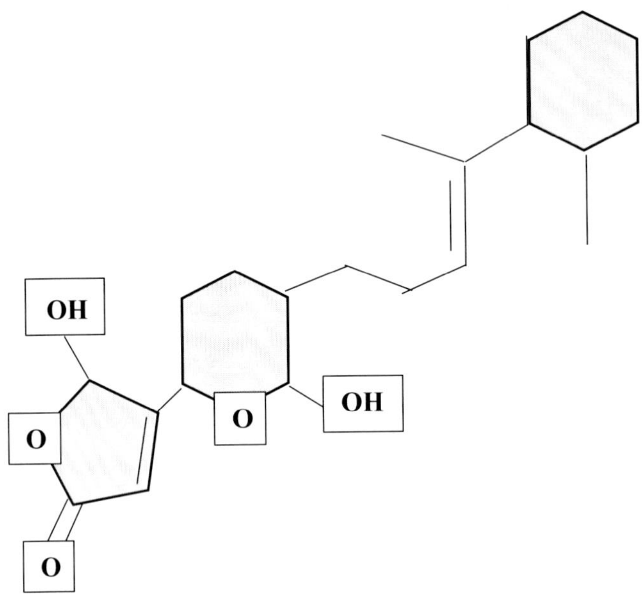

Figure 1. Structure of Manoalide.

3.2. Alkaloids

Alkaloids are a broad class of chemical compounds that exist naturally and have attracted a lot of study since they are powerful anti-infective marine metabolites. These nitrogen-containing substances are frequently seen in terrestrial plants, but they are also made by a range of marine animals, including molluscs, sponges, and tunicates. Marine alkaloids are interesting prospects in the fight against infectious diseases because they display a wide range of biological activities, including antibacterial, antifungal, antiviral, and antiparasitic properties (Adeleye, 2020). For example, the alkaloid manzamine A, which was extracted from a sea sponge, has demonstrated exceptional efficacy against both *Mycobacterium tuberculosis*, the causative agent of tuberculosis, and *Plasmodium falciparum*, the parasite that causes malaria (Abdelmohsen, 2017). Furthermore, the sponge *Latrunculia bocagei* yields the marine alkaloid discorhabdin C (Figure 2), which has shown strong cytotoxic and antibacterial properties and suggests that it could be a promising starting point for the synthesis of novel antibiotics (Zhang, 2018).

Particularly in light of the rising prevalence of antibiotic resistance, the structural diversity and biological potency of marine alkaloids highlight their importance as a rich source of new anti-infective drugs.

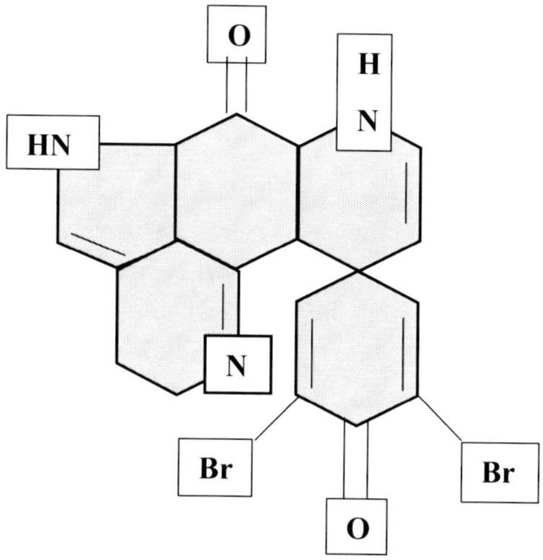

Figure 2. Structure of Discorhabdin C.

3.3. Peptide and Proteins

Marine-derived proteins and peptides have shown great promise as bioactive substances with strong anti-infective qualities. These compounds, which are obtained from marine sources, show a wide range of biological activities, such as antiviral, antifungal, antibacterial, and antiparasitic properties, which makes them interesting candidates for the creation of novel therapeutic medicines. The potential of marine peptides and proteins to target drug-resistant pathogens—a significant issue in contemporary medicine—is one of their main advantages. Antimicrobial peptides (AMPs), for example, have demonstrated broad-spectrum efficacy against a variety of bacteria, including multi-drug-resistant strains like *Pseudomonas aeruginosa* and Methicillin-resistant Staphylococcus aureus (MRSA), when recovered from marine species including fish, sponges, and molluscs (Rajanbabu, 2011).

Furthermore, marine proteins that bind carbohydrates, called lectins, have shown antiviral efficacy by preventing viruses from infecting host cells. As an illustration of its potential as a therapeutic agent, the lectin griffithsin, which is produced from the red algae *Griffithsia*, has demonstrated strong inhibitory effects against HIV and other viruses (Mori, Isolation and characterization of griffithsin, a novel HIV-inactivating protein, from the red alga Griffithsia sp., 2005). Furthermore, it has been shown that marine peptides, such as the cyclodepsipeptide kahalalide F, which is sourced from the marine mollusc *Elysia rufescens*, can demonstrate efficacy against a range of fungal infections and cancer cell lines, hence signifying their diverse therapeutic potential (Hamann, 1993).

These peptides and proteins have specific structural and functional features that are typically enhanced in stability and bioactivity by the peculiar environmental circumstances of the marine ecosystem, which include high pressure, low temperature, and changeable salinity.

The sustainable extraction of these compounds, their intricate structural elucidation, and the requirement for substantial preclinical and clinical testing, however, continue to be major obstacles in the way of their therapeutic application, despite their encouraging potential (Hancock, 2006). However, new peptides and proteins with strong anti-infective activity are still being found via continuous study and advancements in marine biotechnology, providing hope for the creation of innovative medicines against a variety of infectious diseases.

3.4. Polyketides

A broad class of marine metabolites known as polyketides is well known for having strong anti-infective properties, including antimicrobial, antiviral, and antiparasitic actions. Through polyketide synthase pathways, marine microorganisms like bacteria, fungi, and algae synthesise these complex molecules, enabling the development of sophisticated structures with significant biological activity (Rein, 2019). Among the most prominent instances is the macrolide antibiotic erythromycin, which comes from marine *Actinomycetes* and has been used extensively since its discovery to treat bacterial infections (Zhu, 2020). Doxorubicin, a powerful anti-tumour agent generated from terrestrial *Streptomyces*, is another significant polyketide. It has sparked the development of several marine-derived analogues with improved efficacy and fewer side effects (Cimino, 2021). According to a recent study, marine polyketides have the potential to fight multidrug-resistant diseases. One such molecule is salinosporamide A, which was isolated from the marine bacterium *Salinispora tropica* and has shown promising activity against cancer cells and resistant strains of bacteria (Bergmann W. &., 2020). The ongoing investigation of polyketides produced from marine sources is essential to the hunt for novel therapeutic agents, providing hope for the creation of cutting-edge cures for difficult infectious diseases.

4. Marine-Derived Antibiotics

Numerous distinct species found in marine habitats generate a broad range of bioactive substances, including strong antibiotics. These marine-derived antibiotics are frequently more potent and selective than their terrestrial equivalents, opening up new treatment options for infections that are becoming more prevalent and resistant to bacterial strains. The tremendous potential of marine biodiversity in drug discovery is illustrated in this section by highlighting a few noteworthy instances of antibiotics obtained from marine environments and their origins.

4.1. Vancomycin

Vancomycin is a well-known antibiotic that comes from marine sources. It is a glycopeptide antibiotic made by the soil bacterium *Amycolatopsis orientalis*. However, knowledge of its development and mechanics has also benefited from marine sources. Vancomycin is not a marine-derived antibiotic, although marine conditions have sparked the development of analogues and derivatives that increase its effectiveness (Wang, 2014). These discoveries underscore the influence of marine research on the evolution of antibiotic therapies.

4.2. Bryostatin-1

The marine-derived substance bryostatin-1, which is obtained from the bryozoan *Bugula neritina*, demonstrates noteworthy antibacterial and anticancer characteristics. Because of the complex macrocyclic lactone in its structure, which interacts with protein kinase C (PKC) and modifies many biological processes, it has a distinct chemical scaffold. Its antibacterial activity and its therapeutic uses depend on its interaction with PKC (Pettit G. R., 1982). The capacity of bryostatin-1 to modify cell signalling pathways is thought to be responsible for its action, and this ability can be used to create new antibiotics and anticancer drugs.

4.3. Halichondrin B

Derived from the marine sponge *Halichondria okadai*, Halichondrin B is an antibiotic possessing an intricate polyether structure. It is well known for having strong antimicrobial and anticancer properties. The capacity of Halichondrin B to impede microtubule assembly, a critical mechanism in cell division and bacterial development, is attributed to its distinct structure, which consists of numerous polyether rings (Kobayashi, 1993). Halichondrin B's capacity to break cellular microtubules is connected to its efficiency as an antibiotic, according to the structure-activity relationship, which presents a viable option for drug development.

4.4. Fimbrolides

Fimbrolides are a class of antibiotics distinguished by their distinct structure and bicyclic system. They were identified from the marine alga *Fimbriaphyllum fimbriatum*. By inhibiting bacterial protein synthesis, they particularly target ribosomal RNA in their process. Fimbrolides are powerful medicines against a variety of bacterial diseases because of their structure-activity relationship, which shows that the bicyclic structure is essential for binding to the bacterial ribosome and preventing protein synthesis (Sung, 2010).

4.5. Fistularin-3

Fistularin-3, a strong antibacterial compound that exhibits effectiveness against many bacterial and fungal diseases, was extracted from the marine sponge *Ircinia felix*. The molecule is a promising option for additional therapeutic development because of its capacity to effectively suppress microbial growth, which is facilitated by its distinct chemical structure (Kang, 2006).

5. Marine-Derived Antiviral Agents

The wide variety of ecosystems found in marine habitats provides a wealth of bioactive substances, including several antiviral drugs produced from marine life. These antiviral compounds generated from marine sources have attracted a lot of attention because of their distinct modes of action and ability to treat viral infections that are challenging to treat with traditional medications. The origins of various antiviral substances and their particular viral targets are examined in this section, along with important instances and their relevance for antiviral therapy.

5.1. Sources of Marine-Derived Antiviral Compounds

Many marine animals, such as molluscs, sponges, algae, and microbes, are abundant producers of antiviral substances. For example, it is well known that marine sponges produce a variety of secondary metabolites with strong antiviral effects. Didemnin B, a sponge-derived chemical, has demonstrated noteworthy efficacy against many viruses, such as the herpes simplex virus (HSV) and HIV. It was isolated from the marine tunicate *Didemnum molle* (Thompson, 2021). Similarly, by regulating the host's immune response and boosting the efficacy of other antiviral medications, bryostatin-1, which is produced from the marine bryozoan *Bugula neritina*, demonstrates antiviral action against HIV (Tabares, 2020).

Antiviral compounds have also been found in marine algae, especially in the genera Caulerpa and Sargassum. Sulfated polysaccharides derived from *Caulerpa taxifolia*, for instance, have been shown to have inhibitory effects on many enveloped viruses, such as the hepatitis C virus (HCV) and HSV (Jung, 2019). These polysaccharides act by interfering with viral attachment and entry into host cells.

Additionally, marine microorganisms like fungi and bacteria add to the pool of antiviral substances. Actinomycetes produced from marine environments are a source of new antiviral drugs. The marine bacteria *Streptomyces spongiae* produces a chemical called spongistatin-1, which has demonstrated encouraging antiviral efficacy against HSV-1 (Zhao, 2022). Furthermore, terrein, which has been shown to have inhibitory effects against a number of viruses, including the human cytomegalovirus (HCMV), is produced by the marine fungus *Aspergillus terreus* (Guan, 2021).

5.2. Targets of Marine-Derived Antiviral Compounds

Multiple viral targets, including viral entry, replication, and release, are frequently involved in the antiviral actions of substances originating from marine sources. Didemnin B, for example, works by preventing the manufacture of viral proteins and altering the host's immunological response, which hinders the spread of the virus (Thompson, 2021). By stimulating protein kinase C (PKC) pathways, bryostatin-1 targets HIV and can improve immune responses while lowering viral load (Tabares, 2020).

The presence of sulfated polysaccharides in marine algae, such as *Caulerpa taxifolia*, can hinder the attachment of viruses to host cell receptors, hence blocking the entry of viruses and the consequent spread of infection (Jung, 2019). For enveloped viruses, which depend on surface proteins to connect to host cells, this method is essential.

Compounds derived from marine sources also directly impact viral propagation. For instance, Aspergillus terreus produces terrein, which interferes with the synthesis and assembly of viral DNA, preventing the reproduction of viruses like HCMV (Guan, 2021). Spongistatin-1 also lowers the total viral load by inhibiting viral enzymes that are essential for the assembly of viral particles (Zhao, 2022).

6. Clinical and Pre-Clinical Studies of Marine-Derived Antiviral Agents

6.1. Pre-Clinical Studies

Important new information on the mechanisms of action and therapeutic potential of antiviral medicines obtained from marine sources has been made possible by pre-clinical research. Griffithsin, a lectin obtained from the red algae *Griffithsia*, is one of the most researched marine substances. Hepatitis C virus (HCV), SARS-CoV, and HIV are just a few of the viruses, griffinsin has demonstrated strong efficacy against. Griffithsin can bind to viral glycoproteins, which stops the virus from attaching to host cells and inhibits viral reproduction, as shown by in vitro experiments. Further evidence of griffithsin's efficacy has come from tests conducted on animals, which demonstrate that it can lower viral loads and prevent infection (Mori, solation and characterization of griffithsin, a novel HIV-inactivating protein, from the red alga Griffithsia sp., 2005).

Plitidepsin, a cyclic depsipeptide extracted from the marine tunicate *Aplidium albicans*, is another noteworthy molecule produced from marine sources. Plitidepsin has demonstrated potent antiviral efficacy against SARS-CoV-2, the virus that causes COVID-19, according to preclinical research. According to mechanistic research, plitidepsin inhibits host cell elongation factor 1A (eEF1A), which is necessary for the synthesis of viral proteins and reduces viral replication (Schütz, 2021).

6.2. Clinical Studies

Many antiviral medicines originating from marine sources have advanced to clinical trials, indicating their promise as novel therapeutic alternatives. For example, plitidepsin has been the subject of clinical trials to assess its safety and effectiveness in treating COVID-19. Early-stage clinical trials showed that plitidepsin had antiviral action and is well-tolerated by patients, which in some cases improved clinical outcomes and reduced viral load. Because of these results, plitidepsin is now being tested in bigger, more conclusive studies (Varona, 2022).

Likewise, the antiviral qualities of carrageenan, a sulfated polysaccharide obtained from red algae, have been investigated in clinical trials. Nasal sprays containing carrageenan have been studied as a prophylactic against respiratory viral illnesses, such as the flu and the common cold. Carrageenan has been demonstrated in clinical tests to lessen the duration and intensity of symptoms by preventing the virus from adhering to nasal epithelial cells (Eccles, 2010).

7. Marine-Derived Antifungal Agents

The strong action of marine-derived antifungal drugs against a range of fungal diseases, many of which are resistant to standard therapies, has attracted a lot of attention. With its unrivalled variety, the ocean is a rich source of novel metabolites with distinctive structures and modes of action, which gives hope for the advancement of antifungal treatments.

7.1. Halymenol A

Halymenol A, a substance extracted from red algae and generated from *Halymenia sp*., is one of the most well-known marine-derived antifungal drugs. By compromising the integrity of the fungal cell membrane, halymenol A demonstrates potent antifungal action against Candida species, including drug-resistant strains. Halymenol A has the potential to be a novel therapeutic agent due to its unique action, especially in situations where standard antifungals have not worked (Mayer A. M., 2019).

7.2. Bromophycolide A

Bromophycolide A, a marine metabolite of great significance, is sourced from the red alga *Callophycus serratus* in Fiji. Broad-spectrum antifungal activity of bromophycolide A has been established, with specific efficacy against Aspergillus species. It works by preventing the formation of fungal cell walls, which stops the growth of fungi. This mechanism makes it very efficient against fungi that are very resistant to current antifungal medications (Dickey, 2017).

7.3. Tachyplesin

Promising antifungal medicines include marine-derived peptides like Tachyplesin, which was extracted from the horseshoe crab *Tachypleus tridentatus*. Tachyplesin is effective against a variety of fungal infections, such as Candida albicans and Cryptococcus neoformans, by permeabilizing the fungal cell membranes. The peptide's potential as a secure and efficient antifungal therapy is highlighted by its capacity to target fungal membranes without harming mammalian cells.

7.4. Manzamine A

Manzamine A is a significant antifungal agent originating from marine sources, and it was isolated from *Haliclona* sponges. It has been discovered that manzamine A is especially effective against Cryptococcus neoformans, a bacterium that causes serious infections in people with impaired immune systems. By focussing on the mitochondrial electron transport chain, it prevents fungal development by interfering with the fungal cells' ability to produce energy (Ang K. K., 2020).

7.5. Scytonemin

The cyanobacterial metabolite scytonemin has demonstrated antifungal action against a range of pathogenic fungus, including the athlete's foot-causing *Trichophyton rubrum*. Scytonemin works by obstructing the processes that cause fungal oxidative stress, which results in cell death. Because of its distinct mode of action, scytonemin is a good candidate for more research and development as an antifungal medication (Singh, 2021).

8. Marine-Derived Anti-Parasitic Agents

Many different organisms that create distinct metabolites with strong bioactive qualities can be found in marine habitats, which are renowned for their exceptional biodiversity. Marine-derived chemicals have garnered considerable attention due to their efficaciousness in combating a range of parasitic diseases. The production of these compounds by microbes, algae, and marine sponges presents a promising avenue for the discovery of new antiparasitic drugs, particularly given the increasing resistance to traditional therapies (Tasdemir, 2021).

Major global health issues are still posed by parasitic illnesses such trypanosomiasis, leishmaniasis, and malaria, especially in tropical and subtropical countries. Limitations associated with conventional therapy may include toxicity, medication resistance, and poor efficacy. In response, a growing number of scientists are searching marine habitats for novel antiparasitic substances. For example, discodermolide, a strong substance with action against Plasmodium falciparum, the parasite causing malaria, has been demonstrated to be produced by the sea sponge *Discodermia calyx* (Roué, 2012).

In a similar vein, the red algae species Laurencia have produced many halogenated monoterpenes with potent antileishmanial properties, including iso-obtusol and elatol. These substances have shown efficacy against both promastigote and amastigote forms of Leishmania, acting by damaging the parasite's cell membrane integrity and causing the parasite to die (Cavalcante-Silva, 2017). Additionally, marine bacteria such as *Streptomyces spp.* have generated secondary compounds, such as macrolides, that exhibit anti-African sleeping sickness activity against Trypanosoma brucei (Shen, 2019).

Novel mechanisms are frequently involved in the mode of action of marine metabolites against parasites, which makes them attractive targets for therapeutic development. For instance, by targeting particular parasite enzymes and interfering with vital metabolic pathways, manzamine A, a marine alkaloid derived from the sponge *Acanthostrongylophora*, has shown broad-spectrum antiparasitic action, including against leishmaniasis and malaria (Ang K. K., 2018). These substances' distinct chemical structures, which frequently include unusual functional groups and stereochemistry, add to their efficacy and present chances for the creation of novel treatment approaches.

Although marine metabolites have great potential, there are still obstacles in the way of turning these substances into therapeutically effective medicines. It is necessary to handle problems including large-scale synthesis, chemical isolation, and sustainable harvesting. Moreover, comprehensive in vivo and in vitro investigations as well as clinical trials are required to guarantee the security and effectiveness of these medicines (Verma, 2017).

9. Pharmacokinetics and Toxicity of Marine Metabolites

Due to their distinct chemical structures and strong bioactivities, marine metabolites present a viable option for therapeutic development. However, a complete grasp of these substances' pharmacokinetics and toxicity is necessary for the successful transition of these substances from the ocean to clinical application. Adsorption, distribution, metabolism, and excretion (ADME) of these substances are all part of pharmacokinetics, and toxicity evaluations guarantee that they are safe for use in medicine. Determining the viability of marine metabolites as therapeutic possibilities requires consideration of both factors.

9.1. Pharmacokinetics

The intricate and varied structures of marine metabolites frequently impact their pharmacokinetic characteristics, presenting difficulties about absorption, solubility, and stability. For example, a lot of compounds originating from the sea are big, extremely polar, or physically stiff, which can make it harder for the body to absorb and distribute them (Patterson, 2019). Furthermore, these metabolites' distinct chemical surroundings frequently produce molecules with peculiar functions that could compromise their metabolic stability.

The powerful anticancer drug ecteinascidin-743, also known as trobesectin, which is extracted from the sea squirt *Ecteinascidia turbinata*, illustrates the pharmacokinetic difficulties. The intricate structure of trasbectedin, which is typified by several ring systems and various stereocenters, poses considerable challenges to attaining ideal pharmacokinetics. The substance is extensively metabolised in the liver, mostly by cytochrome P450 enzymes, producing both active and inactive metabolites that may have an impact on the toxicity and therapeutic efficacy of the chemical (Le Cesne, 2015).

Furthermore, the distinct modes of action that marine metabolites frequently display may have an impact on their pharmacokinetic profiles. Salinosporamide A, for instance, is a strong proteasome inhibitor that is obtained from the marine bacteria *Salinispora tropica.* It binds to its target quickly and irreversibly, causing biological effects that last long after the chemical is eliminated from the bloodstream (Fenical, 2009). This feature emphasises how important it is for medication developers to have a thorough grasp of the pharmacokinetics and pharmacodynamics of marine metabolites.

9.2. Toxicity

A crucial factor in the creation of medications derived from marine sources is toxicity. Because of their marine origin, marine metabolites frequently display strong biological activity that, if left unchecked, might have hazardous consequences. These substances can become hazardous due to several causes, including as off-target effects, tissue buildup, and the production of toxic metabolites during biotransformation.

For example, one of the most potent known non-protein poisons is the marine toxin palytoxin, which is produced by specific types of marine algae. Palytoxin binds to the Na+/K+-ATPase pump and modifies its function, disrupting cellular ion gradients and ultimately causing cell death (Ciminiello, 2014). Although palytoxin's extraordinary potency makes it an invaluable instrument for researching ion channels, its toxicity prohibits it from being used as a treatment. This illustration shows how carefully efficacy and safety must be balanced when developing medications obtained from marine sources.

The potential for causing hypersensitive reactions is another factor to be taken into account when assessing the toxicity of marine metabolites. This has been noted with marine-derived anticancer drugs like bryostatin-1, which in certain patients might result in severe immune-mediated side effects, such as anaphylaxis (Schaufelberger, 1991). The safe and efficient use of medicines derived from marine sources requires an understanding of the underlying mechanisms of these toxicities and the identification of patient populations that may be at risk.

To forecast potential negative consequences, toxicity assessments usually combine in vivo and in vitro investigations with computational modelling. These analyses aid in determining the marine metabolite therapeutic window—the range of dosages within which the chemical functions well without producing appreciable side effects. Furthermore, continuous developments in formulation methods and drug delivery systems seek to lessen the toxicity of marine metabolites while maintaining their potential for therapeutic use (Gould, 2016).

10. Challenges in Developing Marine Metabolites as Therapeutics

10.1. Extraction and Synthesis Issues

There are various obstacles in the way of developing marine metabolites as medicines, especially during the extraction and synthesis phases. The difficulties stem from the distinct and frequently intricate chemical compositions of marine-derived natural products, in addition to the administrative and environmental limitations linked to obtaining these substances from aquatic settings.

The sustainable gathering of marine organisms is one of the main obstacles to the extraction of marine metabolites. Concerns about ethics and conservation are raised by the fact that many bioactive chemicals are produced from rare or endangered animals. Delicate ecosystems may be disrupted by overharvesting if marine populations are depleted. Thus, to reduce ecological effect, sustainable harvesting methods—such as aquaculture and in vitro growing systems—are crucial (Leal M. C., 2016).

The extraction and purification of marine metabolites are frequently hampered by their intricate chemical structures. Numerous secondary metabolites are produced by marine creatures, including sponges, algae, and microbes, and many of them are only found in tiny levels. Careful methods must be used during the extraction procedure to separate these components without causing any damage. Furthermore, contaminants like salts and organic materials might be introduced by maritime conditions, making the purifying procedure more difficult (Molinski T. F., Drug development from marine natural products, 2009). To increase yield and purity, advanced extraction techniques have been used, such as solid-phase extraction and supercritical fluid extraction (Carballeira, 2008).

Another significant obstacle in the development of marine metabolites as medicines is synthesis. Chemical synthesis is challenging and time-consuming due to the structural complexity of many marine natural products, which are distinguished by distinct stereochemistry and functional groups. The viability of large-scale production, which is essential for clinical and commercial applications, is sometimes restricted by this complexity. To get around these problems, synthetic analogues and derivatives are commonly created, although it might be difficult to replicate the biological action of the original substance (Paterson, 2005).

The "supply problem"—the inability to get adequate amounts of bioactive chemicals for thorough biological testing and medication development—also impedes the discovery of new marine metabolites. The difficulty of cultivating many marine creatures in lab settings due to their slow growth rate exacerbates this problem. Advances in synthetic biology and genomic and proteome techniques are being investigated as potential remedies for this issue by permitting the production of marine biochemical pathways in heterologous hosts (Piel, 2010).

10.2. Regulatory and Market Considerations

While there is great potential in the development of marine metabolite-based therapies, getting these novel chemicals from the ocean to the clinic will need careful navigation of numerous regulatory and commercial obstacles. Market factors, intellectual property concerns, and regulatory frameworks all have a significant impact on how well pharmaceuticals made from marine resources succeed. The timeliness, cost, and viability of turning marine metabolites into commercially viable treatments can all be greatly impacted by these variables (Fleming A. &., 221).

10.3. Regulatory Frameworks

Because marine-derived medicines are innovative and frequently unheard of, the regulatory environment surrounding them is complicated. Before approving new pharmaceuticals, regulatory bodies like the European Medicines Agency (EMA) and the U.S. Food and Drug Administration (FDA) want extensive data on the quality, safety, and efficacy of the drugs. To achieve these strict standards, marine metabolites frequently require considerable preclinical testing due to their distinct chemical structures and bioactivities (Gullo, 2021). Another level of regulatory scrutiny is added when it comes to the extraction and purification processes for marine substances, as they need to follow Good Manufacturing Practices (GMP).

Furthermore, several state laws and international treaties about the preservation and sustainable use of marine biodiversity frequently encompass marine metabolites. One such example is the Nagoya Protocol on Access and Benefit-Sharing, which mandates that the nation of origin get a fair and equitable share of benefits resulting from the exploitation of genetic resources, including marine creatures. It might be difficult to comply with these regulations, especially when sourcing organisms from international oceans or regions outside of national borders (Vierros, 2020).

10.4. Intellectual Property and Patentability

The development of marine metabolite therapies is heavily dependent on IP considerations. Compared to terrestrial medications, patenting chemicals obtained from marine sources might be more complicated because novelty, non-obviousness, and utility must be proven. These compounds' natural origins might occasionally make the patenting process more difficult since courts and patent offices may need proof that the chemical has undergone considerable modification or that its medicinal value was unknown before (Arrieta, 2020).

In addition, concerns about biopiracy and morality surrounding the ownership of marine genetic resources are becoming more and more prominent. Businesses need to overcome these obstacles while making sure that their intellectual property (IP) strategies follow morality and international norms. Maintaining fair access and benefit-sharing while safeguarding innovation is a difficult balance that needs to be carefully considered (Martins A. V., 2014).

10.5. Market Dynamics

Cost-effectiveness, competitive positioning, and production scalability are some of the aspects that need to be taken into account when developing marine metabolite therapies. High manufacturing costs can result from the difficulty of finding and responsibly harvesting marine species as well as the frequently sophisticated and costly extraction and synthesis procedures. These expenses need to be evaluated about prospective market demand and pricing schemes (Rinehart, 2018).

The broader pharmaceutical landscape, which includes competition from synthetic medications and biopharmaceuticals, also has an impact on the market for treatments generated from marine sources. To effectively compete, marine metabolites must exhibit distinct advantages, such as distinct modes of action or better efficacy. Furthermore, the capacity of these medications to meet unmet medical needs or address antibiotic resistance—which is still a major global health concern—often determines how well they do on the market (Jiménez, 2019).

11. Future Perspectives

At the nexus of state-of-the-art biotechnology and the pressing need for new treatments against developing and resistant infections, marine metabolites have a promising future as medicines. The search for novel medications with distinctive modes of action is made possible by the investigation of marine natural products, even in light of the world's ongoing struggles with issues like antibiotic resistance. With its incredible biodiversity, the vast and mostly unexplored marine ecosystem holds the key to a new generation of therapies that have the potential to completely transform current medicine (Molinski T. F., Drug development from marine natural products, 2008).

The development of sophisticated biotechnological techniques for the sustainable synthesis of marine metabolites represents one of the most potential future paths in the field of marine metabolite therapies. Conventional marine creature collecting has sparked environmental issues and is frequently unfeasible for large-scale production in the search for new drugs. But rather than requiring direct extraction from marine sources, recent developments in synthetic biology and metabolic engineering are making it possible for marine metabolites to be biosynthesised in microbial systems (Rocha-Martin, 2020). These technologies enable the optimisation and modification of natural substances to increase their medicinal potential, in addition to making production more sustainable.

Furthermore, the discovery and development of pharmaceuticals originating from marine sources is being revolutionised by the integration of omics technologies, including proteomics, metabolomics, and genomes. Through the deciphering of the genetic and metabolic processes that give rise to bioactive chemicals, researchers can more effectively find novel metabolites and customise them for particular therapeutic uses (Perez-Lopez, 2021). This method also makes it easier to identify marine creatures with great potential for biosynthesis, which opens the door to the discovery of whole new medication classes.

Exciting prospects for the use of marine metabolites in personalised medicine also lie ahead. Marine natural products may be engineered to target certain biological pathways or genetic mutations as our understanding of individual genetic profiles and disease processes advances. As an illustration, marine-derived peptides and alkaloids, which are renowned for their potency and specificity, may be utilised to create tailored treatments for complex disorders like cancer (Mayer A. M.-S., 2010). This personalized approach not only enhances treatment efficacy but also reduces the likelihood of adverse side effects.

Furthermore, a new frontier of bioactive chemicals is anticipated to be discovered through the investigation of deep-sea and extreme marine environments, which are home to creatures with distinctive adaptations. These harsh settings, which include arctic locations and hydrothermal vents, are mainly uncharted territory that may yield new metabolites with hitherto undiscovered chemical structures and biological functions (Martins A. V., Marketed marine natural products in the pharmaceutical and cosmeceutical industries: Tips for success, 2014). To access these isolated habitats and realise their therapeutic potential, advancements in deep-sea exploration equipment and sample methodologies will be essential.

There are still obstacles in the way of getting marine metabolites to the clinical stage, despite the bright potential. There are several obstacles because these chemicals are complicated and because new medication approvals must pass stringent regulatory standards. However, the conversion of marine natural materials into clinically useful medicines is becoming more and more possible with ongoing investments in marine biotechnology, interdisciplinary partnerships, and the creation of novel drug delivery methods (Pérez-Victoria, 2014).

Conclusion

In the battle against infectious diseases, the discovery and application of marine metabolites offer a frontier with enormous potential. Numerous species that produce distinct bioactive chemicals can be found in the vast, largely unexplored waters of the world's seas; many of these compounds are still unknown or poorly understood. The need for innovative antimicrobial medicines has never been greater as infectious illnesses continue to mutate and become resistant to current treatments. Marine environments are a rich, varied, and largely unexplored source of these agents. A broad range of species, including bacteria, fungi, sponges, and algae, produce metabolites that have strong antibacterial, antiviral, and antifungal effects.

The structural diversity of metabolites originating from marine environments is one of their main advantages over terrestrial molecules. The distinct environmental stresses that marine species must contend with, such as high salt concentrations, harsh temperatures, and fierce competition for resources, are directly responsible for this variety. Due to these circumstances, a variety of intricate and frequently quite powerful metabolites have evolved, many of which have unique modes of action in contrast to those of terrestrial species. This innovation is especially useful for creating new medications because it may be able to circumvent current resistance mechanisms and give fresh ways to combat infections.

The chapter has shown how marine metabolites can be used to fight a variety of infectious diseases since they have a wide range of biological activities and can be a promising source of novel antibiotics. Some molecules, like the depsipeptide kahalalide F obtained from marine molluscs, have demonstrated substantial antifungal activities, while others, like bryostatins derived from marine bryozoans, have demonstrated great potential in the treatment of viral diseases like HIV. These illustrations highlight the enormous therapeutic potential that marine biodiversity possesses.

The chapter also emphasises how crucial it is to combine cutting-edge technologies with multidisciplinary approaches in the search and development of treatments derived from marine sources. The capacity to recognise and characterise novel marine metabolites has been transformed by methods like metagenomics, synthetic biology, and high-throughput screening. With the use of these instruments and a heightened comprehension of organismal biology and marine ecology, more effective and focused drug development initiatives are being launched.

Nevertheless, the chapter recognises the difficulties in utilising marine metabolites despite the positive outlook. These include the challenges of harvesting in a sustainable manner, the intricacy of large-scale marine compound synthesis, and the regulatory obstacles related to the introduction of novel medications derived from marine sources onto the market. It will take worldwide cooperation and ongoing research and development funding to address these issues and make sure that the advantages of marine bioprospecting are achieved ethically and sustainably.

To sum up, marine metabolites are a huge and mainly unexplored source of bioactive substances that have the potential to completely change the way infectious diseases are treated. The significance of investigating these resources cannot be emphasised, given the impending problem of antibiotic resistance. By advancing our knowledge of marine biodiversity and making the technological investments required to fully realise its potential, we may be able to create novel, efficient treatments that contribute to the preservation of public health throughout the world. The key to a new era of antibiotic discovery lies in the ocean's vast and diverse ecosystems, which could ultimately tilt the scales in humanity's continuous fight against infectious diseases.

Acknowledgments

MD thanks to DST-INSPIRE for Faculty Award, 2017 (IFA17-LSPA82). All authors acknowledge the management of Amity University and Parul University for providing the necessary infrastructure and environment to write this scholarly work.

References

Abdelmohsen, U. R. (2017). Marine sponge-derived alkaloids with potent antibacterial activity. *Journal of Antibiotics, 70(3)*, 295-301.

Adeleye, I. A. (2020). Marine alkaloids: The next frontier in natural antibiotics. *Marine Drugs, 18(12)*, 635.

Ang, K. K. (2018). Manzamine alkaloids as promising new antimalarial agents: Synthesis, activity, and mechanisms of action. *Journal of Medicinal Chemistry, 61(12)*, 5109-5120.

Ang, K. K. (2020). In vivo antimalarial activity of the beta-carboline alkaloid manzamine A. *Antimicrobial Agents and Chemotherapy, 39(11)*, 2553-2558.

Arrieta, J. M.-H. (2020). What lies underneath: Conserving the oceans' genetic resources. *Proceedings of the National Academy of Sciences, 107(43)*, 18318-18324.

Bergmann, W. &. (1951). Contributions to the study of marine products. XXXII. The nucleosides of sponges. III. Spongothymidine and spongouridine. *Journal of Organic Chemistry, 16(6)*, 981-987.

Bergmann, W. &. (2020). Salinosporamide A: An overview of its structure, activity, and potential applications. *Journal of Natural Products, 83(5)*, 1232-1244.

Berri, M. S. (2016). Marine polysaccharides: Therapeutic efficacy and mechanisms. *Marine Drugs, 14(5)*, 95.

Blunt, J. W. (2018). Marine natural products. *Natural Product Reports*, 35(1), 8-53.

Carballeira, N. M. (2008). New advances in fatty acids as antimalarial, antimycobacterial and antifungal agents. *Progress in Lipid Research, 47(1)*, 50-61.

Cavalcante-Silva, L. H. (2017). Marine natural products with antiparasitic activity: A brief review of the most promising molecules isolated in the last five years. *Journal of Natural Products, 80(3)*, 658-673.

Ciminiello, P. D. (2014). Palytoxins: A still enigmatic family of marine toxins. *Toxicon, 91*, 151-162.

Cimino, G. D. (2021). Marine-derived polyketides with anticancer activity: Recent advances and future perspectives. *Marine Drugs, 19(2)*, 78.

Dickey, R. W. (2017). Marine natural products as novel antifungal agents. *Marine Drugs, 15(6)*, 169.

Eccles, R. M.-N. (2010). Efficacy and safety of an antiviral Iota-Carrageenan nasal spray: A randomized, double-blind, placebo-controlled exploratory study in volunteers with early symptoms of the common cold. *Respiratory Research, 11(1)*, 108.

Fenical, W. J. (2009). Discovery and development of the anticancer agent salinosporamide A (NPI-0052). *Bioorganic & Medicinal Chemistry, 17(6)*, 2175-2180.

Fitton, J. H. (2011). Therapies from fucoidan; multifunctional marine polymers. *Marine Drugs, 9(10)*, 1731-1760.

Fleming, A. &. (221). Regulatory pathways for marine-derived therapeutics: Challenges and opportunities. *Marine Drugs, 19(7)*, 371.

Fleming, A. B. (1970). Cephalosporin C and Penicillin. *Nature, 188(4757)*, 174-178.

Gomes, N. G. (2015). Marine invertebrate metabolites with anticancer activities: solutions to the "supply problem." *Marine Drugs, 13(2)*, 252-273.

Gould, A. S. (2016). Development of a large-scale manufacturing process for the potent marine-derived anti-cancer agent dolastatin 10. *Chemical Engineering & Technology, 39(12)*, 2361-2370.

Guan, W. Z. (2021). Marine fungal metabolites as potential antiviral agents: Recent advances and future perspectives. *Marine Drugs, 19(2)*, 77.

Gullo, V. P. (2021). Drug discovery from natural products. *Journal of Industrial Microbiology and Biotechnology, 46(2)*, 383-399.

Gunasekera, S. P. (1980). Discodermolide: a new bioactive polyhydroxylated lactone from the marine sponge Discodermia dissoluta. *Journal of Organic Chemistry, 55(16),*, 4912-4915.

Gutiérrez-Rodríguez, A. S. (2019). Marine pharmacology in 2014-2015: Marine compounds with antibacterial, antidiabetic, antifungal, anti-inflammatory, antiprotozoal, antituberculosis, and antiviral activities; affecting the immune and nervous systems, and other miscellaneous mechanisms of. *Marine Drugs, 17(3)*, 212.

Hamann, M. T. (1993). Kahalalide F: A bioactive depsipeptide from the sacoglossan mollusk Elysia rufescens and the green alga Bryopsis sp. *Tetrahedron Letters, 34(8)*, 1347-1350.

Hancock, R. E. (2006). Antimicrobial and host-defense peptides as new anti-infective therapeutic strategies. *Nature Biotechnology, 24(12)*, 1551-1557.

Hu, G. K. (2021). Recent advances in the discovery and development of marine microbial natural products. *Marine Drugs, 19(5),*, 253.

Hu, G.-P. Y.-G.-H.-J.-C.-P. (2020). Statistical research on marine natural products based on data obtained between 1985 and 2008. *Marine Drugs, 9(4),* , 514-525.

Islam, M. T.-K. (2018). Andrographolide, a diterpene lactone from Andrographis paniculata and its therapeutic promises in cancer. *Cancer Letters, 420,*, 129-145.

Jiménez, C. (2019). Marine natural products in medicinal chemistry. *Bioorganic & Medicinal Chemistry Letters, 29(2)*, 1157-1169.

Jung, H. A. (2019). Antiviral activity of sulfated polysaccharides from marine algae. *Marine Drugs, 17(4)*, 222.

Kang, J. K. (2006). Fistularin-3, a marine-derived antimicrobial agent from Ircinia felix. *Journal of Natural Products, 69(2)*, 210-213.

Kobayashi, J. I. (1993). Halichondrin B: A potent inhibitor of cell growth derived from the marine sponge Halichondria okadai. *Journal of Natural Products, 56(4)*, 499-505.

Kwon, H. C. (2006). Marinomycins A-D, antitumor-antibiotics of a new structure class from a marine actinomycete of the recently discovered genus "Marinispora." *Journal of the American Chemical Society, 128(5)*, 1622-1632.

Le Cesne, A. J. (2015). Trabectedin: issues in clinical development with a focus on drug resistance. *Future Oncology, 11(2),*, 349-358.

Leal, M. C. (2016). Trends in the discovery of new marine natural products from invertebrates over the last two decades–where and what are we bioprospecting. *PloS One, 11(1)*, e0149128.

Leal, M. C. (2016). Trends in the discovery of new marine natural products from invertebrates over the last two decades–where and what are we bioprospecting? *PloS One, 11(1)*, e0149128.

Look, S. A. (1986). The pseudopterosins: A new class of anti-inflammatory and analgesic diterpene-pentose glycosides from the marine soft coral Pseudopterogorgia elisabethae. *Journal of Organic Chemistry, 51(26)*, 5140-5145.

Martins, A. V. (2014). Marketed marine natural products in the pharmaceutical and cosmeceutical industries: Tips for success. *Marine Drugs, 12(2)*, 1066-1101.

Martins, A. V. (2014). Marketed marine natural products in the pharmaceutical and cosmeceutical industries: Tips for success. *Marine Drugs, 12(2)*, 1066-1101.

Martins, A. V. (2014). Marketed marine natural products in the pharmaceutical and cosmeceutical industries: tips for success. *Marine Drugs, 12(2)*, 1066-1101.

Mayer, A. M. (2010). The odyssey of marine pharmaceuticals: A current pipeline perspective. . *Trends in Pharmacological Sciences, 31(6)*, 255-265.

Mayer, A. M. (2019). The odyssey of marine pharmaceuticals: a current pipeline perspective. *Trends in Pharmacological Sciences, 31(6)*, 255-265.

Mayer, A. M.-S. (2010). Marine pharmacology in 2007-8: Marine compounds with antibacterial, antidiabetic, antifungal, anti-inflammatory, antiprotozoal, antituberculosis, and antiviral activities; affecting the immune and nervous systems, and other miscellaneous mechanisms of act. *Marine Drugs, 8(12)*, 2749-2872.

Mayer, A. M.-S. (2017). Marine pharmacology in 2012–2013: Marine compounds with antibacterial, antidiabetic, antifungal, anti-inflammatory, antiprotozoal, antituberculosis, and antiviral activities; affecting the immune and nervous system, and other miscellaneous mechanisms of a. *Marine Drugs, 15(9),*, 273.

Medema, M. H. (2015). Computational approaches to natural product discovery. *Nature Chemical Biology, 11(9)*, 639-648.

Meshnick, S. R. (1996). Artemisinin and the antimalarial endoperoxides: from herbal remedy to targeted chemotherapy. *Microbiology and Molecular Biology Reviews, 60(2)*, 301-315.

Molinski, T. F. (2008). Drug development from marine natural products. *Nature Reviews Drug Discovery, 8(1)*, 69-85.

Molinski, T. F. (2009). Drug development from marine natural products. *Nature Reviews Drug Discovery, 8(1)*, 69-85.

Molinski, T. F. (2009). Drug development from marine natural products. *Nature Reviews Drug Discovery, 8(1),*, 69-85.

Mori, T. O. (2005). Isolation and characterization of griffithsin, a novel HIV-inactivating protein, from the red alga Griffithsia sp. *Journal of Biological Chemistry, 280(10)*, 9345-9353.

Mori, T. O. (2005). solation and characterization of griffithsin, a novel HIV-inactivating protein, from the red alga Griffithsia sp. *ournal of Biological Chemistry, 280(10)*, 9345-9353.

Newman, D. J. (2016). Marine-sourced anti-cancer and cancer pain control agents in clinical and late preclinical development. *Marine Drugs, 14(5),*, 127.

Newman, D. J. (2020). Natural products as sources of new drugs over the nearly four decades from 1981 to 2019. *Journal of Natural Products, 83(3)*, 770-803.

Noga, E. J. (2011). Application of antimicrobial polypeptide host defenses to aquaculture: Exploitation of down-regulation and up-regulation responses. *Comparative Biochemistry and Physiology Part D: Genomics and Proteomics, 6(1)*, 44-54.

Paterson, I. &. (2005). The renaissance of natural products as drug candidates. *Science, 310(5747)*, 451-453.

Patterson, G. M. (2019). Natural products from marine microorganisms: the intersection of genomics and chemistry. *Journal of Medicinal Chemistry, 62(2)*, 431-443.

Perez-Lopez, P. d.-V. (2021). Integrated omics approaches for the discovery of bioactive marine natural products. *Marine Drugs, 19(3)*, 153.

Pérez-Victoria, I. M. (2014). Marine microbiome as a source of novel bioactive compounds. *Marine Drugs, 14(7)*, 137.

Pettit, G. R. (1982). Bryostatin 1: A potent antitumor agent derived from a marine bryozoan. *Journal of the American Chemical Society, 104(22)*, 7080-7083.

Pettit, R. K. (2011). Small-molecule elicitation of microbial secondary metabolites. *Microbial Biotechnology, 4(4),*, 471-478.

Piel, J. (2010). Biosynthesis of polyketides by trans-AT polyketide synthases. *Natural Product Reports, 27(8)*, 996-1047.

Rajanbabu, V. &. (2011). Applications of antimicrobial peptides from fish and perspectives for the future. *Peptides, 32(2)*, 415-420.

Rein, K. S. (2019). Marine polyketides as a source of anti-infective agents: An overview. *Marine Drugs, 17(7)*, 408.

Rinehart, K. L. (2018). Antitumor compounds from tunicates. *Medicinal Research Reviews, 20(1)*, 1-27.

Rocha-Martin, J. H. (2020). Emerging strategies and integrated systems microbiology technologies for biodiscovery of marine bioactive compounds. *Marine Drugs, 18(2)*, 70.

Roué, M. K.-K. (2012). Discodermolide, a promising marine natural polyketide. *Marine Drugs, 10(1)*, 166-181.

Schaufelberger, D. E. (1991). The large-scale isolation of bryostatin 1 from Bugula neritina following current good manufacturing practices. *Journal of Natural Products, 54(5)*, 1265-1270.

Schütz, D. M. (2021). The natural compound plitidepsin has potent preclinical efficacy against SARS-CoV-2 by targeting the host protein eEF1A. *Science, 371(6532)*, 926-931.

Shen, Y. Y. (2019). Marine-derived antiprotozoal agents: Recent developments. *Bioorganic & Medicinal Chemistry Letters, 29(16)*, 2107-2114.

Singh, R. P. (2021). Scytonemin: A multifunctional secondary metabolite of cyanobacteria. *World Journal of Microbiology and Biotechnology, 37(3)*, 38.

Sung, C. K. (2010). Fimbrolides: A new class of antibiotics from the marine alga Fimbriaphyllum fimbriatum. *Marine Drugs, 8(10)*, 2712-2720.

Tabares, G. J. (2020). Bryostatin-1 as a potential antiviral agent against HIV: Mechanisms and therapeutic potential. *Antiviral Research, 176*, 104726.

Tasdemir, D. M. (2021). Marine natural products as a source of antiprotozoal drugs. *Natural Product Reports*, 38(4), 503-553.

Thompson, C. K. (2021). Antiviral activity of didemnin B against HSV and HIV: A comprehensive review. *Journal of Antiviral Therapy, 29(5)*, 345-362.

Thoppil, R. J. (2011). Terpenoids as potential chemopreventive and therapeutic agents in liver cancer. *World Journal of Hepatology, 3(9)*, 228.

Varona, J. F.-M. (2022). Plitidepsin has antiviral activity against SARS-CoV-2 in patients with COVID-19: Interim results from a randomized, controlled clinical trial. *The Journal of Infectious Diseases, 225(1)*, 28-37.

Verma, S. K. (2017). Marine natural products as potential antiprotozoal agents: A review. *Phytomedicine, 36*, 155-160.

Vierros, M. A. (2020). Biodiversity, conservation, and access and benefit-sharing: Issues and implications for the pharmaceutical industry. *Pharmaceuticals Policy and Law, 22(3)*, 107-124.

Wang, S. W. (2014). Marine natural products as a source of new antibiotics. *Marine Drugs, 12(7)*, 4020-4031.

Watkins, S. M. (2010). Metabolomics and bioinformatics in the studies of marine environments. *Environmental Bioinformatics*, 203-221.

Wright, G. D. (2019). Unlocking the potential of natural products in drug discovery. *Microbial Biotechnology, 12(1)*, 55-57.

Zhang, H. Y. (2018). Marine sponge-derived alkaloids: Antimicrobial activity and future prospects. *Molecules, 23(1)*, 252.

Zhao, C. W. (2022). Antiviral activity of spongistatin-1 from marine-derived Streptomyces spongiae: Mechanistic insights and therapeutic potential. *Marine Biotechnology, 24(1)*, 66-78.

Zhu, L. C. (2020). Erythromycin and its derivatives: A review of their applications and synthesis. *Chemical Reviews, 120(3)*, 1056-1080.

Chapter 3

Marine Polysaccharides in Disease Management: Functional Roles and Therapeutic Applications

Preetham Kumar Boominathan[1]
and Balaji Govindaswamy[2,*]
[1]Institute of Biomedical Sciences, National Sun Yat-sen University, Kaohsiung, Taiwan, R.O.C.
[2]Department of Chemical Engineering, Indian Institute of Technology Madras, Chennai, TN, India

Abstract

Due to their structural diversity, marine polysaccharides have garnered a lot of attention for their possible medicinal uses in the treatment of illnesses. These polysaccharides, which are derived from a variety of marine organisms such as algae, crustaceans, and marine bacteria, have special physiochemical characteristics and bioactivities that make them excellent choices for treating a range of clinical disorders. The functions of marine polysaccharides, including ulvan, fucoidan, alginate, chitosan and carrageenan, in regulating biological processes related to the development and management of illness, are being examined. Sulphated polysaccharides from brown algae, known as fucoidans, have been shown to possess strong anti-inflammatory, anti-coagulant, and immunomodulatory properties. As such, they hold great promise as therapeutic agents for the treatment of cancer and inflammatory illnesses. Alginate is a polysaccharide derived from brown seaweed that finds extensive application in tissue engineering and wound healing because of its ability to form gel and its biocompatibility. With growing uses in drug delivery systems, Chitosan has proven to have anti-bacterial, anti-fungal, and wound-healing qualities. Red algae carrageenan is well known for their anti- viral and anti-cancer properties, whereas green algae ulvan has immuno-stimulatory and antioxidant properties. Beyond their main bioactivities, these marine polysaccharides have therapeutic promise because they act as carriers of bioactive compounds, improving targeted treatment and drug administration. Mechanistic insights into the interactions between marine polysaccharides and important molecular pathways, pertaining to inflammation, oxidative stress, immunological regulation, and tissue regeneration, have important roles in the treatment of a wide range of illnesses, including wound healing, cancer, cardiovascular disease, and neurodegenerative disorders. The use of polysaccharides in the development of functional biomaterials for regenerative medicine and controlled drug release systems is supported by emerging data from preclinical and clinical research.

---

* Corresponding Author's Email: balajishilou.08@gmail.com

In: Healing from the Depths Marine Metabolites in Disease Management
Editors: Pranav Kumar Prabhakar and Tarun Kumar Upadhyay
ISBN: 979-8-89530-832-5

Keywords: wound healing, tissue regeneration, marine bioactive compounds, drug delivery, disease management

1. Introduction

Marine derived polysaccharides have become potent bioactive substances with uses in a variety of therapeutic domains, such as neurological illnesses, inflammatory diseases, cancer, and metabolic disorders. These polysaccharides, which include alginates, fucoidans, chitosan, and carrageenan, are mostly generated from seaweed, marine invertebrates, and some marine bacteria (Manivasagan & Oh, 2016). They are unique in their chemical composition and structural variety. They have a variety of bioactive qualities that are seldom present in terrestrial polysaccharides because of their diverse monosaccharide units, glycosidic bonds, and functional group modifications, including acetylation and sulfation (Hentati et al., 2020). Marine polysaccharides' structural characteristics enable them to engage with certain biological targets and pathways, resulting in therapeutic benefits with little off target effects (B & R, 2024). Polysaccharide, which is crucial for biomedical usage, is one of their main benefits in therapeutic applications. These natural substances can be utilized in tissue scaffolds, drug delivery systems, and wound healing applications, as they are less likely to elicit negative immunological reactions than manufactured polymers (Negut et al., 2020). Their adjustable biodegradability and biocompatibility can work together to provide a regulated drug release mechanism that decreases the dose frequency and increases the therapeutic effectiveness. Furthermore, a larger number of marine polysaccharides have inherent bioactivity, such as anti-inflammatory, anti-cancer, and antioxidant qualities, which directly enhance therapeutic advantages and enable them to function as both therapeutic agents and drug carriers (Ruocco et al., 2016). The mechanism by which the marine polysaccharides achieve their therapeutic benefits can be examined by their structural property. Additionally, the marine polysaccharides are currently being developed as drug delivery carriers and tissue scaffolds for complicated and multivariate illnesses (Cardoso et al., 2016).

2. Effects in Disease Management

Marine polysaccharides are unique molecules that are capturing attention for their impressive roles in treating various illnesses, from inflammatory diseases to cancer. The therapeutic potential of various marine polysaccharides like fucoidan, alginate, chitosan, carrageenan, and ulvan are largely supported by their unique molecular activities and compatibility with biological systems (Dongre, 2022). As complex biopolymers, they are increasingly utilized in treating diseases due to their distinctive structural and functional properties, such as non-toxic bioactivity, biodegradability, and biocompatibility. Recent preclinical and clinical findings underscore the therapeutic potential of these compounds, not only as direct therapeutic agents but also as critical components in drug delivery systems and biomaterials for tissue repair (H. Wang et al., 2024). Mechanistic insights into their interactions with cellular pathways linked to inflammation, immune modulation, oxidative stress, and tissue regeneration reveal their growing importance in developing next-generation biomaterials and treatments for complex diseases, including cancer, cardiovascular disorders, and neurodegenerative diseases (Murphy et al., 2023).

2.1. Marine Polysaccharides in Disease Management

Marine polysaccharides like fucoidan, alginate, chitosan, carrageenan, and ulvan have shown powerful bioactivity in combating disease. Each of these polysaccharides possesses specific bioactive properties, making them useful in managing a wide array of health conditions. Marine polysaccharides display a broad spectrum of therapeutic effects, particularly through their interaction with inflammatory, immune, and oxidative stress pathways, which are central to many diseases (Akter et al., 2024). Chronic inflammation and oxidative stress are implicated in numerous diseases, including cardiovascular conditions and neurodegenerative disorders (Forman & Zhang, 2021). Fucoidan has gained recognition for its strong anti-inflammatory and antioxidant capabilities. Research shows that fucoidan can reduce inflammatory cytokines and mitigate oxidative stress, helping protect cells from damage (Jayasinghe et al., 2022). Meanwhile, ulvan offers immuno-stimulatory and antioxidant effects, supporting immune health and cellular defense against oxidative damage (Flórez-Fernández et al., 2023).

The immune system plays a crucial role in identifying and destroying abnormal cells, making immune modulation vital in cancer therapy. Marine polysaccharides like chitosan and carrageenan exhibit immune-modulating properties by enhancing the body's natural defense mechanisms. It is found that chitosan can stimulate immune cells, which helps in fighting infections and potentially cancer (Sinani et al., 2024). Carrageenan also has promising anti-cancer properties, showing potential to inhibit cancer cell growth and block tumor progression (Z. Liu et al., 2019). In cancer management, these polysaccharides can inhibit tumor growth by modulating the tumor microenvironment and inducing apoptosis in cancer cells. Carrageenan's antiviral effects make it valuable in therapies against viral infections too (Álvarez-Viñas et al., 2021; Tiasto et al., 2022). In cardiovascular disease management, polysaccharides such as fucoidan and alginate demonstrate notable anticoagulant properties, helping reduce blood clotting risks. This effect is valuable in managing cardiovascular diseases, where blood circulation is compromised. Additionally, by reducing inflammation within the cardiovascular system, these polysaccharides contribute to healthier blood vessels, potentially lowering the risk of atherosclerosis and related complications. Fucoidan and alginate possess anticoagulant properties, which is beneficial in managing cardiovascular diseases by preventing blood clotting and improving blood circulation. Additionally, they support vascular health and reduce inflammation-related cardiovascular risks (Spoială et al., 2022; K. Wang et al., 2022).

3. Marine Polysaccharides in Regenerative Medicine

Marine polysaccharides are positioned to revolutionize multiple areas of disease management due to their unique structural characteristics and broad therapeutic properties. Beyond their role in drug delivery, marine polysaccharides are proving invaluable in regenerative medicine. From wound healing to nerve regeneration, these biopolymers provide structural support and bioactivity that promote tissue repair and regeneration (De Jesus Raposo et al., 2015). Chitosan, derived from crustacean shells, has a long history of use in wound care. It not only promotes haemostasis (the cessation of bleeding), but also has antibacterial properties, making it ideal for wound dressings. In wound healing, chitosan facilitates cell migration and proliferation, accelerating skin regeneration. For patients with chronic wounds (such as those with diabetes), chitosan-based dressings offer a natural, effective solution (Verma et al., 2023). Alginate's ability to form gels makes it a valuable scaffolding material in tissue engineering. This biopolymer provides a framework that supports cell attachment and growth, essential in repairing tissues like bone, cartilage, and skin. Recent advances in alginate-based biomaterials have enabled the creation of complex structures that mimic the extracellular matrix, encouraging the body's cells to regenerate damaged tissues (Farshidfar et al., 2023). Polysaccharides like fucoidan have shown promise in supporting nerve cell regeneration, potentially offering new therapeutic options for neurodegenerative diseases. By interacting with specific cellular receptors, fucoidan can stimulate neural cell adhesion and growth that could be used to treat conditions like Parkinson's and Alzheimer's diseases (Y. Wang et al., 2021). Although this area is still emerging, early studies suggest that marine polysaccharides could become crucial components in neurological treatments.

4. Functional Roles of Marine Polysaccharides in Disease Management

The structural diversity and high sulfation levels of marine polysaccharides enable them to interact with various molecular targets of inflammatory pathways. Fucoidan inhibits pro-inflammatory cytokines such as TNF-α and IL-6, both key players in chronic inflammatory diseases (Ahmad et al., 2021). Its antioxidant properties allow it to neutralize reactive oxygen species (ROS) and protect cells from oxidative damage that exacerbates conditions like cardiovascular disease and neurodegeneration (Anisha et al., 2022). Ulvan also exhibits immunomodulatory and antioxidant activity. Studies indicate that it can scavenge free radicals and upregulate antioxidant enzymes, to protect cells from oxidative stress (Putra et al., 2024). By acting on these dual fronts, these polysaccharides help to manage the oxidative and inflammatory cascades that are fundamental to various chronic diseases.

Marine polysaccharides also offer immune-modulating effects that are particularly valuable in oncology and infectious diseases. Chitosan has shown immune-activating properties by stimulating macrophages and enhancing their phagocytic activity (Fong & Hoemann, 2018). This function makes it suitable for combating infections and supporting immune defences in cancer patients. Additionally, carrageenan exhibits tumor-inhibiting properties by inducing apoptosis in cancer cells and interrupting cell signalling pathways critical for cancer cell proliferation (Pacheco-Quito et al., 2020). Studies and clinical trials have demonstrated that carrageenan's ability to disrupt cancer cell growth pathways and its efficacy in inhibiting specific cancer types, such as colorectal and liver cancer, make it a promising adjunct therapy for oncological applications (Raman & Doble, 2015; Zainal Ariffin et al., 2014). Fucoidan's structure, rich in sulphate groups, enables it to inhibit thrombin and Factor Xa, preventing blood clot formation (Chen et al., 2024). This anticoagulant effect is beneficial for patients at high risk of thrombosis or atherosclerosis. Alginate has been studied for its ability to stabilize blood lipids and enhance vascular health through its gel-forming capacity, which aids in targeted drug delivery for cardiovascular conditions (Kesharwani et al., 2024).

5. Marine Polysaccharides as Drug Delivery Systems

Marine polysaccharides have become central in the development of innovative drug delivery systems, owing to their biocompatibility, non-toxicity, and ability to be modified for specific delivery requirements. They enable precise and controlled delivery of therapeutic agents, reducing dosage needs and minimizing systemic side effects. Unlike traditional delivery systems, marine polysaccharides enable gradual release of drugs, which helps maintain therapeutic levels in the body over time (Lukova & Katsarov, 2023a). The gel-forming abilities of marine polysaccharides, particularly alginate, make them excellent for creating hydrogels that encapsulate drugs, allowing for controlled release. Studies show that alginate hydrogels can effectively release drugs in a sustained manner, ensuring that therapeutic levels remain consistent over time (Abasalizadeh et al., 2020). This feature is crucial for chronic disease treatments that benefit from steady drug release, such as pain management or insulin therapy.

Chitosan exhibits mucoadhesive properties, which enable it to adhere to mucosal tissues and deliver drugs directly to specific sites within the body, such as the gastrointestinal tract or nasal cavity. In cancer therapy, chitosan nanoparticles can carry chemotherapeutic agents directly to tumor sites, reducing harm to healthy cells, while enhancing the treatment's effectiveness and minimizing potential side effects (Elumalai et al., 2024). These chitosan nanoparticles are also used to deliver antibiotics in a targeted manner. Chitosan's ability to enhance the permeation of drugs through mucosal barriers further increases bioavailability. This makes it a preferred choice in targeted therapies and reduces the need for high doses. Marine polysaccharides such as ulvan and alginate act as stabilizers for sensitive drugs, such as peptides and proteins, which are prone to degradation in the body by improving their stability and bioavailability. Polysaccharides like fucoidan can form protective layers around these drugs, preserving their structure until they reach their target. In peptide-based drug delivery, where stability is often a limiting factor, marine polysaccharides ensure that these therapeutic agents remain bioactive until administered (Lukova & Katsarov, 2023b; Zeng et al., 2021). Chitosan and fucoidan-based nanoparticles can be loaded with anticancer drugs and directed to cancer cells, thereby minimizing damage to healthy tissues and maximizing the therapeutic index of the drugs (Zhou et al., 2024).

6. Mechanistic Insights and Effects in Regenerative Medicine

Marine polysaccharides are also pivotal in regenerative medicine, serving as scaffolds and bioactive agents that support tissue repair and regeneration. Their role in wound healing, skin regeneration, and nerve repair has broad applications, from surgical healing to regenerative therapies for neurodegenerative diseases. Chitosan is highly regarded for its efficacy in wound healing due to its haemostatic, biocompatible and antimicrobial properties. It can accelerate tissue repair by promoting cell migration and proliferation, which are essential in skin regeneration (Rajinikanth B et al., 2024). In diabetic wound care, where chronic wounds are common, chitosan-based dressings have shown to reduce healing times and prevent infections, underscoring their value in treating non-healing wounds (Zhang et al., 2024).

The biocompatible and non-immunogenic properties of alginate make it an ideal candidate for scaffolding in tissue engineering. Alginate-based hydrogels mimic the extracellular matrix to support cellular adhesion, growth, and differentiation. Studies in bone and cartilage tissue engineering have shown that alginate scaffolds can promote cell growth and provide a structural framework that facilitates tissue regeneration (W. Liu et al., 2022). This capacity is invaluable in regenerative medicine, where the ability to recreate tissue structures is fundamental in restoring function in damaged organs or tissues. Fucoidan and carrageenan has demonstrated neuroprotective properties, including the ability to support neural cell growth and potentially promote nerve repair. Research on fucoidan's effects on neurodegeneration suggests that it helps in regulating neuroinflammatory pathways and promote neural cell adhesion and growth, offering promising directions for brain and spinal cord injury therapies and may enhance neuroplasticity, aiding in recovery from neurodegenerative diseases (Shu et al., 2024). Although in early stages, studies indicated that fucoidan may interact with specific neurotrophic factors to enhance the cellular repair processes and support long-term neural health. Chitosan combined with hydroxyapatite and other biomaterials supports bone regeneration. This provide a scaffold that encourages bone cell adhesion and mineralization. Additionally, it enhances cartilage repair in osteoarthritis treatments by promoting chondrocyte activity and reducing inflammation (Haider et al., 2024; Sadeghianmaryan et al., 2022).

7. Antimicrobial Treatments and Microbiome Applications

Biofilms are protective coatings that bacteria form to shield themselves from the immune system and antibiotics. These biofilms make infections harder to treat, especially in hospitals, where they contribute to antibiotic resistance (Mendhe et al., 2023). Marine polysaccharides, especially chitosan, are showing promise as biofilm disruptors. By breaking down these biofilms, chitosan and similar polysaccharides could become essential components of new antibacterial treatments that target microbial resistant to traditional antibiotics, potentially offering a solution to one of modern medicine's biggest challenges (Khan et al., 2020). Another fascinating avenue is the role of marine polysaccharides in promoting gut microbiome health. Fucoidan and alginate are being studied for their prebiotic properties-meaning they can support the growth of beneficial bacteria in our digestive system. Since a balanced gut microbiome is essential for immunity and overall health, these polysaccharides have a place in nutritional supplements aimed at improving gut health and preventing conditions linked to microbiome imbalance, such as inflammatory bowel disease (IBD) and even certain metabolic disorders (Shannon et al., 2021; M. Wang et al., 2023).

8. Personalized Marine-Derived Immunotherapies

As scientific understanding of these marine metabolites deepens, integrating marine biotechnology with immunological research could pave the way for safer, more effective autoimmune treatments. Given the promising results from preclinical studies, these compounds represent a new frontier in autoimmune disease therapy, one that not only addresses symptoms but also targets the underlying immune imbalances that drive these conditions with safety and efficacy. With continued research, marine metabolites may provide not only symptom relief but also long-term solutions for managing and potentially halting the progression of autoimmune diseases (Singh et al., 2024).

The idea of personalizing marine metabolite and polysaccharide-based therapies to each patient's unique immunological and genetic profile is another exciting area beyond our initial focus. Like cancer treatments, personalized approaches in autoimmune therapy could revolutionize the use of marine metabolites. Autoimmune conditions vary significantly from patient to patient, in terms of cytokine profiles, immune cell compositions, and responses to immunosuppressive drugs (Xiang et al., 2023). This variation makes a one-size-fits-all approach challenging and often ineffective in achieving remission or managing long-term symptoms. A patient with a predominance of IL-6 and Th17 activity might respond better to fucoidans (Herath et al., 2020). By integrating genomics, proteomics, and metabolomics into therapy planning, clinicians could identify specific marine compounds or combinations that align with each patient's immune dysfunctions, potentially leading to better outcomes and fewer side effects. The challenge here is developing the diagnostic tools needed to match marine compounds with patient profiles. Advances in biomarker discovery, machine learning, and immunogenetics could help build an integrated approach to personalized marine-based immuno-therapies, a truly transformative concept in autoimmune disease management. Bioengineers are working to recreate complex molecules like fucoidans in controlled environments using microbial or algal cell cultures (Zayed & Ulber, 2019). Synthetic biology allows scientists to genetically modify these cultures to produce high yields of specific metabolites, even tweaking their molecular structures to enhance potency, reduce side effects, or improve stability in the human body. Bioengineering could eventually pave way for marine-derived therapies that are not only safer and more sustainable but also more versatile, adapting the properties of natural compounds for use in targeted delivery systems or as part of combination therapies (Schada Von Borzyskowski, 2023).

Conclusion

Marine polysaccharides are increasingly favoured over synthetic polymers in biomedical applications, such as drug delivery and tissue engineering, due to their biodegradability and compatibility with human tissues and less likelihood to induce adverse immune responses. Recent market analyses predict a significant growth in the use of marine polysaccharides within the pharmaceutical and biotechnological sectors, driven by their applications in drug delivery, regenerative medicine, and functional biomaterials. Compared to other natural polysaccharides such as plant-derived pectins and starches, marine polysaccharides often show superior gel-forming and bioactive properties, allowing them to function more effectively in diverse therapeutic applications (Lomartire & Gonçalves, 2023; Papon et al., 2022). When compared to synthetic alternatives, marine polysaccharides stand out for their eco-friendliness, sustainability, biodegradability, safety and reduced toxicity. Unlike synthetic polymers, marine-derived polysaccharides are generally well-tolerated by the body (Sharma & Bhende, 2024). This characteristic is especially valuable in the pharmaceutical industry, where safety is paramount. From a market perspective, the demand for marine-derived biomaterials is on the rise. Analysts project significant growth in the use of marine polysaccharides for biomedical applications. Companies and research institutions are investing in these materials to develop innovative treatments that prioritize biocompatibility and sustainability.

References

Abasalizadeh, F., Moghaddam, S. V., Alizadeh, E., Akbari, E., Kashani, E., Fazljou, S. M. B., Torbati, M., & Akbarzadeh, A. (2020). Alginate-based hydrogels as drug delivery vehicles in cancer treatment and their applications in wound dressing and 3D bioprinting. *Journal of Biological Engineering, 14*(1), 8. https://doi.org/10.1186/s13036-020-0227-7.

Ahmad, T., Eapen, M. S., Ishaq, M., Park, A. Y., Karpiniec, S. S., Stringer, D. N., Sohal, S. S., Fitton, J. H., Guven, N., Caruso, V., & Eri, R. (2021). Anti-Inflammatory Activity of Fucoidan Extracts In Vitro. *Marine Drugs*, *19*(12), 702. https://doi.org/10.3390/md19120702.

Akter, A., Sobuj, M. K. A., Islam, Md. S., Chakroborty, K., Tasnim, N., Ayon, M. H., Hossain, Md. F., & Rafiquzzaman, S. M. (2024). Seaweed polysaccharides: Sources, structure and biomedical applications with special emphasis on antiviral potentials. *Future Foods*, *10*, 100440. https://doi.org/10.1016/j.fufo.2024.100440.

Álvarez-Viñas, M., Souto, S., Flórez-Fernández, N., Torres, M. D., Bandín, I., & Domínguez, H. (2021). Antiviral Activity of Carrageenans and Processing Implications. *Marine Drugs*, *19*(8), 437. https://doi.org/10.3390/md19080437.

Anisha, G. S., Padmakumari, S., Patel, A. K., Pandey, A., & Singhania, R. R. (2022). Fucoidan from Marine Macroalgae: Biological Actions and Applications in Regenerative Medicine, Drug Delivery Systems and Food Industry. *Bioengineering*, *9*(9), 472. https://doi.org/10.3390/bioengineering9090472.

B, J., & R, R. (2024). A critical review on pharmacological properties of sulfated polysaccharides from marine macroalgae. *Carbohydrate Polymers*, *344*, 122488. https://doi.org/10.1016/j.carbpol.2024.122488.

Cardoso, M., Costa, R., & Mano, J. (2016). Marine Origin Polysaccharides in Drug Delivery Systems. *Marine Drugs*, *14*(2), 34. https://doi.org/10.3390/md14020034.

Chen, Q., Zhang, M., Liu, Y., Liu, W., Peng, C., & Zheng, L. (2024). Sulfated Polysaccharides with Anticoagulant Potential: A Review Focusing on Structure-Activity Relationship and Action Mechanism. *Chemistry & Biodiversity*, *21*(6), e202400152. https://doi.org/10.1002/cbdv.202400152.

De Jesus Raposo, M., De Morais, A., & De Morais, R. (2015). Marine Polysaccharides from Algae with Potential Biomedical Applications. *Marine Drugs*, *13*(5), 2967–3028. https://doi.org/10.3390/md13052967.

Dongre, R. S. (2022). Marine Polysaccharides in Pharmaceutical Uses: Chitin, Chitosan, Alginate, and Carrageenan. In J. M. Oliveira, H. Radhouani, & R. L. Reis (Eds.), *Polysaccharides of Microbial Origin* (pp. 745–779). Springer International Publishing. https://doi.org/10.1007/978-3-030-42215-8_43.

Elumalai, K., Srinivasan, S., & Shanmugam, A. (2024). Review of the efficacy of nanoparticle-based drug delivery systems for cancer treatment. *Biomedical Technology*, *5*, 109–122. https://doi.org/10.1016/j.bmt.2023.09.001.

Farshidfar, N., Iravani, S., & Varma, R. S. (2023). Alginate-Based Biomaterials in Tissue Engineering and Regenerative Medicine. *Marine Drugs*, *21*(3), 189. https://doi.org/10.3390/md21030189.

Flórez-Fernández, N., Rodríguez-Coello, A., Latire, T., Bourgougnon, N., Torres, M. D., Buján, M., Muíños, A., Muiños, A., Meijide-Faílde, R., Blanco, F. J., Vaamonde-García, C., & Domínguez, H. (2023). Anti-inflammatory potential of ulvan. *International Journal of Biological Macromolecules*, *253*, 126936. https://doi.org/10.1016/j.ijbiomac.2023.126936.

Fong, D., & Hoemann, C. D. (2018). Chitosan Immunomodulatory Properties: Perspectives on the Impact of Structural Properties and Dosage. *Future Science OA*, *4*(1), FSO225. https://doi.org/10.4155/fsoa-2017-0064.

Forman, H. J., & Zhang, H. (2021). Targeting oxidative stress in disease: Promise and limitations of antioxidant therapy. *Nature Reviews Drug Discovery*, *20*(9), 689–709. https://doi.org/10.1038/s41573-021-00233-1.

Haider, A., Khan, S., Iqbal, D. N., Khan, S. U., Haider, S., Mohammad, K., Mustfa, G., Rizwan, M., & Haider, A. (2024). Chitosan as a tool for tissue engineering and rehabilitation: Recent developments and future perspectives – A review. *International Journal of Biological Macromolecules*, *278*, 134172. https://doi.org/10.1016/j.ijbiomac.2024.134172.

Hentati, F., Tounsi, L., Djomdi, D., Pierre, G., Delattre, C., Ursu, A. V., Fendri, I., Abdelkafi, S., & Michaud, P. (2020). Bioactive Polysaccharides from Seaweeds. *Molecules*, *25*(14), 3152. https://doi.org/10.3390/molecules25143152.

Herath, K. H. I. N. M., Kim, H. J., Kim, A., Sook, C. E., Lee, B.-Y., & Jee, Y. (2020). The Role of Fucoidans Isolated from the Sporophylls of Undaria pinnatifida against Particulate-Matter-Induced Allergic Airway Inflammation: Evidence of the Attenuation of Oxidative Stress and Inflammatory Responses. *Molecules*, *25*(12), 2869. https://doi.org/10.3390/molecules25122869.

Jayasinghe, A. M. K., Kirindage, K. G. I. S., Fernando, I. P. S., Han, E. J., Oh, G.-W., Jung, W.-K., & Ahn, G. (2022). Fucoidan Isolated from Sargassum confusum Suppresses Inflammatory Responses and Oxidative Stress in TNF-α/IFN-γ- Stimulated HaCaT Keratinocytes by Activating Nrf2/HO-1 Signaling Pathway. *Marine Drugs*, *20*(2), 117. https://doi.org/10.3390/md20020117.

Kesharwani, P., Sharma, S., Chaudhary, V., Goyal, R., & Gautam, R. K. (2024). Biomedical Applications of Alginates. In Annu (Ed.), *Biopolymers for Biomedical Applications* (1st ed., pp. 53–86). Wiley. https://doi.org/10.1002/9781119865452.ch3.

Khan, F., Pham, D. T. N., Oloketuyi, S. F., Manivasagan, P., Oh, J., & Kim, Y.-M. (2020). Chitosan and their derivatives: Antibiofilm drugs against pathogenic bacteria. *Colloids and Surfaces B: Biointerfaces*, *185*, 110627. https://doi.org/10.1016/j.colsurfb.2019.110627.

Liu, W., Madry, H., & Cucchiarini, M. (2022). Application of Alginate Hydrogels for Next-Generation Articular Cartilage Regeneration. *International Journal of Molecular Sciences*, *23*(3), 1147. https://doi.org/10.3390/ijms23031147.

Liu, Z., Gao, T., Yang, Y., Meng, F., Zhan, F., Jiang, Q., & Sun, X. (2019). Anti-Cancer Activity of Porphyran and Carrageenan from Red Seaweeds. *Molecules*, *24*(23), 4286. https://doi.org/10.3390/molecules24234286.

Lomartire, S., & Gonçalves, A. M. M. (2023). Algal Phycocolloids: Bioactivities and Pharmaceutical Applications. *Marine Drugs*, *21*(7), 384. https://doi.org/10.3390/md21070384.

Lukova, P., & Katsarov, P. (2023a). Contemporary Aspects of Designing Marine Polysaccharide Microparticles as Drug Carriers for Biomedical Application. *Pharmaceutics*, *15*(8), 2126. https://doi.org/10.3390/pharmaceutics15082126.

Lukova, P., & Katsarov, P. (2023b). Contemporary Aspects of Designing Marine Polysaccharide Microparticles as Drug Carriers for Biomedical Application. *Pharmaceutics*, *15*(8), 2126. https://doi.org/10.3390/pharmaceutics15082126.

Manivasagan, P., & Oh, J. (2016). Marine polysaccharide-based nanomaterials as a novel source of nanobiotechnological applications. *International Journal of Biological Macromolecules*, *82*, 315–327. https://doi.org/10.1016/j.ijbiomac.2015.10.081.

Mendhe, S., Badge, A., Ugemuge, S., & Chandi, D. (2023). Impact of Biofilms on Chronic Infections and Medical Challenges. *Cureus*. https://doi.org/10.7759/cureus.48204.

Murphy, E. J., Fehrenbach, G. W., Abidin, I. Z., Buckley, C., Montgomery, T., Pogue, R., Murray, P., Major, I., & Rezoagli, E. (2023). Polysaccharides—Naturally Occurring Immune Modulators. *Polymers*, *15*(10), 2373. https://doi.org/10.3390/polym15102373.

Negut, I., Dorcioman, G., & Grumezescu, V. (2020). Scaffolds for Wound Healing Applications. *Polymers*, *12*(9), 2010. https://doi.org/10.3390/polym12092010.

Pacheco-Quito, E.-M., Ruiz-Caro, R., & Veiga, M.-D. (2020). Carrageenan: Drug Delivery Systems and Other Biomedical Applications. *Marine Drugs*, *18*(11), 583. https://doi.org/10.3390/md18110583.

Papon, N., Copp, B. R., & Courdavault, V. (2022). Marine drugs: Biology, pipelines, current and future prospects for production. *Biotechnology Advances*, *54*, 107871. https://doi.org/10.1016/j.biotechadv.2021.107871.

Putra, N. R., Fajriah, S., Qomariyah, L., Dewi, A. S., Rizkiyah, D. N., Irianto, I., Rusmin, D., Melati, M., Trisnawati, N. W., Darwati, I., & Arya, N. N. (2024). Exploring the potential of Ulva Lactuca: Emerging extraction methods, bioactive compounds, and health applications – A perspective review. *South African Journal of Chemical Engineering*, *47*, 233–245. https://doi.org/10.1016/j.sajce.2023.11.017.

Rajinikanth B, S., Rajkumar, D. S. R., K, K., & Vijayaragavan, V. (2024). Chitosan-Based Biomaterial in Wound Healing: A Review. *Cureus*. https://doi.org/10.7759/cureus.55193.

Raman, M., & Doble, M. (2015). κ-Carrageenan from marine red algae, Kappaphycus alvarezii – A functional food to prevent colon carcinogenesis. *Journal of Functional Foods*, *15*, 354–364. https://doi.org/10.1016/j.jff.2015.03.037.

Ruocco, N., Costantini, S., Guariniello, S., & Costantini, M. (2016). Polysaccharides from the Marine Environment with Pharmacological, Cosmeceutical and Nutraceutical Potential. *Molecules*, *21*(5), 551. https://doi.org/10.3390/molecules21050551.

Sadeghianmaryan, A., Naghieh, S., Yazdanpanah, Z., Alizadeh Sardroud, H., Sharma, N. K., Wilson, L. D., & Chen, X. (2022). Fabrication of chitosan/alginate/hydroxyapatite hybrid scaffolds using 3D printing and impregnating techniques for potential cartilage regeneration. *International Journal of Biological Macromolecules*, *204*, 62–75. https://doi.org/10.1016/j.ijbiomac.2022.01.201.

Schada Von Borzyskowski, L. (2023). Taking Synthetic Biology to the Seas: From Blue Chassis Organisms to Marine Aquaforming. *ChemBioChem*, *24*(13), e202200786. https://doi.org/10.1002/cbic.202200786.

Shannon, E., Conlon, M., & Hayes, M. (2021). Seaweed Components as Potential Modulators of the Gut Microbiota. *Marine Drugs*, *19*(7), 358. https://doi.org/10.3390/md19070358.

Sharma, S., & Bhende, M. (2024). An overview: Non-toxic and eco-friendly polysaccharides—its classification, properties, and diverse applications. *Polymer Bulletin*, *81*(14), 12383–12429. https://doi.org/10.1007/s00289-024-05307-9.

Shu, H., Zhang, X., Pu, Y., Zhang, Y., Huang, S., Ma, J., Cao, L., & Zhou, X. (2024). Fucoidan improving spinal cord injury recovery: Modulating microenvironment and promoting remyelination. *CNS Neuroscience & Therapeutics*, *30*(8), e14903. https://doi.org/10.1111/cns.14903.

Sinani, G., Sessevmez, M., & Şenel, S. (2024). Applications of Chitosan in Prevention and Treatment Strategies of Infectious Diseases. *Pharmaceutics*, *16*(9), 1201. https://doi.org/10.3390/pharmaceutics16091201.

Singh, A. P., Saxena, R., Saxena, S., & Maurya, N. K. (2024). Unveiling the Ocean's Arsenal: Successful Integration of Marine Metabolites into Disease Management. *UTTAR PRADESH JOURNAL OF ZOOLOGY*, *45*(13), 170–188. https://doi.org/10.56557/upjoz/2024/v45i134146.

Spoială, A., Ilie, C.-I., Ficai, D., Ficai, A., & Andronescu, E. (2022). From Biomedical Applications of Alginate towards CVD Implications Linked to COVID-19. *Pharmaceuticals*, *15*(3), 318. https://doi.org/10.3390/ph15030318.

Tiasto, V. A., Goncharov, N. V., Romanishin, A. O., Zhidkov, M. E., & Khotimchenko, Y. S. (2022). κ- and λ-Carrageenans from Marine Alga Chondrus armatus Exhibit Anticancer In Vitro Activity in Human Gastrointestinal Cancers Models. *Marine Drugs*, *20*(12), 741. https://doi.org/10.3390/md20120741.

Verma, D., Okhawilai, M., Goh, K. L., Thakur, V. K., Senthilkumar, N., Sharma, M., & Uyama, H. (2023). Sustainable functionalized chitosan based nano-composites for wound dressings applications: A review. *Environmental Research*, *235*, 116580. https://doi.org/10.1016/j.envres.2023.116580.

Wang, H., Li, X., Xuan, M., Yang, R., Zhang, J., & Chang, J. (2024). Marine biomaterials for sustainable bone regeneration. *Giant*, *19*, 100298. https://doi.org/10.1016/j.giant.2024.100298.

Wang, K., Xu, X., Wei, Q., Yang, Q., Zhao, J., Wang, Y., Li, X., Ji, K., & Song, S. (2022). Application of fucoidan as treatment for cardiovascular and cerebrovascular diseases. *Therapeutic Advances in Chronic Disease*, *13*, 20406223221076891. https://doi.org/10.1177/20406223221076891.

Wang, M., Veeraperumal, S., Zhong, S., & Cheong, K.-L. (2023). Fucoidan-Derived Functional Oligosaccharides: Recent Developments, Preparation, and Potential Applications. *Foods*, *12*(4), 878. https://doi.org/10.3390/foods12040878.

Wang, Y., Wang, Q., Han, X., Ma, Y., Zhang, Z., Zhao, L., Guan, F., & Ma, S. (2021). Fucoidan: A promising agent for brain injury and neurodegenerative disease intervention. *Food & Function*, *12*(9), 3820–3830. https://doi.org/10.1039/D0FO03153D.

Xiang, Y., Zhang, M., Jiang, D., Su, Q., & Shi, J. (2023). The role of inflammation in autoimmune disease: A therapeutic target. *Frontiers in Immunology*, *14*, 1267091. https://doi.org/10.3389/fimmu.2023.1267091.

Zainal Ariffin, S. H., Yeen, W. W., Zainol Abidin, I. Z., Megat Abdul Wahab, R., Zainal Ariffin, Z., & Senafi, S. (2014). Cytotoxicity effect of degraded and undegraded kappa and iota carrageenan in human intestine and liver cell lines. *BMC Complementary and Alternative Medicine*, *14*(1), 508. https://doi.org/10.1186/1472-6882-14-508.

Zayed, A., & Ulber, R. (2019). Fucoidan production: Approval key challenges and opportunities. *Carbohydrate Polymers*, *211*, 289–297. https://doi.org/10.1016/j.carbpol.2019.01.105.

Zeng, Y., Xiang, Y., Sheng, R., Tomás, H., Rodrigues, J., Gu, Z., Zhang, H., Gong, Q., & Luo, K. (2021). Polysaccharide-based nanomedicines for cancer immunotherapy: A review. *Bioactive Materials*, *6*(10), 3358–3382. https://doi.org/10.1016/j.bioactmat.2021.03.008.

Zhang, H., Lin, X., Cao, X., Wang, Y., Wang, J., & Zhao, Y. (2024). Developing natural polymers for skin wound healing. *Bioactive Materials*, *33*, 355–376. https://doi.org/10.1016/j.bioactmat.2023.11.012

Zhou, Y., Liu, J., Ma, S., Yang, X., Zou, Z., Lu, W., Wang, T., Sun, C., & Xing, C. (2024). Fabrication of polymeric sorafenib coated chitosan and fucoidan nanoparticles: Investigation of anticancer activity and apoptosis in colorectal cancer cells. *Heliyon*, *10*(14), e34316. https://doi.org/10.1016/j.heliyon.2024.e34316.

Chapter 4

Marine Metabolites and Inflammation Management

Mirunalini Gobinath[1]
L. Priyanka Dwarampudi[1,*]
R. Shanmugam[2]
Parikshit Roychowdhury[3]
and Naina Mohamed Pakkir Maideen[4,5]
[1]Department of Pharmacognosy and Phytopharmacy, JSS College of Pharmacy,
JSS Academy of Higher Education and Research, Ooty, Tamil Nadu, India
[2]TIFAC CORE HD, JSS College of Pharmacy, JSS Academy of Higher Education and Research,
Ooty, Tamil Nadu, India
[3]Department of Pharmaceutics, JSS College of Pharmacy,
JSS Academy of Higher Education and Research, Ooty, Tamil Nadu, India
[4]College of Medicine, Mohammed Bin Rashid University of Medicine and Health Sciences, Dubai, UAE
[5]Dubai Health, Dubai, UAE

Abstract

Oceans are an abundant source of biomolecules with several anti-inflammatory, antioxidant, and healing properties. Many marine species have unique properties in the ocean compared with terrestrial organisms. For instance, marine algae, seaweed, sponges, corals, snails, fish, and microorganisms produce secondary metabolites to protect themselves from predators. Subsequently, they have all the required characteristics, such as biological activities and structural similarities; marine metabolites are ideal choices for drug discovery and the development of anti-inflammatory moieties. In general, autoimmune disorders represent one of the leading conditions that are not curable up to date. Hence, the need for anti-inflammatory and immunomodulatory agents is in demand in these cases. In addition, marine metabolites tend to interact with molecular mechanisms such as COX-2, MAPK, NFκB, and oxidative pathways. Thus, making them a promising approach for autoinflammatory diseases and chronic inflammatory diseases like diabetes, neurodegenerative disease, cardiovascular diseases, arthritis, and skin disorders like psoriasis. This book chapter emphasizes the significance of marine metabolites in inflammatory management along with their molecular mechanisms.

---

* Corresponding Author's Email: mirunalini.phd@jssuni.edu.in

In: Healing from the Depths Marine Metabolites in Disease Management
Editors: Pranav Kumar Prabhakar and Tarun Kumar Upadhyay
ISBN: 979-8-89530-832-5

Keywords: marine metabolites, inflammation, marine natural products, inflammatory diseases, anti-inflammatory properties

1. Introduction

Inflammation is referred to as the nonspecific defense mechanism of the body to maintain tissue homeostasis, which is induced by various factors, such as pathogens, radiation, chemicals, toxins, and extreme temperatures. These mechanisms lead to several inflammatory diseases and autoimmune disorders, which further cause chronic inflammation. These diseases are classified into chronic inflammatory diseases, neuroinflammatory disorders, metabolic disorders, infectious inflammatory disorders, and autoimmune diseases [1]. Some instances are asthma, COPD, cardiovascular inflammation due to atherosclerosis, neuroinflammation in Alzheimer's, metabolic disorder due to hormonal issues, obesity, inflammation due to infection in tuberculosis, meningitis, and autoimmune inflammation in rheumatoid arthritis, psoriasis, and systemic lupus erythematosus, etc. Other inflammatory diseases are also classified according to the organ system, for example, musculoskeletal system inflammation due to osteoarthritis and inflammation of the gastrointestinal system due to inflammatory bowel diseases and gastritis, etc. These conditions are managed by existing therapeutic options like NSAIDs, corticosteroids, immune modulators, DMARDs, and biologics [2]. In addition, recent insights into the management of inflammation involve various natural products and traditional therapies from India's Ayurveda, Homeopathy, and Chinese herbal medicine, which have been recognized to manage inflammation with fewer side effects. Hence, the perspectives of management of inflammation have shifted towards alternative therapies [3].

Marine natural products are the most abundant natural sources, which have tremendous potential in healing many types of inflammation. Marine organisms such as microalgae, sponges, seaweeds, corals, fish, and microorganisms produce secondary metabolites, which are known as marine metabolites that are produced to protect themselves from predators. Consequently, these metabolites have several potential biological activities, such as antioxidant, immunomodulatory, and anti-inflammatory properties and can be used to treat various inflammatory diseases [4]. Thus, marine metabolites could be a cornerstone in the management of inflammation. This can be achieved by opting for approaches like drug discovery & development and making them a promising strategy for inflammation management. In addition, marine metabolites have unique biological mechanisms and diverse chemical structures that make them novel, effective therapies for a wide range of autoinflammatory disorders [5]. Thus, this chapter emphasizes the significance of marine metabolites in inflammatory management along with their molecular mechanisms such as NFκB, COX, MAPK, and arachidonic acid pathways.

2. Understanding Inflammation

Inflammation is a host defensive response to protect from external insults such as pathogens, heat, chemicals, irritants, radiation, abrasions, trauma, and extreme pressure or temperatures. It is a controlled process that tends to attempt the disposal of microbes or foreign substances at the site of injury and encourages tissue repair so that homeostasis is restored. The 5 main cardinal manifestations of inflammation are coined in Latin as "*calor*," "*dolor*," "*rubor*," "*tumor*," and "*functio laesa*," which means, heat, pain, redness or erythema, swelling or edema, and temporary loss of function [6].

The key steps that are involved in the inflammation process are called inflammatory response and are usually triggered by internal or external factors that release PAMPs (pathogen associated molecular patterns) and DAMPs (damage associated molecular patterns), which are then recognized by pattern-recognition receptors (PRRs), a cell surface protein such as toll-like receptors that are located on immune cells that activate further inflammatory response [7]. These primary steps lead to the following changes:

a) Vascular modifications

Vascular changes due to inflammation generally involve increased permeability of capillaries and arterioles, as well as vasodilation. This increases the size and diameter of arterioles, followed by more blood flow to the injured area. This extra blood flow allows for the removal of any necrotic cells or microbial toxins that have initially caused the inflammatory response. Increased capillary permeability allows the passing of any retained blood components, such as defensive proteins, antibodies, clotting factors, and leukocytes, from the blood plasma to the tissue interstitial spaces [8].

Apart from vasodilation and increased permeability, other inflammatory responses that contribute to the next steps are the release of histamines, kinins, leukotrienes (LTs), prostaglandins (PGs), and activation of the complementary system. The exact process of inflammation commences when the macrophages and mast cells that are found in the tissues respond to the stimuli of PAMPs and DAMPs. Furthermore, it initiates the release of granules from the mast cells to produce inflammatory mediators like histamine, serotonin, chemokines and cytokines that act on endothelial cells to intensify vascular permeability and dilation of blood vessels [9].

b) Recruitment of phagocytes and activation of inflammatory immune cells

In this step, emigration of leukocytes is catalyzed by inflammatory mediators, which begin to adhere to the damaged endothelial walls. The process of recruitment of phagocytes can also be stimulated by pro-inflammatory cytokines such as IL-1, TNF-α, complements, etc., that induce adhesion molecules like integrins and selectins on leukocytes, endothelial, as well as epithelial cells.

Histamines, which are released by mast cells of platelets, basophils, and connective tissues near the damaged area, produce neutrophils and tissue macrophages to further release histamine at the site of injury. These effects not only lead to vascular modifications but also release intracellular adhesion molecules (ICAMs) into the vascular endothelium. Other mediators like bradykinins, which are polypeptides formed from a non-reactive precursor – kininogens that serve as chemotactic molecules for initiating phagocytosis [10]. Other chemotactic agents such as PGs and LTs, produced by damaged tissues, basophils, and mast cells, amplify the functions of bradykinin and histamine, increase permeability, enhance the emigration of phagocytes into the endothelial wall of capillaries, and enhance the adherence of phagocytes to pathogens.

These combined effects cause slow movement of inflammatory mediators and lead to diapedesis.

After chemotaxis causing diapedesis, phagocytic cells that are recruited by chemokines like IL-8, interferon γ, monocyte chemoattractant proteins (MCP), neutrophil chemotactic proteins (NCP), and eosinophil chemotactic proteins (ECP), which is especially related to allergy-mediated inflammation, gather neutrophils, monocytes, macrophages, and cause immune-mediated inflammation, which is further monitored by lymphocytes [11]. Neutrophils are involved in acute inflammatory responses and destroy pathogens by phagocytosis, whereas monocytes get converted to tissue macrophages, which are more potent phagocytes than neutrophils. They are large enough to engulf the damaged tissues, phagocytized membrane ingests the microbes with pseudopods, destroys them inside the vacuole forming phagosome sacs, and digests into the cytoplasm by merging with lysosome forming a complex called phagolysosome. This process produces lysozyme that degrades the pathogen proteins and initiates hydrogen peroxide, superoxide anion, nitric oxide, and hypochlorite anion, which are lethal oxidants causing oxygen-dependent killing or oxidative burst. End products of this process include dead phagocytes, damaged tissues or fluids, and nondegraded materials into residual bodies and pus, indicating the severity of inflammation [12].

c) Tissue repair and regeneration

In the process of repair, macrophages dispose of damaged and necrotized cells and create space for the formation of new cells. Temporary angiogenesis (formation of new blood vessels) is triggered by growth factors and macrophages. Other mediators, including IL-4, IL-10, and transforming growth factors (TGF-β), downregulate the existing inflammatory cytokines that have initially caused inflammation. In addition, fibroblasts that are present at the area of inflammation synthesize collagen, an extracellular matrix, that forms a scar tissue called fibrosis or granulation depending on the severity of the wound. Milder damage restores and regenerates the parenchymal tissue into a healthy state, whereas in severe damage, it forms a non-functional scar. These are the crucial steps involved in tissue repair and regeneration, which include wound healing, and these can be affected by factors like age, nutrition, and blood circulation [13].

In addition to anti-inflammatory moieties, antioxidants also play a prominent role in healing inflammation since there is a strong correlation between inflammation and oxidative stress. It is due to the release of excessive amounts of reactive oxygen species by various macrophages and leukocytes, creating oxidative stress events leading to lipid peroxidation and tissue damage. Apart from these, in autoinflammatory conditions, the involvement of immune cells creates excessive immune reactions that might probably affect the normal cells; thus, this can be overcome by using immunosuppressants and immunomodulating agents [14].

3. Marine Metabolites with Anti-Inflammatory Properties

Oceans are home to abundant and rich marine natural products that have potential to heal various types of inflammation. These include the whole marine biodiversity, such as seaweed, algae, planktons, sponges, corals, microorganisms, and even fish that have metabolites with anti-inflammatory action. The structural key components of these anti-inflammatory metabolites are alkaloids, nitrogen-containing compounds, peptides, polyketides, steroidal compounds, and terpenoids. These components have also been evaluated for their anti-cancer and neuroprotective properties in various recent studies [15].

Marine species are classified into several types. These can be in general or species like bacteria, protozoans, fungi, planktons, invertebrates or vertebrates. Plankton can be further classified as bacterioplankton, which comprises blue-green algae or cyanobacteria and phytoplankton, which consists of diatoms and flagellates, and zooplanktons [16].

Marine invertebrates are a diverse group of organisms that exist in the ocean. Some instances are marine sponges, starfish, jellyfish under the phyla cnidaria, and other species with hard shells. Bryozoan is a phylum of marine invertebrates that live in colonies. Porifera is a phylum that comprises sponges, which are multicellular species, as the name suggests, they have several pores and channels. They have asymmetrical body types and have heterotrophic modes of nutrition. Marine arthropods include more than 1 million species which have jointed appendages as their features. Crustaceans, which are comprised of crabs, lobsters, and shrimps, are some examples of arthropods. The phylum Mollusca contains snails and oysters, etc. Sea cucumbers and sea urchins are examples of the phylum Echinodermata. Marine vertebrates include fish, reptiles and other sea mammals. They produce fatty substances that have various medicinal properties. For instance, fish and cod liver oil have essential fatty acids that are required for our daily needs and used in several health conditions [17].

Most of the anti-inflammatory agents have the mechanism of action of inhibiting cyclooxygenase (COX) enzyme, which is important in arachidonic acid pathways. These agents are called non-steroidal anti-inflammatory drugs, shortly NSAIDs, which exert analgesic, antipyretic, and anti-inflammatory effects. Some instances of synthetic compounds that inhibit COX pathways are aspirin, ibuprofen, indomethacin, naproxen, etc. They work by reducing the sensitivity of peripheral nociceptors and diminishing the production of prostaglandins, thus relieving pain [18]. Generally, prostaglandin protects the mucosal layer of the abdomen and kidneys. Subsequently, blocking prostaglandins might cause damage to the mucosal layer and cause gastric irritation. In COX pathways, COX-1, which is a basal enzyme found in gastric mucosa, platelets, and tissues, and COX-2, which is an inducible enzyme present in the sites of inflammation and cell proliferation. Newer NSAIDs like celecoxib selectively inhibit COX-2, sparing the production of cytoprotective prostaglandins by COX-1, thus leading to fewer adverse effects of gastric irritation and ulceration. Furthermore, thromboxane, which is the product of this pathway, is synthesized from platelets, which have prothrombotic and vasoconstrictive effects, that is the reason why aspirin and other NSAIDs are also used as low-dose antithrombotic prophylaxis. In contrast, in normal physiological conditions, prostacyclin is produced by COX-2 from vascular endothelial cells, which have opposite effects. Consequently, COX-2 selective inhibitors have led to cardiovascular complications due to a decrease in prostacyclin levels without any alterations in thromboxane levels and, thus making selective COX-2 inhibitors an insufficient drug. Also, other classes of drugs, like pseudoephedrine for nasal inflammation or congestion, lead to the risk of various disorders like reversible cerebral vasoconstriction syndrome (RCVS) and posterior reversible encephalopathy syndrome (PRES) [19]. This influences the discovery and the development of potent anti-inflammatory drugs from marine origin due to its fewer risks in side effects. These aspects reveal marine natural products possess antioxidant activities with the tendency to target COX and NFκB pathways. Some of the main active constituents from marine natural products have been elaborated according to their metabolites [20, 21]. The marine-derived drugs that have been approved by FDA are shown in Table 1.

Table 1. Approved marine-derived drugs in the current market

Drug name	Year of approval	Target disease	Proposed mechanism	Source
Cytarabine	1969	Leukemia (Cancer)	By targeting DNA polymerase	Marine sponges
Vidarabine	1976	Antiviral therapy for Herpes simplex	Inhibition of viral DNA polymerase	Marine sponges
Ziconotide	2004	Severe and chronic pain	Targets N-type calcium ion channels	Cone snail
Omega-3-acid ethyl esters	2004	Hypertriglyceridemia	Blocking triglyceride synthesis by targeting synthase enzymes	Fish
Eicosapentanoic acid ethyl esters	2012	Hypertriglyceridemia	Blocking triglyceride synthesis by targeting synthase enzymes	Fish
Eribulin Mesylate	2010	Breast carcinoma	Targeting microtubule formation	Marine sponges
Brentuximab vedotin	2011	Lymphoma and Hodgkin's disorder	Microtubule formation and CD30	Mollusk species or cyanobacterium
Trabectedin (Marine alkaloid metabolite)	2015	Ovarian carcinoma and soft tissue cancer	DNA Structure alteration	Tunicates
Plitidepsin	2018	Lymphoma, Multiple myeloma and leukemia	eukaryotic translation elongation factor 1 alpha 2	Tunicates
Polatuzumab vedotin	2019	Non-Hodgkins and B-cell lymphoma, leukemia, follicular cancer etc.	Microtubule targeted therapy	Cyanobacterium and mollusks
Enfortumab vedotin	2019	Urothelial carcinomas	Nectin-4 targeted and microtubule disturbance	Cyanobacterium and mollusks
Lurbinectedin	2020	Small cell lung cancer	RNA polymerase II inhibition	Tunicates
Distamab vedotin	2021	Endothelial, gastric, advanced breast, urothelial carcinomas	HER2	Cyanobacterium and mollusks

3.1. Marine Alkaloids

Diverse marine organisms produce alkaloids, which have various biological and pharmacological activities. These alkaloids have different structures like indoles, quinolones, etc., that make unique characteristics of anti-inflammatory properties. Alkaloids exhibit immunomodulatory effects in both innate and adaptive immune responses [22].

Indole alkaloids, including Neoechinulin A from the marine fungi *Eurotium sp.* and Asperversiamides B, C, F, and G, from the marine fungi *Aspergillus versicolor*, show decreased nitric oxide by inhibiting iNOS activity in LPS-stimulated macrophage RAW 264.7 cell line studies. This also reduces prostaglandin synthesis PGE_2, and reduces the expression of COX-2, diminishing the production of cytokines like IL-1β, TNF-α, and NFκB via MAPK p38 phosphorylation, thus reducing the proinflammatory response [23].

Penicamide A and their enantiomers, which are quinazoline alkaloids derived from the fungus *Penicillium canescens*, which is isolated from the ascidian *Stylea plicata*, reduces NO production in LPS-mediated macrophage cell lines. Other quinoline alkaloid derivatives like Actiniquinolones A and B, which are derived from the marine fungus Streptomyces spp., inhibit COX-1 and COX-2 activity in cell-line studies. 5-prenyl dihydrorubrumazine and 5-prenyldihydrovariecobrin compounds are diketopiperazine alkaloids from marine fungi *Aspergillus sp.*, which is found in seaweeds, blocks the effects of iNOS and COX-2 expression, and decreases NO and PGE_2 in LPS-induced macrophage cell lines of RAW 264.7 cells [24].

The production of TNF-α, IL-6, MCP-1, IκBα degradation/translocation of NFκB p65 subunit, and p38 phosphorylation of MAPK/JNK are diminished in LPS-mediated macrophage studies with the use of chytocalasan based alkaloid extract, Chaetoglobosin Fex. It also reduces CD14 expression, which is crucial in proinflammatory responses and TLR4 activation [25].

Alkaloids that are obtained from marine sponges include Barettin, which is a brominated alkaloid from the marine sponge *Geodia barretti*, have shown antioxidant and anti-inflammatory activities in in vitro studies. Other derivatives of 6-bromoindole like Geobarettin B and C, exhibited reduced TNF-α, IL-1β, IFNγ and other dendritic cell expressions in LPS-stimulated THP-1 cells and decreased lipid peroxidation in HepG2 cells, which is involved in chronic inflammation [26]. Halichlorine, from *Halichondria okadai*, diminishes the expression of VCAM-1, E-selectin, and ICAM-1, which are involved in the adhesion step of inflammation in LPS-induced aortic endothelial cells. This compound also inhibits macrophage adhesion and NFκB expression, which shows a potent anti-inflammatory effect [27].

Debromohymenialdisine, from *Stylissa spp.*, which is a pyrrole alkaloid, decreased the expressions of IL-6, IL-1β, TNF-α, iNOS, and COX-2. Thus, it further reduces the levels of nitric oxide and prostaglandin synthesis PGE2 in LPS-stimulated Caco-2 and THP-1 cells. In other studies, this compound has also reduced p38, ERK 1 and 2, JNK, and NFκB p65 subunit phosphorylation and increased HO-1 expression, which aligns with the translocation of the transcription factor Nrf2 and other protein factors in oxidative stress and inflammatory pathways. Compounds from *Stylissa flabellata*, and Stylissadines A and B inhibit P2X7 receptors that are involved in ATP receptors in the activation of inflammation, NLRP3 inflammasome, synthesis of proinflammatory cytokines and chemokines, and MAPK activation in THP-1 cells [28].

Extracts such as Tubastribe orthidines A, B, C, E, and F are isolated from ascidian Aplidium orthium, decreasing superoxide synthesis in in vitro studies using PMA-stimulated neutrophils and in an *in-vivo* gouty arthritis model that shows an effect on neutrophil infiltration. Thiazone-containing quinoline quinone alkaloids, Ascidiathiazones A and B obtained from *Aplidium* spp., also have similar effects in neutrophils [29].

Convolutamydine A, which is an oxindole alkaloid from bryozoans, reduced leukocyte migration, IL-6, nitric oxide, PGE_2, and TNF-α in the carrageenan-induced inflammation model. Kottamide D, an imidazole alkaloid extracted from the ascidian *Pycnoclavella kottae,* decreased superoxide anions in cell-line studies. 5 α iodozoanthenamine, which is a zoanthamine alkaloid that is derived from the cnidarian *Zoanthus kuroshio*, shows an anti-inflammatory effect by reducing neutrophils, superoxide anion generation and preventing elastase release. They also decrease reactive oxygen species and NO in LPS-stimulated BV-2 cells [30].

Marine macroalgae, or seaweed, are also a good source of alkaloids. Species like the Caulerpa genus of green algae have shown beneficial effects in cell-line studies used in LPS-mediated macrophages. Aqueous and methanolic extracts of *Caulerpa mexicana* have demonstrated decreased pro-inflammatory cytokines like IL-6, IL-12, and TNF-α production in in vitro and leukocyte migration in air pouch inflammation models, murine zymosan-induced peritonitis and xylene-induced ear edema animal models. The same methanolic extracts have proved activity in a murine model of dextran sulfate sodium-induced ulcerative colitis in which the production of cytokines and interferons IFNγ, IL-12, IL-17A, IL-6 and TNF-α are reduced [31, 32]. The red algae *Gracilaria* species also described reduced inflammation activity, and aqueous extracts of *Gracilaria tenuistipitata* demonstrated promising results in animal studies of Hepatitis C-induced inflammation model. Other species of Caulerpa and *Caulerpa sertularioides* resulted in anti-inflammatory properties by reducing leukocyte migration in the animal model of carrageenan-induced peritonitis. Thus, marine macroalgae have tremendous properties in decreasing inflammation with their structural alkaloids and unique characteristics, which have been demonstrated in various cell-line studies and animal models [33].

3.2. Marine Terpenoids

These are usually located in marine organisms like corals, seaweeds, sponges, and marine microorganisms. They have unique structural diversity due to their harsh environmental pressures and adaptations in ocean life when compared with terrestrial sources. These terpenoids can be divided into various subcategories like halogenated monoterpenes, diterpenes, triterpenoids, and sesquiterpenes based upon their structure and function. Some instances of marine-derived terpenoid compounds and their sources are mentioned below, which have targeting activity on signal transduction pathways, proinflammatory cytokine production, and NFκB activation [34].

Fucoxanthin, a xanthophyll terpenoid compound that can be derived from edible brown seaweed, shows inhibitory action on osteoclast-specific macrophage RAW 264.7 cells in *in vitro* studies. They also increase Nrf2 activation and decrease osteoclast differentiation, cellular infiltration of COX-2, and iNOS expressions in LPS-stimulated murine lung inflammation *in vivo* models. Other evidence from molecular docking states that fucoxanthin blocks the TLR4 receptor, which is required for LPS stimulation [35]. Epiloliolide, a terpenoid from the marine algae *Sargassum horneri,* which is a brown seaweed, decreased the synthesis of TNF-α, IL-6, and IL-1β in human periodontal ligament cells in an *in-vitro* study. Mojabanchromanol, which is also extracted from the same brown algae, diminished ROS and TLR activation in alveolar type 2 epithelial cell line studies [36].

Hirsutanol A, a terpenoid from red algae-derived fungi *Chondrostereum* sp., reduced LPS-induced lung inflammation in *in vivo* studies and STAT3 and MMP-9 production in a mouse endotoxemia model. *Nitschia laevis* extracts also reduced inflammatory response in microglia by increasing Nrf2 and decreasing the expression of pro-inflammatory cytokines and prostaglandins [37].

Diterpenes derived from *Laurencia glandulifera* decreased the synthesis of pro-inflammatory cytokines in cell line studies, and in animal studies, they attenuated the development of dextran sulfate sodium-induced murine colitis [38]. Deacetylphylloketal, which is a terpenoid derived from the marine sponge *Phyllospongia* sp., has shown reduced nitric oxide, PGE_2, and pro-inflammatory cytokines like TNF-α, IL-6, and IL-1β in human epithelial cells and PMA differentiated macrophage cell lines. This is due to blockage of NF-κB translocation and boosting heme oxygenase-1 (HO-1) expression [39].

Monoterpenoids from a marine sponge *Dysidea arenaria* derived compound dysiarenone showed suppression of ROS and LPS-induced inflammatory cytokine mediators by promoting HO-1 expression via an Nrf2-dependent mechanism. Pigments like carotenoids show structural similarities like terpenoids and possess immunomodulatory activity. Carotenoids are also known as tetraterpenes, which are lipophilic and linear conjugated polyenes. Astaxanthin, which is a well-known nutraceutical compound from microalgae, shows modulation of pro-inflammatory cytokines and dendritic cells by increasing expressions of HO-1 and Nrf2 [40].

Polyketides like caniferolide A, derived from the marine actinomycete *Streptomyces caniferus,* diminished NFB, MAPK, JNK, and p38 phosphorylation and increased Nrf2 for the survival of BV2 microglial cells. This states that this marine compound has potential therapeutic avenues in the treatment of Alzheimer disease [41]. Another polyketide, dieckol from the brown seaweed *Ecklonia cava*, reduced the onset and development of NAFLD and decreased NLRP3 inflammasome in a high-fat diet mouse model. 2-epi-jaspine B, a polyketide derived from two marine sponges, *Pachastrissa* sp. and *Jaspis sp.,* respectively, showed SphK1 inhibitoion in an *in-vitro* and Freund's adjuvant rheumatoid arthritis rat model by diminishing IL-6, TNF-α, IL-1β and arthritis score [42]. Tuberatolide B, a polyketide derived from the brown algae *Sargassum macrocarpum*, halts the expression of NFκB/MAPK phosphorylation in cell-lines and zebrafish models in in vivo studies [43].

3.3. Marine Polyphenols

Marine polyphenols have unique structural properties and are obtained naturally from sources like algae, corals, crustaceans, fish, mollusks, and plants. Polyphenols, as the name suggests, that their structure contains multiple phenol groups or hydroxyl groups -OH, which comprise potent antioxidant and anti-inflammatory properties [44]. Some examples of marine polyphenols are phlorotannin, phenolic acids, stilbenes, flavonoids, tannins, and lignans, which assist various marine species to overcome harsh climatic conditions like altered temperature, UV radiation, oxidative stress, and predators. Under these conditions the marine organisms produce metabolites including polyphenols, polysaccharides, xanthophenols, and tocopherols [45].

Phlorofucofuroeckol A, which is a polyketide polyphenol compound extracted from the brown seaweed *Ecklonia cava,* downregulates the expression of SMAD and MAPK signaling in human tracheal fibroblast collagen type 1 protein and promotes the formation of osteoblasts in the bone marrow. Dieckol, which is derived from the same brown algae, has polyphenolic content that shows anti-inflammatory activity [46].

Phlorotannins are most common in macroalgae and are a group of complex polyphenolic compounds with various phloroglucinol units that are categorized based on their linkages. The presence of ether links is represented in phlorethols and fuhalols, dibenzodioxin linkages are eckols, phenyl linkages are fucols, and the presence of both ether and phenyl linkages are fucophlorethols. These compounds have antioxidant properties and are found in brown seaweed species like *Fucus vesiculosus*, *Fucus spiralis*, *Ecklonia cava*, and *Sargassum* species. Phlorotannins have the potential to inhibit LPS inflammation models by altering the NO and PG synthesis, and they also modulate TNF-α, IL-6, and IL-1β cytokines and NFκB p65 subunit expressions [47].

Bromophenols, which are aromatic compounds with hydroxyl groups and bromine ions show anti-inflammatory action by suppressing inflammatory mediators like TNF-α, IL-1β and NFκB signaling [48]. These can be derived from red algae species like *Laurencia* and *Gracilaria* sp. Bromocresol purple, a synthetic dye, which is also isolated from marine red algae *Gracilaria* and other marine sponges. Other compounds that are also isolated from marine *Sargassum* sp., 2,3-Dibromo-1-((2-bromo-4,5-dimethoxybenzyl) oxy) methyl-4,5-dimethoxybenzene, exhibits antioxidant and anti-inflammatory activity [49, 50].

3.4. Marine Polysaccharides

Marine polysaccharides are complex sugars that are mainly derived from marine species like algae, seaweed and other marine invertebrates. These compounds provide energy and structural integrity for the organisms and gained attention due to their several applications in food and pharmaceutical industries. In addition, marine polysaccharides have exceptional characteristics, including high water retention capacity, antioxidant, anti-inflammatory, and antimicrobial properties, it is used in skin care and cosmeceuticals [51].

In marine life, these compounds are involved as structural components, defense mechanisms, energy reservoirs, maintenance of cell communication process, and immune molecular recognition, which is essential for development, growth, and various metabolic activities, microbial interactions and nutrient cycling. These biological macromolecules have several key properties, like anti-inflammatory, and antioxidant, and have shown promising results in inflammatory metabolic disorders, cancer, and diabetes [52].

Fucoidan, which is an overrated sulfated polysaccharide that can be derived from various species of marine brown algae such as *Fucus vesiculosus*, *Sargassum* species, and *Laminaria japonica*. The main structural activity relationship is due to the presence of sulfate group along with fucose sugars, xylose, mannose, and galactose. In recent studies, it is stated that fucoidan is also present in sea cucumbers, *Stichopus chloronotus*, which is a traditional superfood with high nutrition content in Southern China. They have investigated cell-line studies of macrophage RAW 264.7 for their immune modulation activity, where fucoidan has shown promising results in the reduction of NO, IL-6, IL-10, TLR-4, TNF-α levels, and downstream of NFκB signaling pathways [53]. In some studies, fucoidan is often used as a prebiotic agent to enhance gut microbial diversity and promote butyrate production, which assists in the modulation of the immune system. This evidence has also been proved in IBD models, in which fucoidan is extracted from *Undaria pinnatifida*, which has been alleviating the gut health and symptoms of IBD [54].

Other polysaccharides like alginate, carrageenan, chitosan, agar, and laminarian are widely used in the cosmetic, food and pharmaceutical industries for their incredible properties like thickening agents, gelling, stabilizing agents, improving biocompatibility and biodegradability, and in microbial culture media. These polysaccharides can be obtained from species like *Laminaria*, *Kappaphycus alvarezii*, *Gracilaria*, *Ascophyllum nodosum*, and exoskeletons of crustaceans like crabs, shrimps, etc. Some other instances of polysaccharides with anti-inflammatory properties are Ulvan from the green alga *Ulva lactuca*, galactan from *Gelidium sp.*, xyloglucan, porphyran, and sulfated galactans from other red algae species. These compounds have been evaluated for their antiviral, COVID-19, and anticoagulant activity and have further promising therapeutic avenues [55, 56].

Several biomedical applications of marine polysaccharides include tissue engineering, where these polysaccharides are formed into scaffolds for stem cell and regenerative therapy. The reason behind this is that fucoidan and other marine-derived polysaccharides are non-toxic and have more biocompatibility. They could also opt to improve the sensitivity of biosensors. In the case of infection prevention, orthopedic or dental surgeries that use implants also incorporate polysaccharides. They are also used as alternative materials for stent fabrication and reduce restenosis when used as surface coating agents. In peritoneal adhesion, it is used as a physical barrier. To improve cardiac function, polysaccharide-based scaffolds are prepared. They are also incorporated in many drug delivery systems, like stimuli-response, pH response drug release, temperature response, and electric-field responsive drug delivery systems [52, 57].

3.5. Peptides and Proteins

Marine proteins and peptides usually exhibit strong free radical scavenging activities, which are stated in recent research. It is an abundant source of biomolecules and short-chain amino acids, which can be derived from microalgae, seaweeds, sponges, mollusks, tunicates, corals, and marine vertebrates like fish, sharks, etc. These bioactive moieties have a wide range of applications in biomedical, pharmaceutical, nutraceutical and cosmetic industries [58]. Amino acids and glycoproteins are also obtained from these marine sources, and their chemical features have a positive impact on drug and nutritional food development. Edible seaweeds have an enormous amount of nutritional value, and they are consumed as staple food in Southeast Asia. Some instances of primary amino acids present in edible seaweeds are valine, threonine, phenylalanine, lysine, leucine, and arginine [59].

Chlorella 11 peptide which is derived from the green seaweed *Chlorella pyrenoidosa*, shows indirect anti-inflammatory activity in cell-line studies of RAW 264.7 cells by reducing the synthesis of MCP-1 (Monocyte chemoattractant protein-1), which recruits the migration and infiltration of macrophages in the inflammatory process. Furthermore, it also inhibited the formation of VCAM and ICAM-1 proteins in the endothelial cells and blocked the expression of NFκB cell signaling [60]. Other green algae, *Spirulina maxima* and red algae, *Pyropia yezoensis* demonstrated anti-inflammatory properties by suppressing cytokine production in the endothelial cells and direct deactivation of cell-signaling pathways of MAPK, NFκB, p38, JNK and ERK. A bioactive metabolite from the red algae *Ceratodictyon spongiosum*, which is a thiazole-containing cyclic heptapeptide called ceratospongamide, shows anti-inflammatory action by blocking the phospholipase A_2 expression in preclinical studies. Other cyclic heptapeptide cyclomarins, which are isolated from the marine actinomycete *Streptomyces sp.*, have anti-inflammatory potential and have been compared with the standard drug hydrocortisone in an animal model. Thus, marine algae obtained protein hydrolases have promising results in inflammation and molecular targets like TLR4, MAPK, and NFκB [61, 62].

Collagen obtained from marine sources is a good substitute for bovine and porcine derived collagen, which has comparable benefits in retaining the skin texture and enhancing the healing process. Various sources of marine invertebrates and vertebrates, including sponges, cnidarians, starfish, corals, mollusks, and fish, produce high-quality collagen, which can be utilized in various blooming industries. Nevertheless, further investigation is necessary for understanding their physicochemical features [63, 64].

Marine peptides usually have tremendous benefits over other proteins, and their applications include tissue engineering as a regenerative material, nanomaterial for vaccine delivery highlighting the peptide-based vaccines, improving biocompatibility and bioavailability issues, natural preservatives in food processing, and various biomedical and biotechnological applications [65, 66].

3.6. Miscellaneous

As we know, marine metabolites have a potential advantage over terrestrial-derived compounds because of their robust adaptations and modifications due to harsh climatic conditions such as low oxygen, temperature, pH, etc. Some instances that have been used already in the market are essential fatty acids derived from marine sources [38]. These polyunsaturated fatty acids (PUFAs) are usually known as fish oils or cod-liver oils. They are rich sources of omega 3 and 6 fatty acids. Recent evidence states that they are used in various inflammatory conditions like sports-related muscle soreness, rheumatoid arthritis, etc. Thus, incorporating them in the diet for patients with inflammatory conditions has benefits in healing. Eicosanoids like DHA, EPA (docosahexaenoic acid & eicosapentaenoic acid), and arachidonic acid have inhibitory activity on pro-inflammatory cytokines. They can also suppress the migration of leukocytes in the secondary step of inflammation, states some research. Recently, many studies have proved that these essential fatty acids are also obtained from marine algae, which can be an added value for vegans. Extensive research has been conducted in these metabolites and they show indispensable results in improving neurocognitive functions and reducing inflammation via gene expressions [67, 68].

Sterols are lipids that are abundant in microalgae and various brown seaweeds like *Sargassum* and *Undaria* sp. They contain several groups of sterols, such as ergosterol, fucosterol, and poriferasterol, which have anti-inflammatory activity by blocking the production of cytokines including IL-6, TNF-α, and IL-1β in LPS cell-line studies. Moreover, thorough research is required to estimate all the beneficial properties of marine sterols [4]. Phycobiliproteins and phycobilins are the special category of metabolites that are derived from marine algae and cyanobacteria. They have several benefits over cardiovascular diseases due to their cardioprotective nature. They are also used in labs as a fluorescent probe in immunofluorescence assays. They have various uses, like in pharmaceutical industries, nutraceuticals, food colorants, and biomedical applications [69, 70].

4. Challenges and Future Perspectives of Marine Metabolites

Despite their potential and promise in applications in disorders, including various types of inflammation, marine metabolites, or, as a matter of fact, any marine-sourced natural product, has its own set of hurdles in order to be a future game-changer in the healthcare system. The faced challenges could be categorized into two distinct types, namely, discovery challenges and production challenges for marine metabolites [71].

Challenges in the discovery of marine bioactives are particularly vast due to their huge and diverse presence in our oceans. The marine ecosystem being very different from the surface ecosystem gives rise to both structurally and functionally complex metabolites that are majorly still unexplored; furthermore, the expedition into the oceans is costly, and the lack of facilities to facilitate deep-sea exploration limits our exposure to the full potential of the metabolites [72]. There is also a lack of optimized specialized extraction processes for the marine metabolites, as they can be quite different from those of terrestrial metabolites flourishing under deep-sea conditions. Screening techniques such as bioassay-guided fractionation or high-throughput screening are able to solve these issues, but there is a requirement for refinement and improvement [73].

The second obstruction in the discovery phase is a reliable supply and sustainable collection and harvesting; some of the marine metabolites are discovered from sources like sponges and corals. They are very slow-growing and currently are in constant threat from man-made phenomena such as global warming and climate that is constantly affecting their growth, metabolite distinction, and other characters. There is a dire need to develop and strengthen alternative methods like synthetic biology, bioprospecting for marine metabolites, artificial fermentation, and cultivation technology development to create a constant, reliable, and streamlined source for research development and, in the future, to cater the needs of the market demand. Another hurdle in the research process is the lack of marine data. Marine compounds can be quite different from those of terrestrial metabolites, often being more complex and less explored [74].

The synthetic reproduction of metabolites from slow-growing organisms or the highly complex ones can be challenging in the lab. There is a lack of established structures of marine metabolites in comparison to their abundant availability. So, there is a knowledge gap in chemically modifying marine metabolites or in structural activity relation data [30]. There is even a need to develop a reliable structural database for marine metabolite compounds. Apart from these technical issues, the biggest or most significant challenge in real life can be the complex geopolitical nature of the ocean, the lack of inter-governmental collaboration, and the availability of funds to venture into such a study. These types of studies will require a joint effort of governments and academic and private organizations [75, 76].

4.1. Prospects of Marine Metabolites

Unlocking the full potential of marine metabolites in the future will determine advances in biotechnology, synthetic chemistry and environmental conservation. In the future, the supply of marine metabolites, particularly those sourced from various types of microalgae or other microorganisms, can be mass-produced by CRISPR by some simple organisms like yeast or E. coli. In the future, the development of bioreactors and environment friendly synthesis and extraction methods can help in the mass production of macro algae as well as other marine metabolites from sources apart from microorganisms. Culture methods, like aquaculture, marine farming, etc., can be the way to form a sustainable and reliable long-term supply chain of marine metabolites as raw materials [77].

The future of marine metabolites will solely depend on research and development that can establish their significance in a clinical setup or as an adjuvant in therapeutics or cosmetics.

The anti-inflammatory approach of marine metabolites can benefit from further research in-depth diving into the molecular mechanisms establishing their efficacy [78].

Conclusion

In conclusion, we can understand that marine metabolites becoming the emerging therapeutic source of medicines for different disorders, especially those related to inflammation, whether it's chronic or acute, autoimmune or infectious. They may also be a large untapped resource. Marine metabolites will give rise to newer extraction, isolation, and biosynthesis techniques that, coupled with our knowledge of nanotechnology and novel drug delivery systems, can provide better, targeted and side-effect-free natural remedies for various inflammatory diseases. Marine metabolites have been shown to interact with key inflammatory markers like COX, MAPK, NF-Kβ, and others, showing that they have anti-inflammatory potential; however, in-depth research in molecular mechanism studies are required for the establishment of such molecules over the existing conventional drug molecules. These metabolites could help in a range of diseases, such as arthritis, neuroinflammation, and various autoimmune disorders that proceed through inflammation. The challenges facing metabolites lie in the fact that oceans are large vulnerable ecosystems which are constantly being affected by global warming, climate change, and illegal exploitation of oceans through fishing. Harvesting these sources should be a cautious and sustainable approach to keeping the ocean's environmental health in mind. Therefore, the role of artificial ocean farming, aquaculture, and biosynthesis of marine metabolites away from the ocean will be crucial. In the future, with collaborative research, marine metabolites can be the key to bridge the gap between modern medicine and natural biodiversity. The future of humankind could benefit if the horizons of natural remedies for inflammation get expanded from terrestrial to the vast diversity of oceans.

References

[1] Gao C, Wang Q, Ding Y, Kwong CHT, Liu J, Xie B, et al., Targeted therapies of inflammatory diseases with intracellularly gelated macrophages in mice and rats. *Nat Commun.* 2024 Jan 6;15(1):328.

[2] Pisetsky DS. Pathogenesis of autoimmune disease. *Nat Rev Nephrol.* 2023 Aug;19(8):509–24.

[3] Wang Y, Chen S, Du K, Liang C, Wang S, Owusu Boadi E, et al., Traditional herbal medicine: Therapeutic potential in rheumatoid arthritis. *J Ethnopharmacol.* 2021 Oct 28;279:114368.

[4] Ghallab DS, Ibrahim RS, Mohyeldin MM, Shawky E. Marine algae: A treasure trove of bioactive anti-inflammatory compounds. *Mar Pollut Bull.* 2024 Feb 1;199:116023.

[5] Cheung RCF, Ng TB, Wong JH, Chen Y, Chan WY. Marine natural products with anti-inflammatory activity. *Appl Microbiol Biotechnol.* 2016;100:1645–66.

[6] Tortora GJ, Derrickson BH. *Principles of Anatomy and Physiology. John Wiley & Sons*; 2020. 1298 p.

[7] Anderton H, Wicks IP, Silke J. Cell death in chronic inflammation: breaking the cycle to treat rheumatic disease. *Nat Rev Rheumatol.* 2020 Sep;16(9):496–513.

[8] Carestia A, Godin LC, Jenne CN. Step up to the platelet: Role of platelets in inflammation and infection. *Thromb Res.* 2023 Nov 1;231:182–94.

[9] Liao KP, Rist P, Giles J, Santacroce L, Connelly MA, Glynn RJ, et al., Impact of RA treatment strategies on lipids and vascular inflammation in rheumatoid arthritis: a secondary analysis of the TARGET randomized active comparator trial. *Arthritis Res Ther.* 2024 Jun 24;26(1):123.

[10] Hortová-Kohoutková M, Tidu F, De Zuani M, Šrámek V, Helán M, Frič J. Phagocytosis–Inflammation Crosstalk in Sepsis: New Avenues for Therapeutic Intervention. *Shock.* 2020 Nov;54(5):606.
[11] Zheng J, Yao Z, Xue L, Wang D, Tan Z. The role of immune cells in modulating chronic inflammation and osteonecrosis. *Front Immunol* . 2022 Dec 13 ;13.
[12] Ye D, Miyoshi A, Ushitani T, Kadoya M, Igeta M, Konishi K, et al., RAGE in circulating immune cells is fundamental for hippocampal inflammation and cognitive decline in a mouse model of latent chronic inflammation. *Brain Behav Immun.* 2024 Feb 1;116:329–48.
[13] Caballero-Sánchez N, Alonso-Alonso S, Nagy L. Regenerative inflammation: When immune cells help to re-build tissues. *FEBS J.* 2024;291(8):1597–614.
[14] Dhandhi S, Yeshna, Vishal, Monika, Goel B, Chauhan S, et al., The interplay of skin architecture and cellular dynamics in wound healing: Insights and innovations in care strategies. *Tissue Cell.* 2024 Dec 1;91:102578.
[15] Ahmad B, Shah M, Choi S. Oceans as a Source of Immunotherapy. *Mar Drugs.* 2019 May 10;17(5):282.
[16] Sugumaran A, Pandiyan R, Kandasamy P, Antoniraj MG, Navabshan I, Sakthivel B, et al., Marine biome-derived secondary metabolites, a class of promising antineoplastic agents: A systematic review on their classification, mechanism of action and future perspectives. *Sci Total Environ.* 2022 Aug 25;836:155445.
[17] Montefalcone M, Tunesi L, Ouerghi A. A review of the classification systems for marine benthic habitats and the new updated Barcelona Convention classification for the Mediterranean. *Mar Environ Res.* 2021 Jul 1;169:105387.
[18] A. Khalil N, M. Ahmed E, Tharwat T, Mahmoud Z. NSAIDs between past and present; a long journey towards an ideal COX-2 inhibitor lead. *RSC Adv.* 2024;14(42):30647–61.
[19] Maideen NMP, Balasubramanian R, Shanmugam A, Gobinath M, Hussain MHJ. Pseudoephedrine-Induced Risks of Posterior Reversible Encephalopathy Syndrome and Reversible Cerebral Vasoconstriction Syndrome. *Ibnosina J Med Biomed Sci* . 2024 Aug 30
[20] Butranova OI, Zyryanov SK. Selection of NSAIDs for rational pharmacotherapy of chronic musculoskeletal pain: a clinical pharmacologist's perspective. *Neurol Neuropsychiatry Psychosom.* 2024 Apr 24;16(2):87–94.
[21] Ribeiro H, Rodrigues I, Napoleão L, Lira L, Marques D, Veríssimo M, et al., Non-steroidal anti-inflammatory drugs (NSAIDs), pain and aging: Adjusting prescription to patient features. *Biomed Pharmacother.* 2022 Jun 1;150:112958.
[22] Souza CRM, Bezerra WP, Souto JT. Marine Alkaloids with Anti-Inflammatory Activity: Current Knowledge and Future Perspectives. *Mar Drugs.* 2020 Mar;18(3):147.
[23] Islam F, Dehbia Z, Zehravi M, Das R, Sivakumar M, Krishnan K, et al., Indole alkaloids from marine resources: Understandings from therapeutic point of view to treat cancers. *Chem Biol Interact.* 2023 Sep 25;383:110682.
[24] Zhang Z, Sun Y, Li Y, Song X, Wang R, Zhang D. The potential of marine-derived piperazine alkaloids: Sources, structures and bioactivities. *Eur J Med Chem.* 2024 Feb 5;265:116081.
[25] Xu J, Peng WR, Zhang D, Sun HX, Li L, Sun F, et al., Marine sponge-derived alkaloid ameliorates DSS-induced IBD via inhibiting IL-6 expression through modulating JAK2-STAT3-SOCS3 pathway. *Int Immunopharmacol.* 2024 Mar 10;129:111576.
[26] Lind KF, Hansen E, Østerud B, Eilertsen KE, Bayer A, Engqvist M, et al., Antioxidant and Anti-Inflammatory Activities of Barettin. *Mar Drugs.* 2013 Jul;11(7):2655–66.
[27] Lasalo M, Jauffrais T, Georgel P, Matsui M. Marine Microorganism Molecules as Potential Anti-Inflammatory Therapeutics. *Mar Drugs.* 2024 Sep;22(9):405.
[28] Lee SM, Kim NH, Lee S, Kim YN, Heo JD, Rho JR, et al., (10Z)-Debromohymenialdisine from Marine Sponge Stylissa sp. Regulates Intestinal Inflammatory Responses in Co-Culture Model of Epithelial Caco-2 Cells and THP-1 Macrophage Cells. *Molecules.* 2019 Jan;24(18):3394.
[29] Wallen ZD, Demirkan A, Twa G, Cohen G, Dean MN, Standaert DG, et al., Metagenomics of Parkinson's disease implicates the gut microbiome in multiple disease mechanisms. *Nat Commun.* 2022 Nov 15;13(1):6958.

[30] Farag MA, Baky MH, Kühnhold H, Kriege EA, Kunzmann A, Alseekh S, et al., Effects of thermal and UV stress on the polar and non-polar metabolome of photosymbiotic jellyfish and sea anemones. *Mar Pollut Bull.* 2024 Nov 1;208:116983.

[31] Rushdi MI, Abdel-Rahman IAM, Attia EZ, Abdelraheem WM, Saber H, Madkour HA, et al., A review on the diversity, chemical and pharmacological potential of the green algae genus *Caulerpa*. *South Afr J Bot.* 2020 Aug 1;132:226–41.

[32] Palaniyappan S, Sridhar A, Kari ZA, Téllez-Isaías G, Ramasamy T. Evaluation of Phytochemical Screening, Pigment Content, In Vitro Antioxidant, Antibacterial Potential and GC-MS Metabolite Profiling of Green Seaweed Caulerpa racemosa. *Mar Drugs.* 2023 May;21(5):278.

[33] Barbosa J da S, Sabry DA, Silva CHF, Gomes DL, Santana-Filho AP, Sassaki GL, et al., Immunostimulatory Effect of Sulfated Galactans from the Green Seaweed Caulerpa cupressoides var. flabellata. *Mar Drugs.* 2020 May;18(5):234.

[34] Núñez-Pons L, Shilling A, Verde C, Baker BJ, Giordano D. Marine Terpenoids from Polar Latitudes and Their Potential Applications in Biotechnology. *Mar Drugs.* 2020 Aug;18(8):401.

[35] Lee H, Jang H, Heo D, Eom JI, Han CH, Kim SM, et al., Tisochrysis lutea Fucoxanthin Suppresses NF-κB, JNK, and p38-Associated MMP Expression in Arthritis Pathogenesis via Antioxidant Activity. *Antioxidants.* 2024 Aug;13(8):941.

[36] Sanjay, Yoon NY, Park EJ, Lee HJ. Immune-Enhancing Effects of Marine Algae Extracts: Modulation of Macrophage Activation by Sargassum horneri, Sargassum fusiforme, and Undaria pinnatifida. *Appl Sci.* 2024 Jan;14(5):1794.

[37] Tan Y, Deng W, Zhang Y, Ke M, Zou B, Luo X, et al., A marine fungus-derived nitrobenzoyl sesquiterpenoid suppresses receptor activator of NF-κB ligand-induced osteoclastogenesis and inflammatory bone destruction. *Br J Pharmacol.* 2020;177(18):4242–60.

[38] Silva M, Avni D, Varela J, Barreira L. The Ocean's Pharmacy: Health Discoveries in Marine Algae. *Molecules.* 2024 Jan;29(8):1900.

[39] Dai LT, Yang L, Guo JC, Ma QY, Xie QY, Jiang L, et al., Anti-diabetic and anti-inflammatory indole diterpenes from the marine-derived fungus *Penicillium* sp. ZYX-Z-143. *Bioorganic Chem.* 2024 Apr 1;145:107205.

[40] Li JX, Shang RY, Xie DD, Luo XC, Hu TY, Cheng BH, et al., Arenarialins A–F, Anti-inflammatory Meroterpenoids with Rearranged Skeletons from the Marine Sponge Dysidea arenaria. *J Nat Prod.* 2024 Feb 23;87(2):396–403.

[41] Malan A, Choudhary M, Kaur Bamrah P, Kumari D. Potential benefits of marine-derived compounds for slowing the advancement of Alzheimer's disease. *J Asian Nat Prod Res.* 0(0):1–26.

[42] Bu Y, Wu H, Deng R, Wang Y. Therapeutic Potential of SphK1 Inhibitors Based on Abnormal Expression of SphK1 in Inflammatory Immune Related-Diseases. *Front Pharmacol.* 2021 Oct 19 12.

[43] Kim H, Shin HY, Jeong EJ, Lee HD, Hwang KC, Yu KW, et al., Antioxidant and Anti-Inflammatory Activities of Sargassum macrocarpum Extracts. *Antioxidants.* 2022 Dec;11(12):2483.

[44] Nagahawatta DP, Liyanage NM, Jayawardena TU, Jeon YJ. Marine Polyphenols in Cardiovascular Health: Unraveling Structure–Activity Relationships, Mechanisms, and Therapeutic Implications. *Int J Mol Sci.* 2024 Jan;25(15):8419.

[45] Matin M, Koszarska M, Atanasov AG, Król-Szmajda K, Jóźwik A, Stelmasiak A, et al., Bioactive Potential of Algae and Algae-Derived Compounds: Focus on Anti-Inflammatory, Antimicrobial, and Antioxidant Effects. *Molecules.* 2024 Jan;29(19):4695.

[46] Shin HC, Rosenfeld C, Guttendorf RJ, Wade SB, Park YJ, Kim JH, et al., A Pharmacokinetic and Bioavailability Study of Ecklonia cava Phlorotannins Following Intravenous and Oral Administration in Sprague–Dawley Rats. *Mar Drugs.* 2024 Nov;22(11):500.

[47] Carpena M, Pereira CSGP, Silva A, Barciela P, Jorge AOS, Perez-Vazquez A, et al., Metabolite Profiling of Macroalgae: Biosynthesis and Beneficial Biological Properties of Active Compounds. *Mar Drugs.* 2024 Oct;22(10):478.

[48] Dong H, Dong S, Erik Hansen P, Stagos D, Lin X, Liu M. Progress of Bromophenols in Marine Algae from 2011 to 2020: Structure, Bioactivities, and Applications. *Mar Drugs.* 2020 Aug;18(8):411.

[49] Dong H, Wang L, Guo M, Stagos D, Giakountis A, Trachana V, et al., Antioxidant and Anticancer Activities of Synthesized Methylated and Acetylated Derivatives of Natural Bromophenols. *Antioxidants.* 2022 Apr;11(4):786.

[50] Liu M, Hansen PE, Lin X. Bromophenols in Marine Algae and Their Bioactivities. *Mar Drugs.* 2011 Jul;9(7):1273–92.

[51] Kijjoa A, Sawangwong P. Drugs and Cosmetics from the Sea. *Mar Drugs.* 2004 May 25;2(2):73–82.

[52] Wang Y, Guo X, Huang C, Shi C, Xiang X. Biomedical potency and mechanisms of marine polysaccharides and oligosaccharides: A review. *Int J Biol Macromol.* 2024 Apr 1;265:131007.

[53] Jiang S, Yin H, Li R, Shi W, Mou J, Yang J. The activation effects of fucoidan from sea cucumber *Stichopus chloronotus* on RAW264.7 cells *via* TLR2/4-NF-κB pathway and its structure-activity relationship. *Carbohydr Polym.* 2021 Oct 15;270:118353.

[54] Li JK, Veeraperumal S, Aweya JJ, Liu Y, Cheong KL. Fucoidan modulates gut microbiota and immunity in Peyer's patches against inflammatory bowel disease. *Carbohydr Polym.* 2024 Oct 15;342:122421.

[55] Balasubramanian R, Maideen NMP, Muthusamy S, Gobinath M. A Review of Clinical and Preclinical Studies on the Therapeutic Potential of Black Seeds (Nigella sativa) in the Management of Polycystic Ovarian Syndrome (PCOS). *J Pharmacopuncture.* 2023 Mar 31;26(1):1–9.

[56] Pakkir Maideen NM, Balasubramanian R, Gobinath M, Rafiullah M, Muthusamy S. Therapeutic Potentials of Black Seeds (Nigella sativa) in the Management of COVID-19 -A Review of Clinical and In-silico Studies. *Anti-Infect Agents.* 2023 Feb 1;21(1):14–23.

[57] Lee YE, Kim H, Seo C, Park T, Lee KB, Yoo SY, et al., Marine polysaccharides: therapeutic efficacy and biomedical applications. *Arch Pharm Res.* 2017 Sep 1;40(9):1006–20.

[58] Wang YM, Zhang Z, Sheng Y, Chi CF, Wang B. A systematic review on marine umami peptides: Biological sources, preparation methods, structure-umami relationship, mechanism of action and biological activities. *Food Biosci.* 2024 Feb 1;57:103637.

[59] Nagarajan P, Louis LRP, Patil SJ, Adam JK, Krishna SBN. Chapter 13 - Therapeutic potential of biologically active peptides from marine organisms for biomedical applications. In: Atta-Ur-Rahman, editor. *Studies in Natural Products Chemistry.* Elsevier; 2024 p. 467–500.

[60] Xu Y, Tong X, Lu Y, Lu Y, Wang X, Han J, et al., Microalgal proteins: Unveiling sustainable alternatives to address the protein challenge. *Int J Biol Macromol.* 2024 Sep 1;276:133747.

[61] Zhang B, Kawabe R, Kamio M, Uchida H, Matsuura N, Mori-Yasumoto K, et al., Okeaniazole A: Thiazole-containing cyclopeptide from the marine cyanobacterium *Okeania hirsuta* with antileishmanial activity. *Tetrahedron Lett.* 2024 Oct 25;150:155284.

[62] Ribeiro R, Pinto E, Fernandes C, Sousa E. Marine Cyclic Peptides: Antimicrobial Activity and Synthetic Strategies. *Mar Drugs.* 2022 Jun;20(6):397.

[63] Zhang X, Zhuang H, Wu S, Mao C, Dai Y, Yan H. Marine Bioactive Peptides: Anti-Photoaging Mechanisms and Potential Skin Protective Effects. *Curr Issues Mol Biol.* 2024 Feb;46(2):990–1009.

[64] Yao W, Zhang Y, Zhang G. Marine peptides as potential anti-aging agents: Preparation, characterization, mechanisms of action, and future perspectives. *Food Chem.* 2024 Dec 1;460:140413.

[65] Wang Z, Shi Q, Feng Y, Han J, Lu C, Zhou J, et al., Targeted screening of an anti-inflammatory polypeptide from *Rhopilema esculentum* Kishinouye cnidoblasts and elucidation of its mechanism in alleviating ulcerative colitis based on an analysis of the gut microbiota and metabolites. *Food Sci Hum Wellness.* 2024 May 1;13(3):1336–47.

[66] Gobinath M, Ramaswamy S, Roychowdhury P, Dutta S, Shekhar AP, Ambrish V, et al., Nanomaterials for Vaccine Delivery: Innovations and Challenges. In: *Exploring Nanomaterial Synthesis, Characterization, and Applications* . IGI Global; 2025 . p. 267–86.

[67] Wang W, Xu Y, Zhou J, Zang Y. Effects of omega-3 supplementation on lipid metabolism, inflammation, and disease activity in rheumatoid arthritis: a meta-analysis of randomized controlled trials. *Clin Rheumatol.* 2024 Aug 1;43(8):2479–88.

[68] Santalova EA, Kolesnikova SA. 9-Chloro-5,9-dienoic and Other Fatty Acids from Marine Sponge Penares sp. *Russ J Bioorganic Chem.* 2024 Apr 1;50(2):418–31.

[69] Ma J, Hu J, Sha X, Meng D, Yang R. Phycobiliproteins, the pigment-protein complex form of natural food colorants and bioactive ingredients. *Crit Rev Food Sci Nutr.* 2024 Apr 14;64(10):2999–3017.

[70] Ashaolu TJ. The powerful phycobiliproteins-phycocyanin and phycoerythrin: Pleiotropic applications and biofunctional uses. *Algal Res.* 2024 Aug 1;82:103636.

[71] Banday AH, Azha N ul, Farooq R, Sheikh SA, Ganie MA, Parray MN, et al., Exploring the potential of marine natural products in drug development: A comprehensive review. *Phytochem Lett.* 2024 Feb 1;59:124–35.

[72] Zhu Y, Ouyang Z, Du H, Wang M, Wang J, Sun H, et al., New opportunities and challenges of natural products research: When target identification meets single-cell multiomics. *Acta Pharm Sin B.* 2022 Nov 1;12(11):4011–39.

[73] Getachew AT, Jacobsen C, Holdt SL. Emerging Technologies for the Extraction of Marine Phenolics: Opportunities and Challenges. *Mar Drugs.* 2020 Aug;18(8):389.

[74] Stuart KA, Welsh K, Walker MC, Edrada-Ebel R. Metabolomic tools used in marine natural product drug discovery. *Expert Opin Drug Discov.* 2020 Apr 2;15(4):499–522.

[75] Hu D, Jin Y, Hou X, Zhu Y, Chen D, Tai J, et al., Application of Marine Natural Products against Alzheimer's Disease: Past, Present and Future. *Mar Drugs.* 2023 Jan;21(1):43.

[76] seyyedi S reza, Kowsari E, Ramakrishna S, Gheibi M, Chinnappan A. Marine plastics, circular economy, and artificial intelligence: A comprehensive review of challenges, solutions, and policies. *J Environ Manage.* 2023 Nov 1;345:118591.

[77] Liu G, Kong Z, Sun W, Li J, Qi Z, Wu C, et al., Impacts of offshore wind power development on China's marine economy and environment: A study from 2006 to 2019. *J Clean Prod.* 2023 Oct 15;423:138618.

[78] Rigogliuso S, Campora S, Notarbartolo M, Ghersi G. Recovery of Bioactive Compounds from Marine Organisms: Focus on the Future Perspectives for Pharmacological, Biomedical and Regenerative Medicine Applications of Marine Collagen. *Molecules.* 2023 Jan;28(3):1152.

Chapter 5

Marine Metabolites in Autoimmune Disease Therapies: Insights and Opportunities

Preetham Kumar Boominathan[1]
and Balaji Govindaswamy[2,*]
[1]Institute of Biomedical Sciences, National Sun Yat-sen University, Kaohsiung, Taiwan, R.O.C.
[2]Department of Chemical Engineering, Indian Institute of Technology, Madras, Chennai, India

Abstract

Marine metabolites have attracted considerable interest for their potential therapeutic use for treating autoimmune diseases because of their distinct chemical variety and bioactivity. Dysregulated immune responses that assault host tissues cause persistent inflammation and tissue damage in autoimmune illnesses, such as rheumatoid arthritis, multiple sclerosis, inflammatory bowel disease, and systemic lupus erythematosus. Immune suppressants and corticosteroids are common components of traditional therapy. Although they are beneficial, they have significant adverse effects and frequently cause partial remission or recurrence. The exploration of natural chemicals with new methods of action and fewer side effects, especially from marine animals, has been spurred by the need for innovative, focused medicines. Understudied sources of pharmacologically active chemicals can be found in marine environments; metabolites from marine fungi, sponges, algae, cyanobacteria, and mollusks can modulate immune responses. These substances, which are obtained from marine sources, have a variety of bioactivities, including immunomodulatory, antioxidant, and anti-inflammatory properties, making them excellent choices for the treatment of autoimmune diseases. A well-known example is the sesterterpenoid manoalide, which is derived from sea sponges and inhibits phospholipase A2, an essential enzyme in the inflammatory cascade. Fucoidans, which have sulfated polysaccharides derived from brown algae, have also shown a wide range of immunoregulatory properties, such as the ability to suppress complement activation and modify T cell responses, both of which are essential for the advancement of autoimmune diseases. The precise methods via which marine metabolites interact with major pathways linked to autoimmune disorders, including the regulation of cytokine release, mitogen-activated protein kinase (MAPK) pathways, and nuclear factor-kappa B (NF-kB) signalling. Special consideration is given to substances that selectively decrease pro-inflammatory cytokines, which are important players in autoimmune disorders, such as

* Corresponding Author's Email: balajishilou.08@gmail.com

In: Healing from the Depths Marine Metabolites in Disease Management
Editors: Pranav Kumar Prabhakar and Tarun Kumar Upadhyay
ISBN: 979-8-89530-832-5

tumor necrosis factor-alha (TNF-α), interleukin-1β (IL-1 β), and interleukin-6 (IL-6). Additionally, it is described how myeloid-derived suppressor cell (MDSC) recruitment and regulatory T-cell (T reg) activation may be used by marine-derived peptides and alkaloids to restore immunological tolerance. Recent preclinical research has demonstrated that marine metabolites not only target the underlying immunopathogenesis but also reduce the symptoms of autoimmune illnesses in animal models, providing promise for long-term disease management. Future prospects for marine-based chemicals as innovative therapeutics in the management of autoimmune disease appear promising if insights from immunology and marine biotechnology are integrated.

Keywords: marine metabolites, autoimmune disease, immune regulation, cytokine regulation, anti-inflammatory compounds

1. Introduction

Autoimmune diseases are characterized by the failure of the immune system to recognize self-antigens, leading to chronic inflammation and tissue destruction in various organs. Despite advances in understanding the mechanisms of autoimmune disease, conventional treatments remain palliative, focusing on immunosuppression rather than targeted cures (Kim et al., 2024). This often results in significant side effects and diminished quality of life. As the incidence of autoimmune disorders continues to rise globally, there is an urgent need for innovative, effective, and safer therapeutic options (Giacomelli et al., 2021). Marine ecosystems, accounting for over 70% of the Earth's surface, represent a largely untapped reservoir of bioactive compounds (Ghosh et al., 2022). Marine organisms, including sponges, algae, mollusks, and microorganisms, thrive in diverse and often extreme environmental conditions, leading to the evolution of unique chemical defense mechanisms (Putz & Proksch, 2010). These adaptations have provided marine species with a rich array of secondary metabolites that exhibit potent bioactivity against a variety of biological targets (Giordano et al., 2015).

Marine metabolites are a diverse and chemically rich class of natural compounds produced by marine organisms, including sponges, algae, corals, mollusks, and various microorganisms. These compounds are part of the secondary metabolism of marine species, which refers to metabolic processes that produce non-essential yet bioactive compounds critical for survival in challenging ocean environments (Senadheera et al., 2023). Due to unique evolutionary pressures, such as competition for space, defense against predators, and adaptation to extreme conditions, marine organisms have developed a wide array of structurally complex and biologically active molecules (Giordano, 2020). This has resulted in metabolites with potent pharmacological activities, many of which differ from terrestrial natural products in terms of both structure and bioactivity (Gao et al., 2018).

These marine metabolites include alkaloids, peptides, polysaccharides, Terpenoids, and other structurally diverse compounds known for their anti-inflammatory, antioxidant, and immunomodulatory properties (A. Mayer et al., (2024). Recent studies have highlighted the potential of several marine-derived compounds to modulate immune responses, suggesting their potential as leads for autoimmune disease therapies (Montuori et al., 2022). In the realm of autoimmune disease, marine metabolites offer promising opportunities because of their natural diversity and unique mechanisms of action, which often differ from those of traditional small-molecule drugs (Li et al., 2019). Certain marine-derived compounds have been shown to regulate inflammatory pathways or modulate immune cell activity, offering potential as both standalone therapies and as adjuncts to existing treatments (Sen et al., 2024). For example, some compounds from marine sponges and algae have shown the capacity to downregulate pro-inflammatory cytokines or inhibit key enzymes involved in autoimmune pathology, such as cyclooxygenases and lipoxygenase (Meenakshi et al., 2023). Additionally, the immunosuppressive properties of some marine metabolites achieved through T cell modulation or cytokine inhibition; could provide targeted therapeutic effects with reduced systemic toxicity (D'Orazio et al., 2012).

2. Immunomodulatory Properties of Marine Metabolites

Marine metabolites are distinguished by their unique chemical structures, which often result in distinct selective immunological interactions. Unlike synthetic drugs, which can broadly suppress immune function and cause side effects, many marine-derived compounds exhibit more selective immunomodulatory actions that address specific pathways relevant to autoimmune pathogenesis (Tan, 2023). Autoimmune diseases, such as rheumatoid arthritis, multiple sclerosis, inflammatory bowel syndrome, and systemic lupus erythematosus, occur when the immune system mistakenly targets the body's own tissues, leading to chronic inflammation, pain, and tissue damage (Kumar et al., 2024). Although conventional treatments such as immunosuppressants and corticosteroids are widely prescribed, they often lead to partial remission, frequent relapses, and unwanted side effects. This ongoing challenge in autoimmune therapy has prompted scientists to explore the ocean's unique biochemical resources. Marine metabolites exhibit powerful biological activities, including anti-oxidative, anti-inflammatory, and immunomodulatory effects, making them promising candidates for autoimmune therapies (Montuori et al., 2022). Compounds from sea sponges, algae, cyanobacteria, and other marine organisms show potential for targeted immune modulation through unique mechanisms with fewer side effects than conventional treatments. Rather than broadly dampening immune function, these compounds often exhibit fine-tuned effects that target specific inflammatory pathways (Perera et al., 2023). This distinguishes them from traditional anti-inflammatory and immunosuppressive drugs because of their selective, often targeted effects on immune pathways. Their specificity could be critical in mitigating autoimmune inflammation with fewer side effects and greater potential while preserving the immune system's functional capacity.

In autoimmune diseases, the immune system is dysregulated by excessive production of pro-inflammatory cytokines, such as tumor necrosis factor-alpha (TNF-α), interleukin-1 beta (IL-1β), and interleukin-6 (IL-6). These cytokines are pivotal in promoting inflammation, cell recruitment, and tissue damage, thereby contributing to the progression of autoimmune conditions like rheumatoid arthritis, multiple sclerosis, and systemic lupus erythematosus (Xiang et al., 2023). One promising aspect of marine metabolites is their ability to selectively modulate these proinflammatory cytokines. Conventional therapies, while effective in reducing overall inflammation, often lack specificity, leading to broader immunosuppressive effects and associated adverse events. Marine metabolites offer an exciting alternative because of their potential to selectively modulate cytokines, thus achieving a balance between immune suppression and immune activation without compromising the body's entire immune defense (Elbandy, 2022).

Cytokines like TNF-α, IL-1β, and IL-6 play crucial roles in autoimmune disease pathology. TNF-α is a central mediator in inflammatory processes, driving the recruitment and activation of immune cells, such as neutrophils and macrophages, to sites of inflammation. Elevated TNF-α levels are observed in conditions like rheumatoid arthritis, where it stimulates osteoclastogenesis, leading to joint damage (Jang et al., 2021). IL-1β is another pro-inflammatory cytokine that induces fever, activates endothelial cells, and promotes leukocyte adhesion and migration to inflamed tissues. High IL-1β levels contribute to synovitis in arthritis and demyelination in multiple sclerosis (Zhao et al., 2013). IL-6 is involved in T-helper 17 (Th17) cell differentiation, a subset implicated in autoimmune pathology, as well as B cell maturation and antibody production, which can further fuel autoimmune responses by generating self-reactive antibodies (McGeachy et al., 2019). The challenge lies in controlling these cytokines locally within affected tissues without systemically suppressing the immune response, a balance that marine metabolites may help achieve. Fucoidans are sulfated polysaccharides derived from brown algae such as *Sargassum siliquastrum*, and have shown notable promise due to their multi-faceted impact on immune modulation by its ability to down regulate TNF-α, IL-1β, and IL-6 production in preclinical models of inflammation (Jayasinghe et al., 2023). By inhibiting the NF-kB signaling pathway, fucoidan reduce the transcription of these proinflammatory cytokines specifically within the inflamed tissue, helping restore balance without widespread immunosuppression (Olsthoorn et al., 2021). Additionally, fucoidan have been shown to decrease the activity of IL-17 and IL-23 -driven Th17 cells, which are key drivers of autoimmunity, thereby dampening the pathogenic response while leaving protective immune functions intact. They exhibit the capacity to suppress complement activation, a pathway known to intensify tissue damage in conditions like systemic lupus erythematosus and multiple sclerosis. By inhibiting this pathway, fucoidan can prevent tissue damage caused by the immune system (Xue et al., 2020).

Studies on manolide, a sesterterpenoid from sea sponges, have shown that it inhibits phospholipase A2 (PLA2), a key enzyme that catalyzes the release of arachidonic acid and subsequent production a cascade that triggers the production of pro-inflammatory eicosanoids like prostaglandins and leukotrienes. Through this inhibition, manoalide indirectly reduces TNF-α and IL-1β levels that can effectively disrupt inflammatory processes while minimizing broader immunosuppressive effects (Anjum et al., 2016; Moreira et al., 2021). Unlike corticosteroids, which broadly suppress cytokine production, manolide's action on PLA2 limits its effects on eicosanoid-related pathways, offering more focused inflammation control. Manolide's selective action against phospholipase A2 highlights its potential as a targeted anti-inflammatory agent, particularly for diseases like rheumatoid arthritis, where the inflammatory cascade plays a primary role in disease pathology. Preclinical studies in murine models of rheumatoid arthritis have shown that manoalide can significantly reduce joint inflammation and pain, making it a promising alternative to standard anti-inflammatory drugs that often lack specificity (Schrier et al., 1996). Unlike nonsteroidal anti-inflammatory drugs (NSAIDs) that affect a broad spectrum of immune responses, manolide's targeted action against phospholipase A2 could lead to fewer side effects and make it a more sustainable long-term treatment.

Honaucin A, isolated from cyanobacteria, works through an entirely different mechanism by inhibiting the NF-kB signaling pathway, meaning it could help control autoimmune flare-ups without broadly suppressing immune function in conditions such as multiple sclerosis and rheumatoid arthritis (Choi et al., 2012a). NF-kB is a protein complex that controls DNA transcription and is involved in the body's response to infection and inflammation by regulating genes responsible for inflammation and cell survival, making it a central player in autoimmune diseases. In autoimmune diseases, NF-B is often over-activated, leading to excessive production of inflammatory cytokines that damage tissues (Barnabei et al., 2021; Liu et al., 2017). Preliminary studies have shown that nonmucin A reduces the levels of pro-inflammatory cytokines, such as TNF-α and IL-1β, in laboratory models of inflammation with minimal cytotoxicity (Choi et al., 2012b; Salvador-Reyes & Luesch, 2015). By tempering the immune response at a molecular level, nonmucin A offers a targeted approach for managing diseases such as multiple sclerosis and rheumatoid arthritis. Given the direct link between NF-B activation and autoimmune disease progression, nonmucin A's pathway-specific action underscores the appeal of marine metabolites in autoimmune therapies (Mascuch et al., 2018). By selectively targeting these inflammatory mediators, marine metabolites provide a mechanism that dampens immune responses at specific sites of inflammation. This targeted cytokine modulation helps re-establish immune homeostasis without broadly suppressing immunity, which holds promise for the safer long-term management of autoimmune diseases and may predispose patients to infections.

3. Mechanisms of Action of Immune Tolerance Restoration

An interesting area of marine metabolite research is the restoration of immune tolerance—a vital aspect in treating autoimmune diseases, where immune tolerance to self-antigens has been lost. Marine-derived peptides and alkaloids from sponges and cyanobacteria, for instance, have demonstrated potential in modulating regulatory T cells (Tregs) and promoting the activity of myeloid-derived suppressor cells (MDSCs), both crucial in maintaining immune tolerance (Yu et al., 2024). Enhanced Treg activity, facilitated by compounds like fucoidan, can suppress autoreactive immune responses, which is particularly beneficial in autoimmune diseases characterized by overactive immune responses (Xue et al., 2019a).

One of the most significant challenges in treating autoimmune diseases is re-establishing immune tolerance, which involves the body's ability to recognize and ignore self-antigens. In patients with autoimmune disorders, immune tolerance is disrupted, leading to an attack on self-tissues. Regulatory T cells (Tregs) and myeloid-derived suppressor cells (MDSCs) are required for maintaining immune tolerance by suppressing autoreactive T cells and creating an anti-inflammatory environment. However, in autoimmune diseases, Treg function is often impaired, and the population of MDSCs may be insufficient to counterbalance pro-inflammatory immune responses (Tamberi et al., 2024). Marine metabolites may restore immune tolerance by modulating Treg and MDSC function, thereby preventing the immune system from attacking self-antigens without broadly suppressing immune function.

4. Restoration of Treg Function via Marine Metabolites

Regulatory T cells (Tregs), which are characterized by the expression of the transcription factor FoxP3, are essential for immune tolerance. They work by producing anti-inflammatory cytokines like IL-10 and transforming growth factor-beta (TGF-β), which inhibit autoreactive immune cells. In patients with autoimmune disorders, Treg numbers and functionality are often reduced, leading to unchecked inflammatory responses (Rudensky, 2011). Research on marine metabolites has revealed several compounds that enhance Treg function (Abril et al., 2024a; A. Mayer et al., (2024). Fucoidans can increase Treg proliferation and suppress Treg function. This function is crucial in autoimmune diseases, where tolerance is often compromised. By activating the TGF-β pathway, fucoidan promote Treg differentiation and survival, thereby restoring immune tolerance in autoimmune conditions (Y. Xu et al., (2022). In murine models of multiple sclerosis, fucoidan administration was associated with increased Treg counts and a concurrent reduction in inflammation, demonstrating its potential to mitigate disease progression by enhancing Treg-mediated tolerance (Xue et al., 2019b). The gene displays several immunoregulatory functions, including the inhibition of complement activation and modulation of T-cell responses, both critical elements in autoimmune disease progression. These compounds demonstrate that marine metabolites can provide a balanced approach to immune regulation by targeting overactive inflammatory pathways without compromising the immune system as a whole. Carrageenan extracts from red algae also modulate Treg activity through mechanisms involving TGF-β signalling (Besednova et al., 2020). Studies have shown that carrageenan can increase FoxP3 expression in Tregs, thereby making them more effective at controlling inflammatory responses. By fostering Treg activity, carrageenan could theoretically contribute to immune tolerance in diseases like inflammatory bowel disease, where loss of tolerance to gut microbiota plays a significant role (Wu et al., 2016; S.-S. Xu et al., 2017).

5. Enhancing MDSC Activity to Support Immune Suppression

Myeloid-derived suppressor cells (MDSCs) are another essential component of immune response regulation. These cells suppress T cell activation and expansion, thereby controlling immune responses and preventing autoimmune inflammation (Gabrilovich & Nagaraj, 2009). By promoting MDSC recruitment and functionality, marine metabolites may help restore immune homeostasis and prevent autoreactive immune responses in autoimmune conditions. Marine alkaloids from sponges, such as Bromopyrrole alkaloids, recruited MDSCs in animal models of chronic inflammation (Rane et al., 2017). Bromopyrrole alkaloids exert their effects by modulating chemokine production, which attracts MDSCs to sites of inflammation where they can suppress autoreactive T cells (A. M. S. Mayer et al., (2019). This recruitment is particularly valuable in autoimmune diseases, where maintaining local immune suppression is essential for controlling tissue damage without affecting systemic immunity. Marine peptides, particularly those isolated from cyanobacteria, also exhibit MDSC-promoting properties. For example, certain peptides from cyanobacteria activate the STAT3 pathway, which is essential for MDSC expansion and suppressive function. Through STAT3 activation, these peptides could potentially support immune tolerance by bolstering MDSC populations and enhancing their ability to inhibit proinflammatory T cells (Ahmed et al., 2023; Safarzadeh et al., 2021).

6. Potential for Dual Modulation of Tregs and MDSCs

A unique advantage of marine metabolites is their ability to influence both Tregs and MDSCs, offering a multifaceted approach to immune tolerance. By enhancing Treg function and expanding MDSC populations simultaneously, marine metabolites like fucoidan could theoretically re-establish immune tolerance in a comprehensive manner (Haist et al., 2021). This dual modulation is particularly promising for diseases such as systemic lupus erythematosus, in which both Treg insufficiency and impaired MDSC function contribute to disease pathology. The selective cytokine modulation and immune tolerance restoration properties of marine metabolites represent a promising frontier in autoimmune disease therapy. By specifically targeting inflammatory cytokines and bolstering immune tolerance mechanisms, marine-derived compounds like fucoidan, manoalide, bromopyrazole alkaloids, and certain peptides offer a nuanced approach to immune modulation that traditional therapies often lack (Elbandy, 2022; Sanjeewa et al., 2021a). As research advances, the potential for these compounds to achieve disease modification without the broad immunosuppressive effects seen in conventional treatments could revolutionize the management of autoimmune diseases. By selectively targeting the immune responses that drive inflammation, they can mitigate symptoms and alter disease progression without the drawbacks of generalized immunosuppression.

7. Mechanism of Action of Autoimmune Therapy

The human immune system's capacity for self-regulation is typically effective in distinguishing foreign pathogens from host tissues. However, in autoimmune diseases, this regulatory balance falters, resulting in an immune assault on self-cells and tissues. Such dysregulation contributes to chronic conditions characterized by persistent inflammation, pain, and progressive tissue destruction. Standard treatments for autoimmune diseases primarily include broad immunosuppressive agents and corticosteroids. Although beneficial, these therapies are associated with notable adverse effects, such as susceptibility to infections, hepatotoxicity, and potential recurrence of symptoms (Carballido et al., 2020).

Marine ecosystems represent an underexplored reservoir of biologically active compounds. The unique adaptive mechanisms of marine organisms, from sponges to cyanobacteria, have led them to develop metabolites with distinctive chemical structures and functions that may offer novel therapeutic avenues. Notably, these metabolites often exhibit targeted immunomodulatory effects, showing potential to address specific inflammatory pathways involved in autoimmune pathogenesis while avoiding the systemic suppression typical of conventional therapies (Nawaz et al., 2023). The bioactivity of marine metabolites in autoimmune therapies revolves around their unique mechanisms of action, often targeting specific elements within inflammatory pathways. Many marine compounds modulate cytokine release, inhibit proinflammatory signaling cascades, or influence immune cell recruitment, such as myeloid-derived suppressor cells (MDSCs) and Tregs, with precision, a capability that could make them highly effective for conditions where traditional treatments fall short (Florean et al., 2022). One mechanism involves the selective downregulation of pro-inflammatory cytokines such as TNF-α, IL-1β, and IL-6, which are heavily implicated in autoimmune diseases. By selectively downregulating these cytokines, marine metabolites can disrupt the inflammatory cycle that perpetuates tissue damage in autoimmune diseases. For instance, marine-derived peptides can recruit MDSCs and activate Tregs, promoting immune tolerance and potentially restoring the immune balance disrupted in autoimmune conditions (Cheung et al., 2016; Ghallab et al., 2024).

Additionally, some metabolites like fucoidan influence the MAPK and NF-B pathways, reducing the release of inflammatory mediators without compromising overall immune function. This targeted action is particularly valuable for the long-term management of autoimmune diseases, in which minimizing side effects is crucial for maintaining patient quality of life. Marine metabolites thus not only hold promise for symptom management and also address the underlying immunopathology of autoimmune diseases, suggesting the potential for sustained remission and reduced recurrence (Besednova et al., 2022; Sanjeewa et al., 2021b). Marine metabolites such as fucoidan and nonmucin A can inhibit the signaling pathways that lead to cytokine production, thereby reducing inflammation and tissue damage. The unique properties of marine metabolites make them compelling candidates for the treatment of autoimmune diseases. Mayer et al., 2017; Systematic Review and Meta-analysis Expert Group (SRMEG), Universal Scientific Education and Research Network (USERN), Tehran, Iran & Elahe, 2022).

Other metabolites directly influence immune cell behavior directly. Marine peptides and alkaloids can stimulate the recruitment of myeloid-derived suppressor cells (MDSCs) and regulatory T cells (Tregs), which are both important for maintaining immune tolerance (Florean et al., 2022; Proksch, 1994). This immune tolerance effect is particularly valuable in autoimmune conditions, where the immune system's failure to recognize "self" tissues leads to chronic inflammation and tissue damage. By restoring tolerance, these compounds may not only relieve symptoms but also alter the course of the disease itself. Their selective immune modulation; ability to restore immune tolerance, and targeted cytokine regulation offer potential advantages over broad-spectrum immunosuppressant drugs, which often come with significant side effects. Compounds such as manoalide, fucoidan, and nonmucin A demonstrate the range of bioactivities present in marine ecosystems, with each metabolite offering unique mechanisms of action that either prevent immune cells from overreacting or encourage immune tolerance, reducing the chances of the body attacking itself. By modulating key inflammatory pathways and promoting immune tolerance, these metabolites demonstrate a novel approach to treating autoimmune diseases, with early studies indicating their effectiveness in preclinical models (Abril et al., 2024b; Anisha et al., 2022; Pradhan et al., 2020).

Marine metabolites show promise beyond autoimmune diseases, potentially offering insights into regenerative medicine and other chronic conditions. For instance, the immunomodulatory and anti-inflammatory properties that make these compounds effective in autoimmune diseases could also benefit treatments for neurodegenerative disorders like Alzheimer's disease and Parkinson's disease, where inflammation and immune dysfunction play significant roles (Naik et al., 2024). Similarly, the tissue-repair properties found in some marine compounds may lend themselves to wound healing and recovery therapies for patients undergoing intensive treatments like chemotherapy or radiation (Shanthi Kumari et al., 2020; Wang et al., 2024).

Conclusion

The potential of marine metabolites in the development of autoimmune disease therapies is increasingly evident given their unique chemical diversity, bioactivity, and often selective immunomodulatory properties. Marine-derived compounds, including alkaloids, peptides, polysaccharides, and Terpenoids, have demonstrated remarkable capacity to target inflammatory and immune pathways implicated in autoimmune pathology. Through mechanisms such as cytokine modulation, the inhibition of inflammatory enzymes, and the regulation of T cell activity, these compounds offer promising routes to conventional immunosuppressive drugs.

References

Abril, A. G., Carrera M, Pazos M. (2024a). Marine Bioactive Compounds with Functional Role in Immunity and Food Allergy. *Nutrients*, *16*(16), 2592. https://doi.org/10.3390/nu16162592.

Abril, A. G., Carrera M, Pazos M. (2024b). Marine Bioactive Compounds with Functional Role in Immunity and Food Allergy. *Nutrients*, *16*(16), 2592. https://doi.org/10.3390/nu16162592.

Ahmed, S., Alam, W., Aschner, M., Filosa, R., Cheang, W. S., Jeandet, P., Saso, L., & Khan, H. (2023). Marine Cyanobacterial Peptides in Neuroblastoma: Search for Better Therapeutic Options. *Cancers*, *15*(9), 2515. https://doi.org/10.3390/cancers15092515.

Anisha, G. S., Padmakumari, S., Patel, A. K., Pandey, A., & Singhania, R. R. (2022). Fucoidan from Marine Macroalgae: Biological Actions and Applications in Regenerative Medicine, Drug Delivery Systems and Food Industry. *Bioengineering*, *9*(9), 472. https://doi.org/10.3390/bioengineering9090472.

Anjum, K., Abbas, S. Q., Shah, S. A. A., Akhter, N., Batool, S., and Hassan, S. S. U. (2016). Marine Sponges as a Drug Treasure. *Biomolecules & Therapeutics*, *24*(4), 347–362. https://doi.org/10.4062/biomolther.2016.067.

Barnabei, L., Laplantine, E., Mbongo, W., Rieux-Laucat, F., & Weil, R. (2021). NF-κB: At the Borders of Autoimmunity and Inflammation. *Frontiers in Immunology* 12, 716469. https://doi.org/10.3389/fimmu.2021.716469.

Besednova, N. N., Andryukov, B. G., Zaporozhets, T. S., Kuznetsova, T. A., Kryzhanovsky, S. P., Ermakova, S. P., Galkina, I. V., & Shchelkanov, M. Yu. (2022). Molecular Targets of Brown Algae Phlorotannins for the Treatment of Inflammatory Processes of Various Origins. *Marine Drugs*, *20*(4), 243. https://doi.org/10.3390/md20040243.

Besednova, N. N., Zaporozhets, T. S., Kuznetsova, T. A., Makarenkova, I. D., Kryzhanovsky, S. P., Fedyanina, L. N., and Ermakova, S. P. (2020). Extracts and Marine Algae Polysaccharides for the Treatment and Prevention of Inflammatory Diseases of the Intestine. *Marine Drugs*, *18*(6), 289. https://doi.org/10.3390/md18060289.

Carballido, J. M., Regairaz, C., Rauld, C., Raad, L., Picard, D., & Kammüller, M. (2020). Emerging Jamboree of Transformative Therapies for Autoimmune Diseases. *Frontiers in Immunology*, *11*, 472. https://doi.org/10.3389/fimmu.2020.00472.

Cheung, R. C. F., Ng, T. B., Wong, J. H., Chen, Y., and Chan, W. Y. (2016). Marine natural products with anti-inflammatory activity. *Applied Microbiology and Biotechnology*, *100*(4), 1645–1666. https://doi.org/10.1007/s00253-015-7244-3.

Choi, H., Mascuch, S. J., Villa, F. A., Byrum, T., Teasdale, M. E., Smith, J. E., Preskitt, L. B., Rowley, D. C., Gerwick, L., & Gerwick, W. H. (2012a). Honaucins A–C, Potent Inhibitors of Inflammation and Bacterial Quorum Sensing: Synthetic Derivatives and Structure-Activity Relationships. *Chemistry & Biology*, *19*(5), 589–598. https://doi.org/10.1016/j.chembiol.2012.03.014.

Choi, H., Mascuch, S. J., Villa, F. A., Byrum, T., Teasdale, M. E., Smith, J. E., Preskitt, L. B., Rowley, D. C., Gerwick, L., & Gerwick, W. H. (2012b). Honaucins A–C, Potent Inhibitors of Inflammation and Bacterial

Quorum Sensing: Synthetic Derivatives and Structure-Activity Relationships. *Chemistry & Biology*, *19*(5), 589–598. https://doi.org/10.1016/j.chembiol.2012.03.014.

D'Orazio, N., Gammone, M. A., Gemello, E., De Girolamo, M., Cusenza S, Riccioni G. (2012). Marine Bioactives: Pharmacological Properties and Potential Applications against Inflammatory Diseases. *Marine Drugs*, *10*(4), 812–833. https://doi.org/10.3390/md10040812.

Elbandy, M. (2022). Anti-Inflammatory Effects of Marine Bioactive Compounds and Their Potential as Functional Food Ingredients in the Prevention and Treatment of Neuroinflammatory Disorders. *Molecules*, *28*(1), 2. https://doi.org/10.3390/molecules28010002.

Florean, C., Dicato, M., & Diederich, M. (2022). Immune-modulating and anti-inflammatory marine compounds against cancer. *Seminars in Cancer Biology*, *80*, 58–72. https://doi.org/10.1016/j.semcancer.2020.02.008.

Gabrilovich, D. I., & Nagaraj, S. (2009). Myeloid-derived suppressor cells as regulators of the immune system. *Nature Reviews Immunology*, *9*(3), 162–174. https://doi.org/10.1038/nri2506.

Gao, H., Li, G., & Lou, H.-X. (2018). Structural Diversity and Biological Activities of Novel Secondary Metabolites from Endophytes. *Molecules*, *23*(3), 646. https://doi.org/10.3390/molecules23030646.

Ghallab, D. S., Ibrahim, R. S., Mohyeldin, M. M., and Shawky, E. (2024). Marine algae: A treasure trove of bioactive anti-inflammatory compounds. *Marine Pollution Bulletin*, *199*, 116023. https://doi.org/10.1016/j.marpolbul.2023.116023.

Ghosh, S., Sarkar, T., Pati, S., Kari, Z. A., Edinur, H. A., and Chakraborty, R. (2022). Novel Bioactive Compounds From Marine Sources as a Tool for Functional Food Development. *Frontiers in Marine Science*, *9*, 832957. https://doi.org/10.3389/fmars.2022.832957.

Giacomelli, R., Afeltra, A., Bartoloni, E., Berardicurti, O., Bombardieri, M., Bortoluzzi, A., Carubbi, F., Caso, F., Cervera, R., Ciccia, F., Cipriani, P., Coloma-Bazán, E., Conti, F., Costa, L., D'Angelo, S., Distler, O., Feist, E., Foulquier, N., Gabini, M., … Tzioufas, A. (2021). The growing role of precision medicine in the treatment of autoimmune diseases; results of a systematic review of literature and Experts' Consensus. *Autoimmunity Reviews*, *20*(2), 102738. https://doi.org/10.1016/j.autrev.2020.102738.

Giordano, D. (2020). Bioactive Molecules from Extreme Environments. *Marine Drugs*, *18*(12), 640. https://doi.org/10.3390/md18120640.

Giordano, D., Coppola, D., Russo, R., Denaro, R., Giuliano, L., Lauro, F. M., Di Prisco, G., Verde, C. (2015). Marine Microbial Secondary Metabolites. In *Advances in Microbial Physiology* (Vol. 66, pp. 357–428). Elsevier. https://doi.org/10.1016/bs.ampbs.2015.04.001.

Haist, M., Stege, H., Grabbe, S., & Bros, M. (2021). The Functional Crosstalk between Myeloid-Derived Suppressor Cells and Regulatory T Cells within the Immunosuppressive Tumor Microenvironment. *Cancers*, *13*(2), 210. https://doi.org/10.3390/cancers13020210.

Jang, D., Lee, A.-H., Shin, H.-Y., Song, H.-R., Park, J.-H., Kang, T.-B., Lee, S.-R., and Yang, S.-H. (2021). The Role of Tumor Necrosis Factor Alpha (TNF-α) in Autoimmune Disease and Current TNF-α Inhibitors in Therapeutics. *International Journal of Molecular Sciences*, *22*(5), 2719. https://doi.org/10.3390/ijms22052719.

Jayasinghe, A. M. K., Kirindage, K. G. I. S., Fernando, I. P. S., Kim, K.-N., Oh, J.-Y., and Ahn, G. (2023). The Anti-Inflammatory Effect of Low-molecular-weight Fucoidan from Sargassum siliquastrum in Lipopolysaccharide-Stimulated RAW 264.7 Macrophages by Inhibiting NF-κB/MAPK Signaling Pathways. *Marine Drugs*, *21*(6), 347. https://doi.org/10.3390/md21060347.

Kim, K. M., D'Elia, A. M., & Rodell, C. B. (2024). Hydrogel-based approaches to target hypersensitivity mechanisms underlying autoimmune disease. *Advanced Drug Delivery Reviews*, *212*, 115395. https://doi.org/10.1016/j.addr.2024.115395.

Kumar, S., Kaushik, D., & Sharma, S. K. (2024). Autoimmune Disorders: Types, Symptoms, and Risk Factors. In K. Raza and S. Singh (Eds.). *Artificial Intelligence and Autoimmune Diseases* (Vol. 1133, pp. 3–31). Springer Nature Singapore. https://doi.org/10.1007/978-981-99-9029-0_1.

Li, R., Gu, Y.-C., & Zhang, W. (2019). Emerging Marine Immunomodulatory Small-molecules (2010-Present). *Current Chemical Biology*, *13*(3), 187–196. https://doi.org/10.2174/2212796813666190716102614.

Liu, T., Zhang, L., Joo, D., & Sun, S.-C. (2017). NF-κB signaling in inflammation. *Signal Transduction and Targeted Therapy*, *2*(1), 17023. https://doi.org/10.1038/sigtrans.2017.23.

Mascuch, S. J., Boudreau, P. D., Carland, T. M., Pierce, N. T., Olson, J., Hensler, M. E., Choi, H., Campanale, J., Hamdoun, A., Nizet, V., Gerwick, W. H., Gaasterland, T., & Gerwick, L. (2018). Marine Natural Product Honaucin A Attenuates Inflammation by Activating the Nrf2-ARE Pathway. *Journal of Natural Products*, *81*(3), 506–514. https://doi.org/10.1021/acs.jnatprod.7b00734.

Mayer, A. M. S., Guerrero, A. J., Rodríguez, A. D., Taglialatela-Scafati, O., Nakamura, F., Fusetani, N. (2019). Marine Pharmacology in 2014–2015: Marine Compounds with Antibacterial, Antidiabetic, Antifungal, Anti-Inflammatory, Antiprotozoal, Antituberculosis, Antiviral, and Anthelmintic Activities; Affecting the Immune and Nervous Systems, and Other Miscellaneous Mechanisms of Action. *Marine Drugs*, *18*(1), 5. https://doi.org/10.3390/md18010005.

Mayer, A., Mayer, V., Swanson-Mungerson, M., Pierce, M., Rodríguez, A., Nakamura, F., & Taglialatela-Scafati, O. (2024). Marine Pharmacology in 2019–2021: Marine Compounds with Antibacterial, Antidiabetic, Antifungal, Anti-Inflammatory, Antiprotozoal, Antituberculosis, and Antiviral Activities; Affecting the Immune and Nervous Systems, and Other Miscellaneous Mechanisms of Action. *Marine Drugs*, *22*(7), 309. https://doi.org/10.3390/md22070309.

Mayer, A., Rodríguez, A., Taglialatela-Scafati, O., & Fusetani, N. (2017). Marine Pharmacology in 2012–2013: Marine Compounds with Antibacterial, Antidiabetic, Antifungal, Anti-Inflammatory, Antiprotozoal, Antituberculosis, and Antiviral Activities; Affecting the Immune and Nervous Systems, and Other Miscellaneous Mechanisms of Action. *Marine Drugs*, *15*(9), 273. https://doi.org/10.3390/md15090273.

McGeachy, M. J., Cua, D. J., & Gaffen, S. L. (2019). The IL-17 Family of Cytokines in Health and Disease. *Immunity*, *50*(4), 892–906. https://doi.org/10.1016/j.immuni.2019.03.021.

Meenakshi, D. U., Narde, G. K., Siddiqui, S., and Ahuja, A. (2023). Marine Bioactive Phytoconstituents in Autoimmune Disorders: Role and Mechanism - A Review. *Anti-Inflammatory & Anti-Allergy Agents in Medicinal Chemistry*, *22*(1), 10–29. https://doi.org/10.2174/1871523022666230731104529.

Montuori, E., De Pascale, D., & Lauritano, C. (2022). Recent Discoveries on Marine Organism Immunomodulatory Activities. *Marine Drugs*, *20*(7), 422. https://doi.org/10.3390/md20070422.

Moreira, V., E. Leiguez, P. M. Janovits, P. M., Maia-Marques, C. M. Fernandes, C. M., & Teixeira, C. (2021). Inflammatory Effects of Bothrops Phospholipases A2: Mechanisms Involved in Biosynthesis of Lipid Mediators and Lipid Accumulation. *Toxins*, *13*(12), 868. https://doi.org/10.3390/toxins13120868.

Naik, B., Richa, S., Bharadwaj, S., Mishra, S., Kumar, V., Kumar, V., Saris, P. E. J., Gupta, A. K., Mishra, R., Gupta, U., Rustagi, S., and Preet, M. S. (2024). Utilization of marine algal metabolites to fight neurodegenerative diseases. *Frontiers in Marine Science*, *11*, 1370839. https://doi.org/10.3389/fmars.2024.1370839.

Nawaz, T., Gu, L., Fahad, S., Saud, S., Jiang, Z., Hassan, S., Harrison, M. T., Liu, K., Khan, M. A., Liu, H., El-Kahtany, K., Wu, C., Zhu, M., and Zhou, R. (2023). A comprehensive review of the therapeutic potential of cyanobacterial marine bioactives: Uncovering the hidden treasures of the sea. *Food and Energy Security*, *12*(5), e495. https://doi.org/10.1002/fes3.495.

Olsthoorn, S. E. M., Wang, X., Tillema, B., Vanmierlo, T., Kraan, S., Leenen, P. J. M., and Mulder, M. T. (2021). Brown Seaweed Food Supplementation: Effects on Allergy and Inflammation and Its Consequences. *Nutrients*, *13*(8), 2613. https://doi.org/10.3390/nu13082613.

Perera, R. M. T. D., Herath, K. H. I. N. M., Sanjeewa, K. K. A., and Jayawardena, T. U. (2023). Recent Reports on the Bioactive Compounds of Marine Cyanobacteria in Relation to Human Health Applications. *Life*, *13*(6), 1411. https://doi.org/10.3390/life13061411.

Pradhan, B., Nayak, R., Patra, S., Jit, B. P., Ragusa, A., and Jena, M. (2020). Bioactive Metabolites from Marine Algae as Potent Pharmacophores against Oxidative Stress-Associated Human Diseases: A Comprehensive Review. *Molecules*, *26*(1), 37. https://doi.org/10.3390/molecules26010037.

Proksch, P. (1994). Defensive roles for secondary metabolites from marine sponges and sponge-feeding nudibranchs. *Toxicon*, *32*(6), 639–655. https://doi.org/10.1016/0041-0101(94)90334-4.

Putz, A., and Proksch, P. (2010). Chemical Defense in Marine Ecosystems. In M. Wink (Ed.), *Functions and Biotechnology of Plant Secondary Metabolites* (pp. 162–213). Wiley-Blackwell. https://doi.org/10.1002/9781444318876.ch3.

Rane, R. A., Nandave, M., Nayak, S., Naik, A., Shah, D., Alwan, W. S., Sahu, N. U., Naphade, S. S., Palkar, M. B., Karunanidhi, S., Thapliyal, N., & Karpoormath, R. (2017). Synthesis and pharmacological evaluation of marine bromopyrazole alkaloid-based hybrids with anti-inflammatory activity. *Arabian Journal of Chemistry*, *10*(4), 458–464. https://doi.org/10.1016/j.arabjc.2014.06.004.

Rudensky, A. Y. (2011). Regulatory T cells and Foxp3. *Immunological Reviews*, *241*(1), 260–268. https://doi.org/10.1111/j.1600-065X.2011.01018.x.

Safarzadeh, E., Mohammadi, A., Mansoori, B., Duijf, P. H. G., Hashemzadeh, S., Khaze, V., Kazemi, T., Derakhshani, A., Silvestris, N., & Baradaran, B. (2021). STAT3 Silencing and TLR7/8 Pathway Activation Repolarize and Suppress Myeloid-Derived Suppressor Cells From Breast Cancer Patients. *Frontiers in Immunology* 11, 613215. https://doi.org/10.3389/fimmu.2020.613215.

Salvador-Reyes, L. A., & Luesch, H. (2015). Biological targets and mechanisms of action of natural products from marine cyanobacteria. *Natural Product Reports*, *32*(3), 478–503. https://doi.org/10.1039/C4NP00104D.

Sanjeewa, K. K. A., Herath, K. H. I. N. M., Yang, H. W., Choi, C. S., & Jeon, Y. J. (2021a). Anti-Inflammatory Mechanisms of Fucoidans for the Treatment of Inflammatory Diseases: A Review. *Marine Drugs*, *19*(12), 678. https://doi.org/10.3390/md19120678.

Sanjeewa, K. K. A., Herath, K. H. I. N. M., Yang, H. W., Choi, C. S., & Jeon, Y. J. (2021b). Anti-Inflammatory Mechanisms of Fucoidans for the Treatment of Inflammatory Diseases: A Review. *Marine Drugs*, *19*(12), 678. https://doi.org/10.3390/md19120678.

Schrier, D. J., Flory, C. M., Finkel, M., Kuchera, S. L., Lesch, M. E., and Jacobson, P. B. (1996). Effects of the phospholipase a_2 inhibitor, manoalide, on cartilage degradation, stromelysin expression, and synovial fluid cell counts induced by intraarticular injection of human recombinant interleukin-1α in the rabbit. *Arthritis & Rheumatism*, *39*(8), 1292–1299. https://doi.org/10.1002/art.1780390805.

Sen, K., Kumar Das, S., Ghosh, N., Sinha, K., & Sil, P. C. (2024). Lupeol: A dietary and medicinal triterpene with therapeutic potential. *Biochemical Pharmacology*, *229*, 116545. https://doi.org/10.1016/j.bcp.2024.116545.

Senadheera, T. R. L., Hossain, A., & Shahidi, F. (2023). Marine Bioactives and Their Application in the Food Industry: A Review. *Applied Sciences*, *13*(21), pp. 12088. https://doi.org/10.3390/app132112088.

Shanthi Kumari, K.; Shivakrishna, P.; Ganduri, V. S. R. (2020). Wound-healing activities of bioactive compounds isolated from Micrococcus sp. OUS9 isolated from marine water. *Saudi Journal of Biological Sciences*, *27*(9), 2398–2402. https://doi.org/10.1016/j.sjbs.2020.05.007.

Systematic Review and Meta-analysis Expert Group (SRMEG), Universal Scientific Education and Research Network (USERN), Tehran, Iran, & Elahe, A.-D. (2022). Possible Therapeutic Targets of Natural Marine Products Based on PI3K/AKT Dependent Inhibitors in Viral Infection COVID-19. *Cellular Physiology and Biochemistry*, *56*(6), 707–729. https://doi.org/10.33594/000000595.

Tamberi, L., Belloni, A., Pugnaloni, A., Rippo, M. R., Olivieri, F., Procopio, A. D., & Bronte, G. (2024). The Influence of Myeloid-Derived Suppressor Cell Expansion in Neuroinflammation and Neurodegenerative Diseases. *Cells*, *13*(7), 643. https://doi.org/10.3390/cells13070643.

Tan, L. T. (2023). The Impact of Marine Chemical Ecology Research on the Discovery and Development of New Pharmaceuticals. *Marine Drugs*, *21*(3), 174. https://doi.org/10.3390/md21030174.

Wang, H., Li, X., Xuan, M., Yang, R., Zhang, J., & Chang, J. (2024). Marine biomaterials for sustainable bone regeneration. *Giant*, *19*, 100298. https://doi.org/10.1016/j.giant.2024.100298.

Wu, W., Wang, F., Gao, X., Niu, T., Zhu, X., Yan, X., & Chen, H. (2016). Synergistic effect of κ-carrageenan on oxazolone-induced inflammation in BALB/c mice. *BMC Gastroenterology*, *16*(1), 41. https://doi.org/10.1186/s12876-016-0459-7.

Xiang, Y., Zhang, M., Jiang, D., Su, Q., & Shi, J. (2023). The role of inflammation in autoimmune disease: A therapeutic target. *Frontiers in Immunology* 14, 1267091. https://doi.org/10.3389/fimmu.2023.1267091.

Xu, S.-S., Liu, Q.-M., Xiao, A.-F., Maleki, S. J., Alcocer, M., Gao, Y.-Y., Cao, M.-J., & Liu, G.-M. (2017). *Eucheuma cottonii* Sulfated Oligosaccharides Decrease Food-allergic Responses in Animal Models by Up-regulating Regulatory T (Treg) Cells. *Journal of Agricultural and Food Chemistry*, *65*(15), 3212–3222. https://doi.org/10.1021/acs.jafc.7b00389.

Xu, Y., Xiong, J., Sun, X., and Gao, H. (2022). Targeted nanomedicine-based remodeling of the immunosuppressive tumor microenvironment for enhanced cancer immunotherapy. *Acta Pharmaceutica Sinica B*, *12*(12), 4327–4347. https://doi.org/10.1016/j.apsb.2022.11.001.

Xue, M., Liang, H., Ji, X., Liu, Y., Ge, Y., Hou, L., & Sun, T. (2019a). Fucoidan prevents murine autoimmune diabetes by suppressing TLR4 signaling pathways, regulating DC/Treg-induced immune tolerance, and improving gut microecology. *Nutrition & Metabolism*, *16*(1), 87. https://doi.org/10.1186/s12986-019-0392-1.

Xue, M., Liang, H., Ji, X., Liu, Y., Ge, Y., Hou, L., & Sun, T. (2019b). Fucoidan prevents murine autoimmune diabetes by suppressing TLR4 signaling pathways, regulating DC/Treg-induced immune tolerance, and improving gut microecology. *Nutrition & Metabolism*, *16*(1), 87. https://doi.org/10.1186/s12986-019-0392-1.

Xue, M., Liang, H., Ji, X., Zhou, Z., Liu, Y., Sun, T., & Zhang, L. (2020). Effects of fucoidan on gut flora and the prevention of 1,2-dimethylhydrazine-induced colorectal carcinogenesis. *The Journal of Nutritional Biochemistry*, *82*, 108396. https://doi.org/10.1016/j.jnutbio.2020.108396.

Yu, J., Wang, X., Du, P., & Shi, H. (2024). The therapeutic potential of marine alkaloids in treating breast cancer. *Frontiers in Marine Science*, *11*, 1440928. https://doi.org/10.3389/fmars.2024.1440928.

Zhao, R., Zhou, H., & Su, S. B. (2013). The critical role for interleukin-1β in the progression of autoimmune diseases. *International Immunopharmacology*, *17*(3), 658–669. https://doi.org/10.1016/j.intimp.2013.08.012.

Chapter 6

Marine-Derived Compounds as Promising Agents for Managing Cardiovascular Disorders

Nagaraju Bandaru[1,*]
Nalla Swathi[2]
Prashik B. Dudhe[1]
Mohan Gandhi Bonthu[1]
and Makarand Suresh Gambhire[1]
[1]School of Pharmaceutical Sciences (SOPS), Sandip University, Nasik, Maharashtra, India
[2]Department of Pharmacology, Malla Reddy Pharmacy College, Maisammaguda, Hyderabad, India

Abstract

Marine organisms have long been recognized as prolific sources of bioactive compounds with diverse pharmacological properties. In recent times, there has been a rising interest in marine metabolites and their possible cardiovascular benefits. This abstract delves into the emerging field of marine metabolites and its implications for cardiovascular health. Marine metabolites encompass a wide array of compounds, including peptides, polyphenols, terpenes, and alkaloids, among others. These molecules exhibit a remarkable spectrum of biological activities, including antioxidant, anti-inflammatory, anti-thrombotic, and vasodilator effects. These properties hold promise for preventing and treating cardiovascular disorders (CVDs), a leading cause of global morbidity and mortality. Several marine metabolites have demonstrated promising cardiovascular effects in preclinical studies. For instance, marine-derived peptides have shown potential in regulating blood pressure and lipid metabolism, while polyphenols have exhibited anti-atherosclerotic and endothelial protective properties. Furthermore, marine terpenes and alkaloids have displayed antiplatelet and anticoagulant activities, thus mitigating the risk of thrombotic events. Despite these encouraging findings, translating marine metabolites into clinically viable therapeutics poses several challenges. Issues such as compound isolation, structural elucidation, bioavailability, and safety profiles warrant further investigation. Furthermore, extensive clinical trials are needed to demonstrate the effectiveness and safety of marine-derived substances in humans. In conclusion, marine metabolites represent a rich source of bioactive molecules with immense potential for

* Corresponding Author's Email: bnrajupharma@gmail.com

In: Healing from the Depths Marine Metabolites in Disease Management
Editors: Pranav Kumar Prabhakar and Tarun Kumar Upadhyay
ISBN: 979-8-89530-832-5

combating Cardiovascular diseases CVDs. Continued research efforts are needed to harness the therapeutic potential of these compounds, leading to the development of novel cardiovascular drugs and preventive strategies.

Keywords: marine; metabolites; health; cardiovascular; research

1. Introduction

1.1. Overview of Cardiovascular Disorders

Cardiovascular diseases refer to a wide spectrum of ailments affecting the heart and blood arteries. These illnesses are major primary reasons of global mortality and morbidity, affecting millions of individuals and posing significant public health challenges. Understanding the various types of cardiovascular disorders, their causes, symptoms, and treatments is crucial for prevention and management [1].

Types of Cardiovascular Disorders:

- Coronary artery disease: It is known as Ischemia, and it occurs when the coronary arteries supplying blood to muscle of the heart becomes thin or obstruct owing to atherosclerosis. Atherosclerosis is the formation of plaque on walls of arteries. This results in angina (chest discomfort) as well as heart attacks (myocardial infarction) [2].
- Heart failure: This occurs when the heart cannot pump enough blood due to various disorders like hypertension, cardiomyopathy, and coronary artery disease. Symptoms may include retention of fluid, tiredness, and shortness of breath [3].
- Arrhythmias: These are abnormal cardiac rhythms in which the heart beats excessively quickly, too slowly, or irregularly. Common kinds include atrial fibrillation (AFib), ventricular tachycardia, and bradycardia. Arrhythmias can cause palpitations, disorientation, and, in extreme situations, abrupt cardiac arrest [4].
- Stroke: It happens when the blood flow to a brain portion ceases or diminishes, where the brain tissue is deprived of oxygen and nutrients. Hypertension, commonly known as high blood pressure, is marked by excessive force of blood against arterial walls. It causes increase in the risk of multiple cardiovascular diseases like heart failure, coronary artery disease and stroke. Hypertension frequently has no symptoms, making regular monitoring critical.
- Congenital heart defects: These are structural defects of the heart that appear at birth. They can vary from basic problems such as tiny holes in the heart walls to more complex issues like malformed heart valves or chambers. Treatment may involve surgery or other medical interventions [5].
- Cardiomyopathy: The condition of heart muscle that hinders the blood pumping capacity properly. Cardiomyopathy is classified into three types: hypertrophic, dilated, and restrictive. Symptoms may include shortness of breath, limb edema, and arrhythmias [6].

- Periphery artery disease: It develops when the arteries supplying blood to the limbs, typically, legs become constricted or obstructed. It can cause discomfort, cramps, and paralysis, as well as limb amputation in extreme cases [7].
- Stroke: It occurs when the flow of blood to a brain portion is restricted depriving of nutrients and oxygen in brain tissue. Strokes can be ischemic (due to blocked arteries) or hemorrhagic (due to bleeding in the brain). Prompt treatment is essential to reduce brain damage and improve outcomes [8].

1.1.1. Causes and Risk Factors

The causes of cardiovascular disorders are multifactorial, involving genetic, lifestyle, and environmental factors. Common risk factors include [9]:

- Age: Risk rises as per age
- Gender: Risk in men is higher whereas in women it increases after menopause
- History of cardiovascular diseases in family rises the risk
- Smoking significantly increases the risk of cardiovascular disease
- High saturated fat, trans fat, and cholesterol diets can increase the risk of developing atherosclerosis
- Physical inactivity is connected to several cardiovascular problems
- Obesity raises the risk of hypertension, diabetes, and coronary artery disease
- Diabetes considerably raises the risk of cardiovascular disease
- Controlling high blood pressure is crucial to prevent heart disease and stroke
- High LDL cholesterol levels can cause arterial plaque formation.

Cardiovascular disorders are complex and varied, requiring a multifaceted approach for prevention, diagnosis, and treatment. Awareness and management of risk factors, along with advances in medical and surgical treatments, can significantly improve results for the persons with cardiovascular disease. Regular check-ups and a proactive approach to heart health are essential in mitigating the impact of these potentially life-threatening conditions [10].

1.2. Importance of Marine-Derived Compounds in Medicine

Marine-derived compounds have emerged as a significant source of novel pharmaceuticals and therapeutic agents, demonstrating the immense potential of the marine environment for medical advancements. The oceans, occupying over70% surface of the earth, harbour a diverse array of life forms, many of which produce unique bioactive compounds with extraordinary medicinal properties. These compounds have contributed to the development of various drugs and therapies, underscoring the importance of marine biodiversity in the field of medicine [11]. One of the key reasons marine-derived compounds are invaluable in medicine is their structural diversity and unique chemical properties. Marine organisms such as sponges, corals, algae, and microorganisms produce a wider range of secondary metabolites that are not found in terrestrial organisms. These compounds have evolved to help marine organisms survive in extreme and competitive environments, leading to the development of unique mechanisms of action that can be harnessed for therapeutic purposes. For instance, compounds like bryostatin, derived from marine bryozoans, and ecteinascidin, from sea squirts, exhibit potent anti-cancer properties and are being investigated for their potential to treat various cancers (Figure 1) [12].

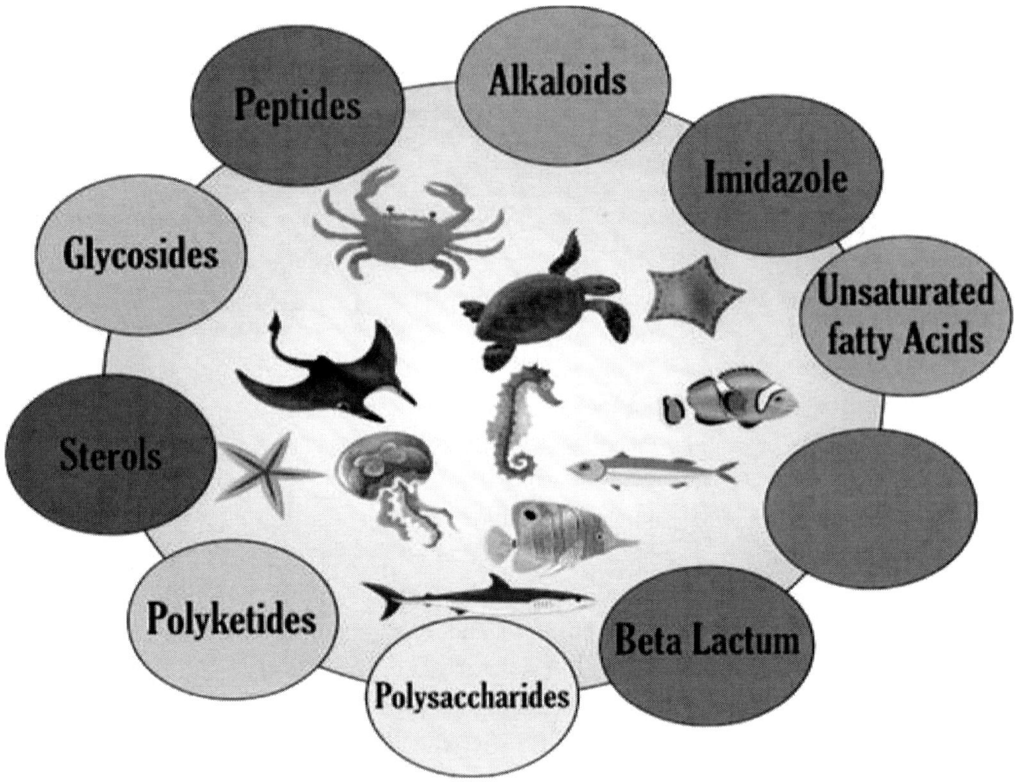

Figure 1. Role of different marine products in research.

Moreover, marine-derived compounds have shown significant potential in addressing antibiotic resistance, a growing global health concern. The discovery of new antibiotics is crucial as many pathogens have developed resistance to existing drugs. Marine environments, with their unique microbial ecosystems, offer a promising reservoir of novel antimicrobial agents. For example, compounds like marinopyrroles and salinamides, isolated from marine bacteria, have demonstrated strong antibacterial activity against resistant strains. These discoveries highlight the oceans as a critical source of new antibiotics that could combat resistant infections and save lives [13].

In addition to anti-cancer and antimicrobial properties, marine-derived compounds have shown promise in treating a variety of other diseases and conditions. For instance, the cone snail venom peptide, ziconotide, is used as a powerful analgesic for chronic pain management, offering an alternative to opioid-based painkillers without the associated risk of addiction. Similarly, omega-3 fatty acids obtained from marine fish oils are well known for their cardiovascular advantages, including as lowering the risk of heart disease and cholesterol levels. The versatility and efficacy of marine-derived compounds make them indispensable in the development of new and improved therapeutic agents (Figure 2) [14].

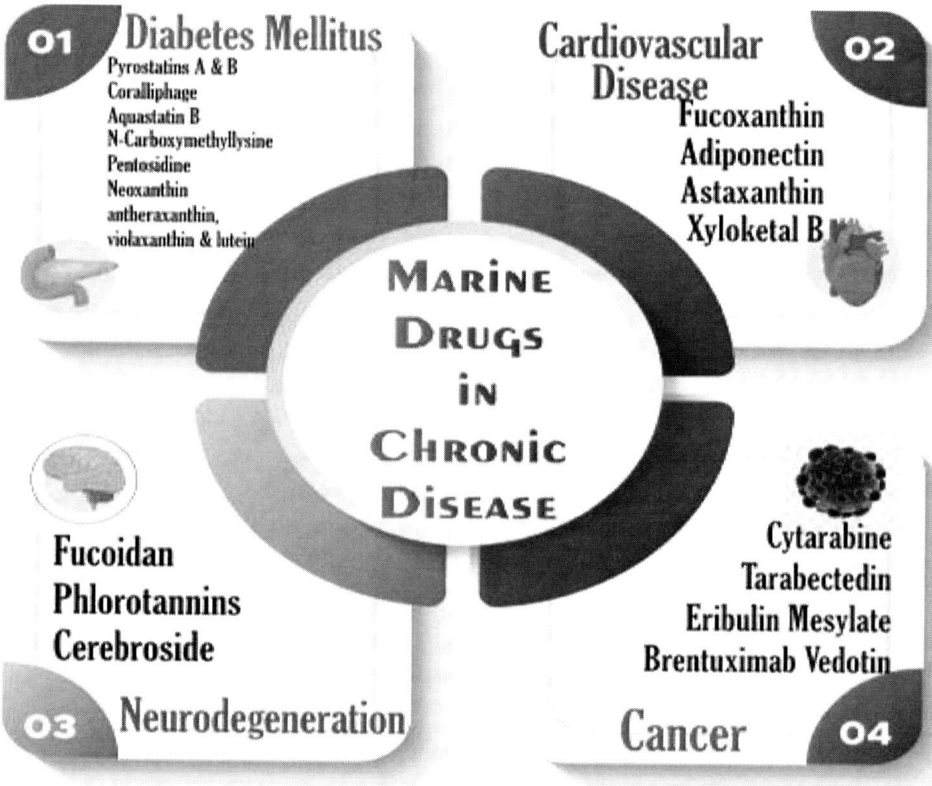

Figure 2. Role of different marine drugs in various pathological conditions.

Furthermore, marine biotechnology plays a vital role in drug discovery, and development. Advances in genomics, metagenomics, and bioinformatics have enabled researchers to explore marine biodiversity more efficiently and identify promising bioactive compounds. Techniques such as genome mining and synthetic biology are being employed to enhance the production of these compounds and modify their structures for improved efficacy and safety. This interdisciplinary approach is accelerating the pace of marine drug discovery and opening new avenues for medical innovation [15].

Despite the significant potential of marine-derived compounds, challenges remain in their exploration and development. The vastness of the oceans and the complexity of marine ecosystems make it difficult to fully tap into their potential. Additionally, sustainable harvesting and environmental conservation are critical considerations to ensure that the exploitation of marine resources does not harm marine biodiversity. Collaboration among the researchers, pharmaceutical companies, and conservationists are important to balance the exploration of marine compounds with the preservation of marine ecosystems. Finally, marine-derived compounds hold immense promise for advancing medicine and addressing some of the most pressing health challenges of our time. Their unique chemical properties, coupled with the advancements in marine biotechnology, have set up the way for the discovery of novel drugs and therapies. As we continue to explore the vast and largely untapped marine environment, the potential for ground-breaking medical innovations remains vast, reinforcing the importance of marine biodiversity in the future of medicine (Figure 3) [16].

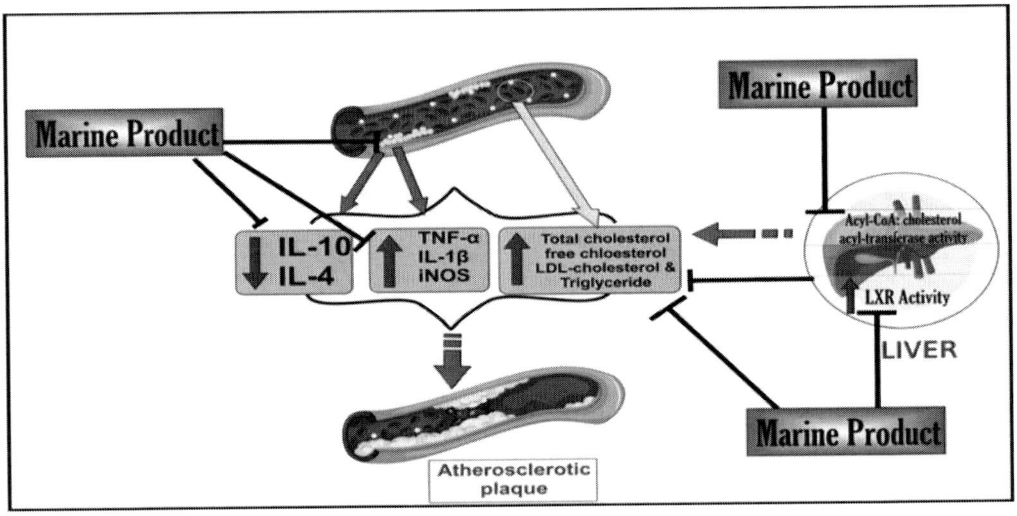

Figure 3. Role of marine drugs in Hyperlipidemia.

2. Bioactive Marine Compounds and Their Mechanisms of Action

2.1. Marine Products from Algae

Algal marine products refer to a range of products derived from various types of algae found in marine environments. Algae are a broad class of photosynthetic organisms that occur in both marine and fresh water environments. They can be as small as unicellular species or as huge as multicellular seaweeds. Algal marine products acquired more attention in recent times for their potential as sustainable sources of food, feed, biofuels, pharmaceuticals, and other valuable compounds [17].

2.1.1. Fucoidan

Fucoidan, a Sulphated polysaccharide is present in several brown seaweed species, including bladder wrack, wakame, and kombu. Because of its possible health benefits, especially its potential function in cardiovascular illnesses, it has garnered great interest in biomedical research. Fucoidan has potential cardiovascular benefits, but further studies are needed to fully understand its mechanism of action [18].

It has been proven that fucoidan possesses anticoagulant and antiplatelet features. These characteristics may contribute to a decreased risk comprised of thrombotic events such as strokes and heart attacks, and the prevention of blood clot formation. It exhibits antioxidant activity, which can help neutralize free radicals and reduce oxidative stress in cardiovascular system. Oxidative stress is incriminated in the development of atherosclerosis along with other cardiovascular disorders. Chronic inflammation contributes significantly to the development and evolution of cardiovascular illnesses. Fucoidan is reported to possess anti-inflammatory properties that help to mitigate inflammation within blood vessels and cardiac tissue. Some studies suggest that fucoidan may have direct cardio protective effects, including the ability to protect cardiac cells from damage caused by ischemia-reperfusion injury, a common phenomenon in conditions like heart attacks and cerebrovascular accidents. Fucoidan was investigated for its potential to modulate lipid metabolism, including the regulation of cholesterol levels. High levels of cholesterol are the risk factor for atherosclerosis and cardiovascular disease, compounds that can help maintain healthy lipid profiles may contribute to cardiovascular health [19].

Endothelial dysfunction, referring to the impaired function of the cells that line blood vessels, is an important hallmark of many cardiovascular diseases. Some research suggests that fucoidan may help improve endothelial function, thereby supporting overall cardiovascular health. Preliminary studies indicate that fucoidan may have hypotensive (blood pressure-lowering) effects, which could be beneficial for individuals with hypertension, A significant risk factor for heart diseases [20].

2.1.2. Alginate

Alginate is a polymer produced from brown seaweeds, including kelp and brown algae. It has a widespread use in various industries, including food, pharmaceuticals, and biomedical applications, due to its distinctive properties, especially biocompatibility, biodegradability, as well as gel-forming ability. While alginate's specific role in cardiovascular diseases is not as extensively studied as other marine-derived compounds like fucoidan, there are potential applications and mechanisms that suggest its relevance in cardiovascular health [21]:

- Blood Lipid Regulation: Some research suggests that alginate may have cholesterol-lowering effects, potentially through mechanisms such as binding to bile acids in the gastrointestinal system, which may lead to increased excretion of cholesterol. By helping to regulate blood lipid levels, alginate may contribute to minimizing the risk of atherosclerosis and cardiovascular illnesses linked dyslipidemia [22].
- Blood Pressure Management: Alginate has been investigated for its ability to regulate blood pressure levels. Animal studies have demonstrated that alginate supplementation can lower blood pressure, possibly via processes that involve inhibition of Renin Angiotensin Aldosterone System (RAAS) or improved endothelial function. Elevated blood pressure is the predominant risk factor for strokes, and heart attacks, and compounds that lower blood pressure may protect the cardiovascular system [23].
- Antioxidant Activity: Alginate's antioxidant properties can help neutralize free radicals and mitigate the oxidative stress. Oxidative stress is linked to the development of cardiovascular illnesses such as atherosclerosis, hypertension, and heart failure. By scavenging free radicals and preventing oxidative damage to blood vessels and cardiac tissues, alginate may contribute to cardiovascular health [24].
- Anti-inflammatory properties: Long-term inflammation is the key mark for the progression of cardiovascular diseases. Alginate is reported to have anti-inflammatory activity, which may help mitigate inflammatory processes within blood vessels and cardiac tissues, potentially lowering the risk of atherosclerosis, impaired endothelial function, and other inflammatory cardiovascular conditions [25].
- Blood Clot Prevention: Alginate has the capacity to form gels in presence of divalent cations like calcium ions and is used in a variety of medical applications, including wound care and haemostatic agents. While its role in preventing blood clot formation specifically in cardiovascular diseases is not well-established, alginate's hemostatic

properties suggest potential utility in managing thrombotic events associated with conditions such as the heart attacks and strokes [26].

2.1.3. Ulvan

Ulvan, a sulfated polysaccharide that exists in green seaweeds, specifically Ulva spp. (commonly called as Sea lettuce It is structurally similar to other sulfated polysaccharides such as fucoidan and carrageenan, but it comes from different types of seaweed. Ulvan has gained attention in biomedical research due to its diverse biological activities and potential health benefits. While research specifically on Ulvan's role in cardiovascular diseases is limited compared to other marine polysaccharides, there are potential applications and mechanisms that suggest its relevance in cardiovascular health [27].

- Anticoagulant and Antiplatelet Activity: Ulvan, like other sulfated polysaccharides, has anticoagulant and antiplatelet properties that may help prevent blood clotting and lessen the risk of thrombotic events like heart attacks and strokes [28].
- Antioxidant Effects: Ulvan has antioxidant properties, which may assist to neutralize free radicals and reduce oxidative stress in the cardiovascular system. Oxidative stress is linked to the onset of heart diseases like atherosclerosis and hypertension. By eating free radicals and preventing oxidative damage to blood vessels and cardiac tissues, ulvan may contribute to cardiovascular health [29].
- Anti-inflammatory properties: The chronic inflammation, important criteria in the development of cardiovascular diseases. Ulvan has anti-inflammatory properties that may help reduce inflammation in blood vessels and cardiac tissues, potentially lowering the risk of atherosclerosis, endothelial dysfunction, and other cardiovascular diseases [30].
- Endothelial function improvement: Endothelial dysfunction, characterized by impaired function of the cells lining blood vessels, is a hallmark of many cardiovascular diseases. Some research suggests that ulvan may help improve endothelial function, thereby supporting overall cardiovascular health. [31]
- Blood pressure regulation: Although limited, some studies have suggested that ulvan may have hypotensive (blood pressure-lowering) effects, which could be beneficial for individuals with hypertension that is the big risk factor of cardiovascular diseases [32].
- Lipid metabolism regulation: Ulvan has been investigated for its potential to modulate lipid metabolism, including the regulation of cholesterol levels. High levels of cholesterol raises the atherosclerosis and heart attack risk, and compounds that can help maintain healthy lipid profiles may contribute to cardiovascular health [33].

2.1.4. Carrageenan

Carrageenanis the sulphated polysaccharide derived from red seaweed, such as Irish moss (Chondrus crispus) and Gigartina spp. It is widely used in food industry as thickening, stabilizing, gelling agent, as well as in pharmaceuticals and cosmetics. While carrageenan is primarily known for its applications in food and other industries, there is limited research on its specific role in cardiovascular diseases. However, some potential mechanisms and roles have been suggested [34-35].

- Anticoagulant Properties: Carrageenan, particularly low-molecular-weight carrageenan's, has been reported to exhibit anticoagulant activity by inhibiting blood coagulation factors. This anticoagulant activity assists in preventing the blood clots (thrombi) and reduces the risk of thrombotic events like heart strokes, and attacks [36].
- Anti-inflammatory Effects: Chronic inflammation contributes considerably to the development and progression of cardiovascular disorders such atherosclerosis and hypertension. According to some study, carrageenan may have anti-inflammatory characteristics, which might help decrease inflammation in blood vessels and cardiac tissues potentially lowering the risk of cardiovascular complications associated with inflammation [37].
- Antioxidant Activity: Carrageenan has been reported to exhibit antioxidant activity, which can help neutralize free radicals and lower the oxidative stress in the cardiovascular system. Oxidative stress is linked to the pathophysiology of cardiovascular illnesses, and substances with antioxidant properties may help protect against oxidative damage to blood vessels and cardiac tissues [38].
- Endothelial Function Improvement: Endothelial dysfunction, characterized by impaired function of cells lining the blood vessels, is a hallmark of many cardiovascular diseases. Some research suggests that carrageenan may help improve endothelial function, thereby supporting overall cardiovascular health.
- Blood Pressure Regulation: Limited evidence suggests that carrageenan may have hypotensive (blood pressure-lowering) effects, which could be beneficial for individuals with hypertension which is the major risk factor for cardiovascular diseases [39].

2.1.5. Fucoxanthin

Fucoxanthin, natural carotenoid pigment found in brown seaweeds, such as Kelp, Wakame, and Hijiki. It is responsible for the brown or olive-green color of these seaweeds. Fucoxanthin has gained attention in biomedical research due to its various potential health benefits, including its role in cardiovascular health. While research on fucoxanthin's specific effects on cardiovascular diseases is ongoing and evolving, several potential mechanisms and roles have been proposed [40].

- Anti-inflammatory Effects: Chronic inflammation mainly contributes for the development and progression of cardiovascular disorders including atherosclerosis and hypertension. Fucoxanthin has been shown to have anti-inflammatory characteristics, which may help decrease inflammation inside blood vessels and cardiac tissues, thereby reducing the risk of cardiovascular problems linked with inflammation [41].
- Antioxidant Activity: Fucoxanthin exhibits strong antioxidant qualities can help neutralize free radicals and minimize oxidative stress in the cardiovascular system. Oxidative stress is linked to the pathophysiology of cardiovascular disorders and compounds with antioxidant properties may help to protect against oxidative damage to blood vessels and cardiac tissues [42].
- Regulation of Lipid Metabolism: Fucoxanthin has been investigated for its potential to modulate lipid metabolism, including the regulation of cholesterol levels. Elevated cholesterol levels elevate atherosclerosis and cardiovascular disease risk, and

compounds that help to retain healthy lipid profiles may benefit cardiovascular health. Fucoxanthin may promote the breakdown of fat stores and inhibit the deposition of triglycerides in adipose tissue, potentially leading to improvements in lipid profiles [43].

- Anti-obesity Effects: Obesity, the primary risk factor for coronary artery disease, hypertension, and stroke. Fucoxanthin has been reported to have anti-obese effects by promoting the oxidation of fat and inhibiting the differentiation of preadipocytes into mature adipocytes. By reducing adiposity and improving metabolic parameters, fucoxanthin may indirectly contribute to cardiovascular health [44].
- Anti-hypertensive Effects: Studies suggest that fucoxanthin may have hypotensive (blood pressure-lowering) effects, which could be beneficial for individuals with hypertension. High blood pressure is the important risk factor for cardiovascular illnesses, and substances that may assist manage blood pressure levels may have beneficial effects on the cardiovascular system [45].

2.2. Animal Origin Marine Products

Animal origin marine products, often referred to as seafood, encompass a wide variety of food items derived from animals living in the ocean, seas, and other saltwater bodies. These products are a significant source of nutrition for people around the world, providing essential proteins, fats, minerals, and vitamins [46]. Several marine animal origin products have been investigated for their potential benefits in treating cardiovascular diseases. Here are some examples:

- Fish Oil: Fish oil is prevalent in omega-3 fatty acids, especially EPA and DHA, and has been widely investigated for cardiovascular effects. It can decrease triglyceride levels and reduce inflammation, decrease blood pressure, and prevent the accumulation of plaque in arteries region, all of which aid in lowering the risk of heart disease [47].
- Omega-3 Supplements: Along with fish oil, nutrients containing omega-3 fatty acids obtained from marine sources such as krill oil or algae oil have also shown promise in improving cardiovascular health [48].
- Fish Collagen: Collagen derived from the skin and scales of fish has been investigated for its potential to improve cardiovascular health. Collagen peptides derived from fish collagen may help improve arterial stiffness and endothelial function, which are important factors in preventing cardiovascular diseases [49].
- Shellfish: Certain shellfish, such as oysters and mussels, have rich content of minerals, vitamins, and omega-3 fatty acids, and these help in maintaining heart health. Additionally, shellfish contain bioactive compounds like taurine and glycogen, which may have cardio protective effects [50].
- Seafood with high levels of Taurine: It is an amino acid found in higher concentrations in certain marine animals like shellfish and fish. Studies implicated that taurine supplementation may possesses beneficial effects on heart's health, including lowering blood pressure and improving cholesterol levels [51].

- Marine peptides: Peptides derived from the sea sources, such as fish or shellfish have been investigated for their potential cardiovascular benefits. These peptides may exert antioxidant, anti-inflammatory, and antithrombotic effects, which could help prevent or manage cardiovascular diseases [52].
- Marine Antioxidants: Certain marine organisms produce antioxidants to protect themselves from oxidative stress in their environment. Extracts from these organisms, such as algae or marine microorganisms, may contain antioxidants that could benefit cardiovascular health by reducing oxidative damage and inflammation While research on the cardiovascular benefits of marine animal origin products is promising, it's critical to consult with a medical professional before commencing a new supplement regimen, especially for individuals with pre-existing cardiovascular concerns or those utilizing drugs [53].

2.3. Marine-Derived Drugs

Several drugs have been developed from marine organisms or their metabolites. For example [54]:

- Ziconotide (Prialt): Derived from the venom of the cone snail Conus magus, ziconotide is used as a painkiller for severe chronic pain.
- Yondelis (Trabectedin): Originally isolated from a marine tunicate, Yondelis is used in the soft stissue sarcoma and ovarian cancer treatment.
- Eribulin (Halaven): Originally derived from a marine sponge, Eribulin is used in the metastatic breast cancer treatment.

Marine-Derived Anticancer Compounds: Many marine organisms produce compounds with anticancer properties. These include substances like cyanobacteria metabolites, marine sponges-derived compounds (e.g., Discodermolide), and compounds from marine algae (e.g., Fucoidan) [55].

Antibiotics and Antimicrobial Compounds: Marine organisms produce a variety of compounds with antimicrobial properties, which can potentially be developed into antibiotics or antifungal agents. For instance, marine bacteria have been a source of novel antibiotics, such as salinosporamide A [56].

Antioxidants and Anti-Inflammatory Compounds: Some of the marine compounds possess anti-inflammatory and antioxidant properties can be advantageous for several health issues, notably cardiovascular disease, neurodegenerative disorders, and arthritis [57].

Cosmetic Ingredients: Marine-derived compounds like marine collagen, seaweed extracts, and marine-derived enzymes are used in cosmetics and skincare products provide hydrating, anti-aging, and skin-regenerating characteristics [58].

Nutraceuticals and Dietary Supplements: Substances derived from marine organisms, i.e., omega-3 fatty acids obtained from fish oil, chitosan from shellfish, and astaxanthin from microalgae, are used in dietary supplements for their potential health benefits [59].

Table 1. Catalogue of algae-based products with lipid-lowering properties

Name	Source	Specimen	Implication	Action	Reference
Porphyran	Pyropiayezoensis	Hyperlipidemic mice	Low density and high density lipoprotein cholesterol, total cholesterol, triglycerides	To be known	[65]
Sulphated polysaccharides	Pedinatestrastromatica	The rat model of isoproterenol-induced myocardial infraction	Low density, and high density lipo protein cholesterol, triglycerides and total cholesterol	Controls the level of expression of SREBP-2, and LDL-R	[66]
Fucoidan A3	Ascophyllum nodosum	Hyperlipidemic mice	Total Cholesterol, Tri glycerides, fat pad index	Improve the expression of RCT-related gene	[67]
Fucoidan	Ascophyllum nodosum	Hyperlipidemic mice	Total Cholesterol, Tri glycerides, Fat pad index	Improve the expression of proteins, and RCT-related genes	[68]
Fucoidan	Fucus vesiculosus	Hyperlipidemic mice	Total Cholesterol, Tri glycerides, High and low density lipoprotein cholesterol	Manage the key enzymes expression that are involved in cholesterol and triglyceride synthesis	[69]
Fucoidan A2	Ascophyllum nodosum	Hyperlipidemic mice	Triglycerides, Fat pad index, total cholesterol and high density lipoprotein cholesterol	Adjusts the expression of RCT-related protein	[70]
Fucoidan	Kjellmaniella crassifola	Hyperlipidemic rats	Tri glycerides	Inhibits aliphatic acid and cholesterol synthesis; quickens aliphatic acid oxidation	[71]
Fucoidan	Sargassum wightii	Hyperlipidemic mice	Total Cholesterol	Inhibits activity of HMGCoA reductase and improves HL, LCAT, and LPL activity	[72]
Fucoidan	Brown algae	Mice lacking apo lipoprotein E	Low and high density lipoprotein cholesterol, total Cholesterol, Tri glycerides	Induce faecal cholesterol elimination	[73]
Sodium alginate	Turbinaria ornata J. Agardh (turner)	Diabetic rats	Low density and high density lipoprotein cholesterol and total cholesterol	Induce faecal elimination of cholesterol	[74]
Sodium Alginate and amide form of Sodium Alginate	Brown algae	Hyperlipidemic rats	Total Cholesterol, Tri glycerides	Induce faecal cholesterol and and coprostanol elimination	[75]
Clcium alginate	Brown algae	Hyperlipidemic rats	Total Cholesterol	Increase bile acid's faecalexcretion	[76]
Ulvan	Ulva pertusa	Hyperlipidemic mice	Low and High density lipoprotein cholesterol, total cholesterol, triglycerides	Safeguard against liver damage triggered by oxidative stress	[77]
Ulvan	Ulvapertusa	Hyperlipidemic mice	Low density and High density lipoprotein cholesterol, Total cholesterol, Triglycerides,	Enhance hypolipidemic activities	[78]
Carrageenan	Kappaphycus alvarezii	Hyperlipidemic mice	Total Cholesterol	Regulates the genes involved in lipid metabolism	[79]

Name	Source	Specimen	Implication	Action	Reference
Carrageenan	Echeuma spinosum	Hyperlipidemic mice	Total cholesterol, Triglycerides, Low density and High density lipoprotein cholesterol	Enhance intestinal content, restrict mucosal absorption, and promotes peristaltic movement in small intestine	[80]
Fucoxanthin	Laminaria japonica	Hyperlipidemic mice	Total Cholesterol, Tri glycerides	Controls IRS-1/PI3KT expression and AMPK signalling pathway	[81]
Fucoxanthin	Phaeodactylum tricornutum	Hyperlipidemic mice	Tri glycerides	Improve the activity of key enzymes associated with lipoprotein metabolism.	[82]
Fucoxanthin	Algae	Obese people	Tri glycerides	Mitigate the accumulation of lipids	[83]
Phloro tannins	Ecklonia cava	hypocholesterolemic people	Triglycerides, High density and low density lipoprotein cholesterol	Mitigates the lipid accumulation	[84]
Diphlorethohydroxycarmalol (DPHC)	Ishige okamurae	Hyperlipidemic mice	Triglycerides, Low density lipoprotein cholesterol, High density lipoprotein cholesterol	Mitigate the bio-synthesis of lipids	[85]

Bioremediation and Biotechnology: Certain marine-derived enzymes and microorganisms are being explored for their potential applications in bioremediation (cleaning up pollutants from the environment) and biotechnology processes [60].

Marine Biopolymers: Polysaccharides are extracted from marine algae, comprising Alginate, and Carrageenan, have Several industrial applications, such as thickening, stabilizing, and gelling agents in food, medicines, and cosmetics [61].

3. Marine-Derived Compounds in Cardiovascular Disease Management

3.1. Marine Products in CAD and Hypertension

Marine products, particularly those derived from algae and fish, have shown significant promise in managing coronary artery disease (CAD), and hypertension. These health conditions are major contributors to cardiovascular disease, a major source of illness and mortality globally.

3.1.1. Omega-3 Fatty Acids

One of the primary marine products beneficial for CAD and hypertension is Omega-3 fatty acids. These are mostly present in salmon, mackerel, and sardines like fatty fishes, and in certain algae. These Omega-3 fatty acids, notably EPA and DHA, contain anti-inflammatory characteristics and help reduce blood triglyceride levels, lowering of blood pressure, and reducing the risk of heart arrhythmias. Regular consumption of omega-3-rich marine products is accompanied with decreased risk of heart attacks and strokes [62].

3.1.2. Bioactive Compounds in Algae

Marine algae are rich in bioactive substances, like fucoidans, phlorotannins, and carotenoids, which possess anti-inflammatory and antioxidant nature. These compounds helps to improve endothelial function, lowering arterial stiffness, and decreasing blood pressure, which are crucial in managing hypertension and preventing the progression of CAD [63].

3.1.3. Algal Supplements

Algal supplements are available as oils or capsules that provide a plant-based source containing omega-3 fatty acids, making them suitable for vegetarians and those allergic to fish. These supplements are effective in promoting cardiovascular health and managing hypertension (Table 1). Incorporating marine products into a proper balanced diet plays a vital role in prevention and management of CAD and hypertension, supporting overall cardiovascular health [64].

3.2. Marine Products on Arrhythmias

Marine products, like omega-3 fatty acids in fish oil, have been investigated for their potential effects on arrhythmias. Arrhythmias, which are irregular heartbeats, can range from benign to life-threatening conditions. Omega-3 fatty acids, namely eicosapentaenoic acid (EPA) and docosahexaenoic acid (DHA), are believed to have cardioprotective properties. Research suggests that these marine-derived compounds may help stabilize the heart's electrical activity, reducing the likelihood of arrhythmias. They are thought to modulate ion channels, decrease inflammation, and improve autonomic tone, all of which contribute to maintaining a regular heart rhythm. Some studies have indicated that regular consumption of fish or fish oil supplements can lower the risk of sudden cardiac death, which is often triggered by ventricular arrhythmias [86].

However, the evidence is not entirely consistent. While some clinical trials have shown beneficial effects, others have reported no significant impact on arrhythmias. The variability in results may be due to differences in study design, dosages, and populations studied. Overall, incorporating marine products rich in omega-3 fatty acids in the diet appears to offer some potential benefits for heart health and arrhythmia prevention, but more research is required to completely understand their role and efficacy in this area [87].

3.3. Marine Products on Heart Failure

Several studies have suggested that regular consumption of fish or fish oil supplements may help improve clinical outcomes in heart failure patients. For instance, supplementation of omega-3 has been linked with a reduction in mortality and hospitalization rates among these individuals. Additionally, the anti-inflammatory properties of omega-3 fatty acids may help mitigate the chronic inflammatory state often observed in heart failure (Table 2).

Table 2. Marine derived products on Heart failure

Name	Source	Specimen	Implication	Action	Ref.
Furanone	Fungi Setophaeriasp.	RAW 264.7 macrophage cells	Total cholesterol, Triglycerides, low-density lipoprotein cholesterol	Up regulate PPAR∞	[88]
Xyloketal B	Fungi xylariasp	Hyperlipidemic mice	Triglycerides, low density lipoprotein cholesterol, Total Cholesterol	Reduces the accumulation of lipids	[89]
Cube natural sea salt	Marine	3T3-L1 adipocytes, Hyperlipidemic mice	Total cholesterol, triglycerides, and low density lipoprotein cholesterol	Reduces the accumulation of lipids, regulates lipolysis and beta-oxidation	[90]
Deep sea water	Marine	3T3-L1 preadipocytes	Low density lipoprotein cholesterol, triglycerides, and total cholesterol	Enhances the metabolism of lipids	[91]
Deep sea water	Marine	3T3-L1 pre-adipocytes, Hyperlipidemia mice	Triglycerides, low density lipoprotein cholesterol, total cholesterol	Enhances lipolysis and oxidation of fatty acids	[92]
Deep sea water	Marine	HepG2 cells	Total Cholesterol, triglycerides, Low-density lipoprotein cholesterol	Induce transcription of the LDLR gene and ApoA1, while inhibiting the production of PCSK9 mRNA.	[93]
Deep sea water	Marine	Hyperlipidemic rats	High density and low density lipoprotein cholesterol and total cholesterol	Enhance LDLR gene expression, supress synthesis of fatty acid and SREBP-1c expression	[94]

4. Challenges and Future Prospects

4.1. Present Challenges Encountered during Drug Development from Marine Sources

Drug development from the marine sources presents unique challenges, despite the rich biodiversity and potential for novel compounds found in oceans. Firstly, the collection and supply of marine organisms can be logistically difficult and environmentally disruptive. Many promising marine species inhabit deep or remote locations, making their harvest costly and technically challenging. Secondly, marine organisms often produce compounds in minute quantities, necessitating advanced extraction and synthesis techniques to obtain sufficient amounts for study and development. This scarcity complicates the scaling up necessary for drug production [95].

Additionally, the complex structures of marine-derived compounds pose significant challenges for chemical synthesis and modification, essential steps in drug development. The intricate molecular architecture often requires sophisticated and expensive methods to replicate or alter for therapeutic use. Furthermore, marine environments are dynamic and sensitive, meaning that the impact of bio-prospecting activities on ecosystems must be carefully managed to avoid ecological damage. Regulatory hurdles also exist, as the approval process for new drugs derived from marine sources is rigorous, requiring extensive testing for safety and efficacy. Finally, there is often a lack of comprehensive knowledge about marine species and their biochemical pathways, leading to difficulties in understanding the mechanisms of action and potential side effects of marine-derived compounds. Overcoming these challenges requires interdisciplinary collaboration, advanced technologies, and sustainable practices [96].

4.2. Regulatory and Safety Considerations

Developing drugs from marine sources involves navigating complex regulatory and safety considerations to ensure efficacy and safety. Initially, bioactive compounds must be identified and extracted sustainably, with clear documentation of their sources. Preclinical testing involves toxicological studies, pharmacokinetics, pharmacodynamics, and understanding the compound's mechanism of action to establish safe dosage levels and potential side effects [97].

Clinical trials, conducted in multiple phases, assess the safety and efficacy of drug in humans. Phase I examines safety in healthy volunteers, Phase II for efficacy and further safety in patients and Phase III finalizes the effectiveness and monitors side effects in large-scale trials. Successful completion of these phases is required for regulatory approval, which involves submitting comprehensive data to agencies like the FDA or EMA. Environmental and ethical considerations are crucial, ensuring that extraction and production do not harm marine ecosystems and that indigenous knowledge and rights are respected. Quality control and manufacturing must adhere to Good Manufacturing Practices (GMP) to ensure the drug's purity and consistency [98].

By meeting these stringent regulatory and safety standards, the pharmaceutical industry can effectively harness the therapeutic potential of marine-derived compounds while safeguarding patient health and marine biodiversity.

5. Future Directions and Research Needs

Marine-derived compounds have shown considerable promise in managing cardiovascular disorders, offering a rich source of novel therapeutic agents with unique mechanisms of action. However, several future directions and research needs must be addressed to fully realize their potential [99].

5.1. Future Directions

Leveraging HTS to rapidly identify potential cardiovascular agents from marine sources. This technology can screen thousands of compounds efficiently, accelerating the discovery process. Utilizing genomics, proteomics, and metabolomics to understand the mechanisms of molecules that are underlying the cardiovascular benefits of marine compounds. These approaches can help identify biomarkers for efficacy and safety [100]. Employing computational models to predict the pharmacokinetics, pharmacodynamics, and potential side effects of marine compounds. This can streamline the drug development process and reduce reliance on animal testing. Integrating large datasets from marine bioprospecting with clinical data to identify promising compounds and predict their therapeutic potential [101]. Developing sustainable cultivation methods for marine organisms to ensure a stable supply of bioactive compounds without depleting natural resources. This approach supports environmental conservation while providing a reliable source for drug development. Engineering microorganisms to produce marine-derived compounds. This innovative approach can enhance the production of scarce or hard-to-harvest compounds [102].

Conducting clinical trials of phase I, and II for analysing the safety and efficacy of promising marine-derived cardiovascular agents. This step is crucial for translating preclinical findings into clinical applications. Investigating how marine-derived compounds can be tailored to each patient based on genetic, epigenetic, and metabolic profiles. Personalized approaches can enhance the efficacy and reduce adverse effects of treatments [103].

5.2. Research Needs

Identifying specific molecular targets of marine-derived compounds in cardiovascular pathways. Understanding these interactions can reveal new therapeutic targets and mechanisms of action. Developing advanced in-vitro, and in-vivo models of cardiovascular diseases in studying the effects of marine compounds. These models should mimic human disease conditions closely [104]. Conducting long-term toxicity studies to ensure that marine-derived compounds are safe for chronic use in managing cardiovascular disorders. Comprehensive profiling of potential side effects to identify and mitigate adverse reactions [105].

Establishing international guidelines for the development and approval of marine-derived drugs. Harmonized regulations can facilitate global research collaborations and streamline the approval process. Ensuring ethical practices in bioprospecting and benefit-sharing with local communities who contribute traditional knowledge [106]. Encouraging collaboration between marine biologists, pharmacologists, cardiologists, and bioinformaticians to integrate diverse expertise in drug development. Fostering partnerships between academic institutions, pharmaceutical companies, and governmental agencies to fund and advance research on marine-derived cardiovascular therapies [107].

Finally, while marine-derived compounds hold great promise for managing cardiovascular disorders, their full potential can only be realized through a concerted effort in advanced research, sustainable practices, and robust regulatory frameworks. Addressing these future directions and research needs will pave the way for innovative and effective cardiovascular treatments derived from the ocean's vast and largely unexplored biodiversity.

Conclusion

Exploration of marine-derived compounds as potential agents for managing cardiovascular disorders represents a compelling avenue of research with promising therapeutic implications. The rich biodiversity of marine environments offers a vast array of bioactive molecules that exhibit diverse pharmacological properties, ranging from anti-inflammatory and antioxidant effects to modulation of lipid metabolism and vascular function. These compounds have demonstrated significant potential in preclinical studies, showing efficacy in mitigating cardiovascular risk factors i.e., hypertension, atherosclerosis, and thrombosis.

However, translating these promising findings into clinically approved therapies requires overcoming several challenges. These include rigorous safety assessments, comprehensive understanding of pharmacokinetics and pharmacodynamics, and scalability of production methods. Collaborative efforts across disciplines—spanning marine biology, pharmacology, medicinal chemistry, and clinical cardiology—are crucial for advancing research and development in this field. Moreover, adherence to sustainable harvesting practices and ethical considerations in bio prospecting ensure the preservation of marine ecosystems and respect for indigenous knowledge.

As research progresses, continued investment in innovative screening technologies, mechanistic studies, and clinical trials will be pivotal in unlocking the total therapeutic potential of marine-derived compounds. Ultimately, integrating these novel agents into cardiovascular care has the potential to revolutionize treatment strategies and improve outcomes for patients worldwide.

References

[1] World Health Organization. *World Health Statistics 2020*, World Health Organization: Geneva, Switzerland, 2020, p. 12.

[2] Mayakrishnan V, Kannappan P, Abdullah N, Ahmed ABA. Cardioprotective activity of polysaccharides derived from marine algae: An overview. *Trends Food Sci Technol.* 2013,30:98–104.

[3] Kobiyama K, Ley K. Atherosclerosis. *Circ Res.* 2018,123(10):1118-1120. Woollard KJ, Geissmann F. Monocytes in atherosclerosis: subsets and functions. *Nat Rev Cardiol.* 2010,7(2):77-86.

[4] Borén J, Williams KJ. The central role of arterial retention of cholesterol-rich apolipoprotein-B-containing lipoproteins in the pathogenesis of atherosclerosis: a triumph of simplicity. *Curr Opin Lipidol.* 2016,27(5):473-83.

[5] Navar-Boggan AM, Peterson ED, D'Agostino RB Sr, Neely B, Sniderman AD, Pencina MJ. Hyperlipidemia in early adulthood increases long-term risk of coronary heart disease. *Circulation.* 2015,131:451–458.

[6] Albany CJ, Trevelin SC, Giganti G, Lombardi G, Scotta C. Getting to the Heart of the Matter: The Role of Regulatory T-Cells (Tregs) in Cardiovascular Disease (CVD) and Atherosclerosis. *Front Immunol.* 2019,10:2795.

[7] Kim YG, Cho YR, Park GM, et al., High-density lipoprotein cholesterol and the risk of obstructive coronary artery disease beyond low-density lipoprotein cholesterol in non-diabetic individuals. *Eur J Prev Cardiol.* 2020,27:706–714.

[8] Maffeis C, Surano MG, Cordioli S, et al., A High-fat vs. a Moderate-fat Meal in Obese Boys: Nutrient Balance, Appetite, and Gastrointestinal Hormone Changes. *Obesity (Silver Spring).* 2010,18:449–455.

[9] Fisher J, Liu Y, Birch LL, Rolls BJ. Effects of portion size and energy density on young children's intake at a meal. *Am J Clin Nutr.* 2007,86:174–179. *Mar Drugs.* 2020,18:390.

[10] Cotton JR, Burley VJ, Weststrate JA, Blundell JE. Dietary fat and appetite: Similarities and differences in the satiating effect of meals supplemented with either fat or carbohydrate. *J Hum Nutr Diet.* 2007,20:186–199.

[11] Luo J, Yang H, Song BL. Mechanisms and regulation of cholesterol homeostasis. *Nat Rev Mol Cell Biol.* 2020,21:225–245.

[12] Pedersen TR, Kjekshus J, Berg K, et al., Randomized Trial of Cholesterol-Lowering in 4444 Patients with Coronary-Heart-Disease—The Scandinavian Simvastatin Survival Study (4S). *Lancet.* 1994,344:1383–1389.

[13] Brown MS, Goldstein JL. A Receptor-Mediated Pathway for Cholesterol Homeostasis. *Science.* 1986,232:34–47.

[14] Shepherd J, Cobbe SM, Ford I, et al., Prevention of Coronary Heart-Disease with Pravastatin in Men with Hypercholesterolemia. *N Engl J Med.* 1995,333:1301–1307.

[15] Collins R, Armitage J, Parish S, Sleight P, Peto R, Heart Protection Study Collaborative Group. MRC/BHF Heart Protection Study of cholesterol lowering with simvastatin in 20,536 high-risk individuals: a randomized placebo-controlled trial. *Lancet.* 2002,360:7–22.

[16] Cardoso SM, Pereira OR, Seca AML, Pinto DCGA, Silva AMS. Seaweeds as Preventive Agents for Cardiovascular Diseases: From Nutrients to Functional Foods. *Mar Drugs.* 2015,13:6838–6865.

[17] Seca AML, Pinto DCGA. Overview on the Antihypertensive and Anti-Obesity Effects of Secondary Metabolites from Seaweeds. *Mar Drugs.* 2018,16:237.

[18] Lee HG, Lu YA, Li XN, et al., Anti-Obesity Effects of *Grateloupia elliptica*, a Red Seaweed, in Mice with High-Fat Diet-Induced Obesity via Suppression of Adipogenic Factors in White Adipose Tissue and Increased Thermogenic Factors in Brown Adipose Tissue. *Nutrients.* 2020,12:308.

[19] Tapia-Martinez J, Hernandez-Cruz K, Franco-Colin M, et al., Safety evaluation and antiobesogenic effect of *Sargassum liebmannii* J. Agardh (Fucales: Phaeophyceae) in rodents. *J Appl Phycol.* 2019,31:2597–2607.

[20] Xu SY, Huang XS, Cheong KL. Recent Advances in Marine Algae Polysaccharides: Isolation, Structure, and Activities. *Mar Drugs.* 2017,15:388.

[21] Manlusoc JKT, Hsieh CL, Hsieh CY, Salac ESN, Lee YT, Tsai PW. Pharmacologic application potentials of sulfated polysaccharide from marine algae. *Polymers.* 2019,11:1163.

[22] Ale MT, Mikkelsen JD, Meyer AS. Important determinants for fucoidan bioactivity: A critical review of structure-function relations and extraction methods for fucose-containing sulfated polysaccharides from brown seaweeds. *Mar Drugs.* 2011,9:2106–2130.

[23] Cao J, Wang SC, Yao CW, Xu Z, Xu XM. Hypolipidemic effect of porphyran extracted from *Pyropia yezoensis* in ICR mice with high fatty diet. *J Appl Phycol.* 2016,28:1315–1322.

[24] Lekshmi VS, Kurup GM. Sulfated polysaccharides from the edible marine algae *Padina tetrastromatica* protects heart by ameliorating hyperlipidemia, endothelial dysfunction and inflammation in isoproterenol induced experimental myocardial infarction. *J Funct Foods.* 2019,54:22–31.

[25] Li B, Lu F, Wei XJ, Zhao RX. Fucoidan: Structure and bioactivity. *Molecules.* 2008,13:1671–1695.

[26] Laurienzo P. Marine polysaccharides in pharmaceutical applications: An overview. *Mar Drugs.* 2010,8:2435–2465.

[27] Bai X, Hu B, Song S, Liang H, Ji A. Research progress of fucoidan in anti-tumor mechanism. *Chem Life.* 2019,39:998–1003.

[28] Kim J, Kim DH, Chun SH, Park HY, Lee KW. Enhancement natural killer cell activity of fucoidan in lung metastasis in vivo model. *FASEB J.* 2018,32:1.

[29] Li X, Zhang S. Study on the effect of anti-aging and structure-function relationship of fucoidan from kelp. *Sci Technol Food Ind.* 2015,36:117–121.

[30] Tomori M, Nagamine T, Miyamoto T, Iha M. Evaluation of the immunomodulatory effects of fucoidan derived from *Cladosiphon okamuranus* Tokida in mice. *Mar Drugs.* 2019,17:547.

[31] Daub CD, Mabate B, Malgas S, Pletschke BI. Fucoidan from *Ecklonia maxima* is a powerful inhibitor of the diabetes-related enzyme, alpha-glucosidase. *Int J Biol Macromol.* 2020,151:412–420.

[32] Yang ZX, Yin JY, Wang YF, et al., The fucoidan A3 from the seaweed *Ascophyllum nodosum* enhances RCT-related genes expression in hyperlipidemic C57BL/6J mice. *Int J Biol Macromol.* 2019,134:759–769.

[33] Yin JY, Wang J, Li FH, et al., The fucoidan from the brown seaweed *Ascophyllum nodosum* ameliorates atherosclerosis in apolipoprotein E-deficient mice. *Food Funct.* 2019,10:5124–5139.

[34] Krizshanovsky SP, Kuznetsova TA, Geltser BI, et al., Fucoidan from brown algae *Fucus evanescens*: New perspectives in the treatment of atherosclerosis. *Russ J Biother.* 2017,16:82–87.

[35] Park J, Yeom M, Hahm DH. Fucoidan improves serum lipid levels and atherosclerosis through hepatic SREBP-2-mediated regulation. *J Pharmacol Sci.* 2016,131:84–92.

[36] Husni A, Pawestri S, Isnansetyo A. Blood glucose level and lipid profile of alloxan-induced diabetic rats treated with Na-alginate from seaweed *Turbinaria ornata* (Turner) J. agardh. *J Teknol.* 2016,78:7–14.

[37] Marounek M, Volek Z, Skrivanova E, et al., Comparative study of the hypocholesterolemic and hypolipidemic activity of alginate and amidated alginate in rats. *Int J Biol Macromol.* 2017,105:620–624.

[38] Idota Y, Kogure Y, Kato T, et al., Cholesterol-lowering effect of calcium alginate in rats. *Biol Pharm Bull.* 2016,39:62–67.

[39] Xue YT, Li S, Liu WJ, et al., The mechanisms of sulfated polysaccharide drug of propylene glycol alginate sodium sulfate (PSS) on bleeding side effect. *Carbohydr Polym.* 2018,194:365–374.

[40] Collins R, Armitage J, Parish S, Sleight P, Peto R, Heart Protection Study Collaborative Group. MRC/BHF Heart Protection Study of cholesterol lowering with simvastatin in 20536 high-risk individuals: A randomised placebo-controlled trial. *Lancet.* 2002,360:7–22.

[41] Preiss D, Tobert JA, Hovingh GK, Reith C. Lipid-modifying agents, from statins to PCSK9 inhibitors. *J Am Coll Cardiol.* 2020,75:1945–1955.

[42] Nikolic D, Banach M, Chianetta R, et al., An overview of statin-induced myopathy and perspectives for the future. *Expert Opin Drug Saf.* 2020,19:601–615.

[43] Hamed I, Ozogul F, Ozogul Y, Regenstein JM. Marine bioactive compounds and their health benefits: A review. *Compr Rev Food Sci Food Saf.* 2015,14:446–465.

[44] Hu XQ, Tao NP, Wang XC, Xiao JB, Wang MF. Marine-derived bioactive compounds with anti-obesity effect: A review. *J Funct Foods.* 2016,21:372–387.

[45] Yokota T, Nomura K, Nagashima M, Kamimura N. Fucoidan alleviates high-fat diet-induced dyslipidemia and atherosclerosis in ApoE(Shl) mice deficient in apolipoprotein E expression. *J Nutr Biochem.* 2016,32:46–54.

[46] Nasri M. Protein hydrolysates and biopeptides: Production, biological activities, and applications in foods and health benefits. A review. *Adv Food Nutr Res.* 2017,81:109–159.

[47] Han Y, Gao ZG, Chen LQ, et al., Multifunctional oral delivery systems for enhanced bioavailability of therapeutic peptides/proteins. *Acta Pharm Sin B.* 2019,9:902–922.

[48] Renukuntla J, Vadlapudi AD, Patel A, Boddu SHS, Mitra AK. Approaches for enhancing oral bioavailability of peptides and proteins. *Int J Pharm.* 2013,447:75–93.

[49] Salamat-Miller N, Johnston TP. Current strategies used to enhance the paracellular transport of therapeutic polypeptides across the intestinal epithelium. *Int J Pharm.* 2005,294:201–216.

[50] Zupancic O, Bernkop-Schnurch A. Lipophilic peptide character: What oral barriers fear the most. *J Control Release.* 2017,255:242–257.

[51] Lee, E. J., Hur, J., Ham, S. A., Jo, Y., Lee, S., Choi, M. J., Seo, H. G. Fish collagen peptide inhibits the adipogenic differentiation of preadipocytes and ameliorates obesity in high fat diet-fed mice. *Int. J. Biol. Macromol.* 2017, *104*, 281–286.

[52] Nasri, R., Abdelhedi, O., Jemil, I., Ben Amor, I., Elfeki, A., Gargouri, J., Boualga, A., Karra-Chaabouni, M., Nasri, M. Preventive effect of goby fish protein hydrolysates on hyperlipidemia and cardiovascular disease in Wistar rats fed a high-fat/fructose diet. *RSC Adv.* 2018, *8*, 9383–9393.

[53] Benaicheta, N., Labbaci, F. Z., Bouchenak, M., Boukortt, F. O. Effect of sardine proteins on hyperglycaemia, hyperlipidaemia and lecithin: Cholesterol acyltransferase activity, in high-fat diet-induced type 2 diabetic rats. *Br. J. Nutr.* 2016, *115*, 6–13.

[54] Jemil, I., Abdelhedi, O., Nasri, R., Mora, L., Marrekchi, R., Jamoussi, K., ElFeki, A., Hajji, M., Toldra, F., Nasri, M. Hypolipidemic, antiobesity and cardioprotective effects of sardinelle meat flour and its hydrolysates in high-fat and fructose diet fed Wistar rats. *Life Sci.* 2017, *176*, 54–66.

[55] Abdelhedi, O., Khemakhem, H., Nasri, R., Jridi, M., Mora, L., Ben Amor, I., Jamoussi, K., Toldra, F., Gargouri, J., Nasri, M. Assessment of cholesterol, glycemia control and short- and long-term antihypertensive effects of smooth hound viscera peptides in high-salt and fructose diet-fed Wistar rats. *Mar. Drugs* 2019, *17*, 194.

[56] Ben Slama-Ben Salem, R., Ktari, N., Bkhairia, I., Nasri, R., Mora, L., Kallel, R., Hamdi, S., Jamoussi, K., Boudaouara, T., El-Feki, A., et al., In vitro and in vivo anti-diabetic and anti-hyperlipidemic effects of protein hydrolysates from *Octopus vulgaris* in alloxanic rats. *Food Res. Int.* 2018, *106*, 952–963.

[57] Saravanan, P., Davidson, N. C., Schmidt, E. B., Calder, P. C. Cardiovascular effects of marine omega-3 fatty acids. *Lancet* 2010, *376*, 540–550.

[58] Bhatt, D. L., Steg, P. G., Miller, M., Brinton, E. A., Jacobson, T. A., Ketchum, S. B., Doyle, R. T., Juliano, R. A., Jiao, L. X., Granowitz, C., et al., Cardiovascular risk reduction with icosapent ethyl for hypertriglyceridemia. *N. Engl. J. Med.* 2019, *380*, 11–22.

[59] Miller, M. Icosapent ethyl for hypertriglyceridemia: Insights from the REDUCE-IT trial. *Future Cardiol.* 2019, *15*, 391–394.

[60] Mason, R. P., Dawoud, H., Jacob, R. F., Sherratt, S. C. R., Malinski, T. Eicosapentaenoic acid improves endothelial function and nitric oxide bioavailability in a manner that is enhanced in combination with a statin. *Biomed. Pharmacother.* 2018, *103*, 1231–1237.

[61] Albracht-Schulte, K., Gonzalez, S., Jackson, A., Wilson, S., Ramalingam, L., Kalupahana, N. S., Moustaid-Moussa, N. Eicosapentaenoic acid improves hepatic metabolism and reduces inflammation independent of obesity in high-fat-fed mice and in HepG2 cells. *Nutrients* 2019, *11*, 599.

[62] Zuniga, J., Cancino, M., Medina, F., Varela, P., Vargas, R., Tapia, G., Videla, L. A., Fernandez, V. N-3 PUFA supplementation triggers PPAR-alpha activation and PPAR-alpha/NF-kappa B interaction: Anti-inflammatory implications in liver ischemia-reperfusion injury. *PLoS ONE* 2011, *6*, 9.

[63] Tanaka, N., Irino, Y., Shinohara, M., Tsuda, S., Mori, T., Nagao, M., Oshita, T., Mori, K., Hara, T., Toh, R., et al., Eicosapentaenoic acid-enriched high-density lipoproteins exhibit anti-atherogenic properties. *Circ. J.* 2018, *82*, 596–601.

[64] Tang, Y., Lu, F., Lin, H., Fu, L. Physico-chemical property and hypoglycemic effect in vitro of soluble dietary fiber from *Ulva lactuca*. *J. Fujian Agric. For. Univ. Nat. Sci. Ed.* 2017, *46*, 702–707.

[65] Li, W. D., Jiang, N. F., Li, B. X., Wan, M. H., Chang, X. T., Liu, H., Zhang, L., Yin, S. P., Qi, H. M., Liu, S. M. Antioxidant activity of purified ulvan in hyperlipidemic mice. *Int. J. Biol. Macromol.* 2018, *113*, 971–975.

[66] Jiang, N.F., Li, B. X., Wang, X. Q., Xu, X. N., Liu, X. L., Li, W. D., Chang, X. T., Li, H., Qi, H. M. The antioxidant and antihyperlipidemic activities of phosphorylated polysaccharide from *Ulva pertusa*. *Int. J. Biol. Macromol.* 2020, *145*, 1059–1065.

[67] Chin, Y. X., Mi, Y., Cao, W. X., Lim, P. E., Xue, C. H., Tang, Q. J. A pilot study on anti-obesity mechanisms of *Kappaphycus alvarezii*: The role of native carrageenan and the leftover sans-carrageenan fraction. *Nutrients* 2019, *11*, 1133.

[68] Qiu, X., Zhong, W. W. Antihyperglycemic and antihyperlipidemic effects of low-molecular-weight carrageenan in rats. *Open Life Sci.* 2018, *13*, 379–384.

[69] Petryk, N., Shevchenko, O. Anti-inflammatory activity of mesenchymal stem cells in lambda-carrageenan-induced chronic inflammation in rats: Reactions of the blood system, leukocyte-monocyte ratio. *Inflammation* 2020.

[70] de Campos Facchin, B. M., da Rosa, J. S., Luz, A. B. G., Moon, Y. J. K., de Lima, T. C., Casoti, R., Biavatti, M. W., Dalmarco, E. M., Frode, T. S. Systemic administration of *Calea pinnatifida* inhibits inflammation induced by carrageenan in a murine model of pulmonary neutrophilia. *Mediat. Inflamm.* 2020, *2020*.

[71] Tkachenko, A. S., Gubina-Vakulyck, G. I., Klochkov, V. K., Kavok, N. S., Onishchenko, A. I., Gorbach, T. V., Nakonechna, O. A. Experimental evaluation of the impact of gadolinium orthovanadate $GdVO_4:Eu^{3+}$ nanoparticles on the carrageenan-induced intestinal inflammation. *Acta Medica Hradec Kralove* 2020, *63*, 18–24.

[72] Maeda, H., Fukuda, S., Izumi, H., Saga, N. Anti-oxidant and fucoxanthin contents of brown alga *Ishimozuku* (*Sphaerotrichia divaricata*) from the west coast of Aomori, Japan. *Mar. Drugs* 2018, *16*, 255.

[73] Zhang, Y. P., Xu, W., Huang, X. Q., Zhao, Y. Q., Ren, Q. Q., Hong, Z. A., Huang, M. Q., Xing, X. Fucoxanthin ameliorates hyperglycemia, hyperlipidemia and insulin resistance in diabetic mice partially through IRS-1/PI3K/Akt and AMPK pathways. *J. Funct. Foods* 2018, *48*, 515–524.

[74] Chang, Y. H., Chen, Y. L., Huang, W. C., Liou, C. J. Fucoxanthin attenuates fatty acid-induced lipid accumulation in FL83B hepatocytes through regulated Sirtl/AMPK signaling pathway. *Biochem. Bioph. Res. Commun.* 2018, *495*, 197–203.

[75] Cui, Y., Tang, N., Niu, J., Zhu, M., Cheng, M., Zhao, C., Qi, H., Liu, S. Study on the anti-aging effect of fucoxanthin on H_2O_2-induced premature senescence in WI-38 human diploid fibroblast cells. *Chin. J. Mar. Drugs* 2018, *37*, 48–52.

[76] Grasa-Lopez, A., Miliar-Garcia, A., Quevedo-Corona, L., Paniagua-Castro, N., Escalona-Cardoso, G., Reyes-Maldonado, E., Jaramillo-Flores, M. E. *Undaria pinnatifida* and fucoxanthin ameliorate lipogenesis and markers of both inflammation and cardiovascular dysfunction in an animal model of diet-induced obesity. *Mar. Drugs* 2016, *14*, 148.

[77] Gille, A., Stojnic, B., Derwenskus, F., Trautmann, A., Schmid-Staiger, U., Posten, C., Briviba, K., Palou, A., Bonet, M. L., Ribot, J. A lipophilic fucoxanthin-rich *Phaeodactylum tricornutum* extract ameliorates effects of diet-induced obesity in C57BL/6J mice. *Nutrients* 2019, *11*, 796.

[78] Hitoe, S., Shimoda, H. Seaweed fucoxanthin supplementation improves obesity parameters in mildly obese Japanese subjects. *Funct. Foods Health D* 2017, *7*, 246–262.

[79] Qin, Y. Applications of bioactive seaweed substances in functional food products. *J. Food Sci. Technol. China* 2019, *37*, 18–23.

[80] Choi, E. K., Park, S. H., Ha, K. C., Noh, S. O., Jung, S. J., Chae, H. J., Chae, S. W., Park, T. S. Clinical trial of the hypolipidemic effects of a brown alga *Ecklonia cava* extract in patients with hypercholesterolemia. *Int. J. Pharmacol.* 2015, *11*, 798–805.

[81] Ding, Y., Wang, L., Im, S., Hwang, O., Kim, H. S., Kang, M. C., Lee, S. H. Anti-obesity effect of diphlorethohydroxycarmalol isolated from brown alga *Ishige okamurae* in high-fat diet-induced obese mice. *Mar. Drugs* 2019, *17*, 637.

[82] Yang, Z. X., Liu, G. J., Wang, Y. F., Yin, J. Y., Wang, J., Xia, B., Li, T., Yang, X. Q., Hou, P. B., Hu, S. M., et al., Fucoidan A2 from the brown seaweed *Ascophyllum nodosum* lowers lipid by improving reverse cholesterol transport in C57BL/6J mice fed a high-fat diet. *J. Agric. Food Chem.* 2019, *67*, 5782–5791.

[83] Peng, Y. B., Wang, Y. F., Wang, Q. K., Luo, X., He, Y. H., Song, Y. F. Hypolipidemic effects of sulfated fucoidan from *Kjellmaniella crassifolia* through modulating the cholesterol and aliphatic metabolic pathways. *J. Funct. Foods* 2018, *51*, 8–15.

[84] Chen, S., Wang, W., Cai, L., Zhong, S., Zhong, S., Xie, E. The regulatory effect of *Sargassum* fucoidan on the lipid metabolism related enzymes and cholesterol synthesis key enzyme of hyperlipidemic mice. *J. Chin. Inst. Food Sci. Technol.* 2017, *17*, 10–16.

[85] Xie, Z., Kang, H., He, L., Kuang, R. Effects and mechanism of deep-sea fish oil on blood lipid in hyperlipidemia rats. *Chin. J. Clin. Pharmacol. Ther.* 2017, *22*, 870–874.

[86] Beppu, F., Li, H., Yoshinaga, K., Nagai, T., Yoshinda, A., Kubo, A., Kanda, J., Gotoh, N. Dietary starfish oil prevents hepatic steatosis and hyperlipidemia in C57BLI6N mice fed a high-fat diet. *J. Oleo Sci.* 2017, *66*, 761–769.

[87] Wu, D., Xu, H., Chen, J. Y., Zhang, L. S. Effects of astaxanthin supplementation on oxidative stress: A systematic review and meta-analysis of randomized controlled trials. *Int. J. Vitam. Nutr. Res.* 2020, *90*, 179–194.

[88] Zhang, K., Guo, J., Dai, R. Effects of astaxanthin on lipid-lowering, anti-oxidation and anti-inflammation of atherosclerosis in ApoE knockout mice. *Chin. J. Immunol.* 2020, *36*, 794–798.

[89] Zou, T., Yang, H., Tang, X., Ling, W. Protection of astaxanthin against high-fat diet-induced non-alcoholic fatty liver disease in mice. *Acta Nutr. Sin.* 2016, *38*, 386–390.

[90] Zuo, Z., Shao, Y., Liu, Y., Bu, Y., Wang, H., Li, N., Liu, Z. Astaxanthin regulates liver lipid metabolism and circadian gene expression. *Food Sci.* 2019, *40*, 165–172.

[91] Jia, Y., Wu, C., Kim, J., Kim, B., Lee, S. J. Astaxanthin reduces hepatic lipid accumulations in high-fat-fed C57BL/6J mice via activation of peroxisome proliferator-activated receptor (PPAR) alpha and inhibition of PPAR gamma and Akt. *J. Nutr. Biochem.* 2016, *28*, 9–18.

[92] Faraone, I., Sinisgalli, C., Ostuni, A., Armentano, M. F., Carmosino, M., Milella, L., Russo, D., Labanca, F., Khan, H. Astaxanthin anticancer effects are mediated through multiple molecular mechanisms: A systematic review. *Pharmacol. Res.* 2020, *155*, 104689.

[93] Gunasinghe, M. A., Kim, A. T., Kim, S. M. Inhibitory effects of vanadium-binding proteins purified from the sea squirt *Halocynthia roretzi* on adipogenesis in 3T3-L1 adipocytes. *Appl. Biochem. Biotech.* 2019, *189*, 49–64.

[94] Lindequist, U. Marine-derived pharmaceuticals—challenges and opportunities. *Biomol. Ther. (Seoul)* 2016, *24*(6), 561–571. https://doi.org/10.4062/biomolther.2016.181.

[95] Malve, H. Exploring the ocean for new drug developments: Marine pharmacology. *J. Pharm. Bioallied Sci.* 2016, *8*(2), 83–91. https://doi.org/10.4103/0975-7406.171700.

[96] Atanasov, A. G., Zotchev, S. B., Dirsch, V. M., et al., Natural products in drug discovery: advances and opportunities. *Nat. Rev. Drug Discov.* 2021, *20*, 200–216. https://doi.org/10.1038/s41573-020-00114-z.

[97] Supanekar, S. P., Naik, M. S., Sarwade, V. V. Marine resources for marine-derived drugs: A review. *Uttar Pradesh J. Zool.* 2024, *45*(5), 175–186.

[98] Xiang, Z., Han, M., Zhang, H. Nanomaterials based flexible devices for monitoring and treatment of cardiovascular diseases (CVDs). *Nano Res.* 2022, *89*, 248.

[99] Abd, M., Mohammed, E., Badawy, D., Naing, L., Johar, S., Ong, S., Rahman, H. A., Lin, C., Raja, C., Pengiran, I., et al., Scoping review: Are CVDs risk calculators using the digital platform beneficial for CVDs prevention and management? *Res. Sq.* 2022. https://doi.org/10.21203/rs.3.rs-XXXXX/v1. (Preprint)

[100] Vogel, B., Acevedo, M., Appelman, Y., Bairey Merz, C. N., Chieffo, A., Figtree, G. A., Guerrero, M., Kunadian, V., Lam, C. S. P., Maas, A. H. E. M., et al., The Lancet Women and Cardiovascular Disease Commission: Reducing the global burden by 2030. *Lancet* 2021, *397*, 2385–2438.

[101] Jankowski, J., Floege, J., Fliser, D., Böhm, M., Marx, N. Cardiovascular disease in chronic kidney disease: Pathophysiological insights and therapeutic options. *Circulation* 2021, *143*, 1157–1172.

[102] Ferraz, C. A. A., Grougnet, R., Nicolau, E., Picot, L., de Oliveira Junior, R. G. Carotenoids from marine microalgae as antimelanoma agents. *Mar. Drugs* 2022, *20*, 618.

[103] Liang, B., Cai, X.-Y., Gu, N. Marine natural products and coronary artery disease. *Front. Cardiovasc. Med.* 2021, *8*, 739932.

[104] Ahmed, I., Asgher, M., Sher, F., Hussain, S. M., Nazish, N., Joshi, N., Sharma, A., Parra-Saldívar, R., Bilal, M., Iqbal, H. M. N. Exploring marine as a rich source of bioactive peptides: Challenges and opportunities from marine pharmacology. *Mar. Drugs* 2022, *20*, 208.

[105] Nigam, M., Suleria, H. A. R., Farzaei, M. H., Mishra, A. P. Marine anticancer drugs and their relevant targets: A treasure from the ocean. *DARU J. Pharm. Sci.* 2019, *27*, 491–515.

[106] Kwon, O. S., Ahn, S., Jeon, J. E., Park, I. G., Won, T. H., Sim, C. J., Park, H. G., Oh, D. C., Oh, K. B., Noh, M., et al., Psammocindoles A–C: Isolation, synthesis, and bioactivity of indole-γ-lactams from the sponge *Psammocinia vermis*. *Org. Lett.* 2021, *23*, 4667–4671.

[107] Kleks, G., Holland, D. C., Kennedy, E. K., Avery, V. M., Carroll, A. R. Antiplasmodial alkaloids from the Australian bryozoan *Amathia lamourouxi*. *J. Nat. Prod.* 2020, *83*, 3435–3444.

Chapter 7

Beyond the Shore: Marine Metabolites in Cancer Therapy

Adil Ali[1]
Ankit Paul[1]
Tarun Kumar Upadhyay[1,*]
and Pranav Kumar Prabhakar[2]
[1]Department of Life Sciences, Parul Institute of Applied Sciences and Research and Development Cell, Parul University, Vadodara, Gujarat, India
[2]Department of Biotechnology, School of Engineering and Technology, Nagaland University, Meriema, Kohima, Nagaland, India

Abstract

Cancer is one of the leading causes of death globally. It is one of the deadliest diseases, causing multiple metabolic and physiological alterations leading to an uncontrolled proliferation of cells due to transformation, disorders of apoptosis, proliferation, invasion, angiogenesis, and metastasis. Conservative chemotherapy is currently utilized to treat cancer because there has been a lack of effective targeted drugs. Several studies have been performed to find novel anticancer drugs from natural sources because the existing therapies are either very expensive or non-efficient with side effects. As a result, there has been an urgent requirement for novel drugs with distinctive modes of action. This book chapter discusses the effects of numerous bioactive compounds derived from marine sources, focusing on marine plants, algae, bacteria, actinomycetes, fungi, sponges, and soft corals. Marine-derived secondary metabolites exhibit a diverse array of biological activities. Anticancer effects of marine natural products through *in vitro* and *in vivo* studies were first introduced; their antiproliferative effects or the induction of cell cycle arrest, apoptosis, or autophagy induction activity of marine metabolites in preventing tumorigenesis has sparked interest in exploring this field of study. *Aspergillus spp.*, *Penicillium spp.*, *Neosartorya spp.*, *Dichotomomyces spp.*, *Paradendryphiella spp.*, *Westerdykella spp.*, *Streptomyces spp.*, and *Nocardiopsis spp.* are some of the marine-based fungi species which are reported to possess antitumorigenic activity.

---

* Corresponding Author's Email: tarun_bioinfo@yahoo.co.in

In: Healing from the Depths Marine Metabolites in Disease Management
Editors: Pranav Kumar Prabhakar and Tarun Kumar Upadhyay
ISBN: 979-8-89530-832-5

Keywords: marine, secondary metabolite, cancer, anticancer drug, microorganism, plants

1. Introduction

Cancer has been a global disease with high mortality and morbidity. According to the World Health Organization (WHO), it has been projected that by 2030, there will be a total of 21 million newly diagnosed incidences of cancer diseases and 13 million death tolls attributed to this illness [1]. At present, 13% of all death tolls around the globe are engendered by carcinoma, as well as it is appraised that 37% of such mortalities may be evaded by altering or averting the main etiological factors, such as smoking, tobacco chewing, exposure to radiation, liquor consumption, infections, and by some uncertain, unknown factors. The majority of existing anticancer treatments have adverse effects. Subsequently, there is a compelling need for the hour for novel and safer drugs to combat cancer. Pharmaceutical firms have shown decreased interest in medications generated from natural products. However, it remains the most superior platform for delivering innovative, efficient, and distinctive chemical structures that could have significant potential in the treatment or preclusion of carcinoma. They can also act as scaffolds or lead molecules to explore more potent anticancer medications. Approximately 1355 medicines received approval for therapeutic use between 1981 to 2010. Out of them, 128 were anti-oncogenic drugs, with roughly thirty-five percent derived from organic products or composites isolated from natural organic sources [2].

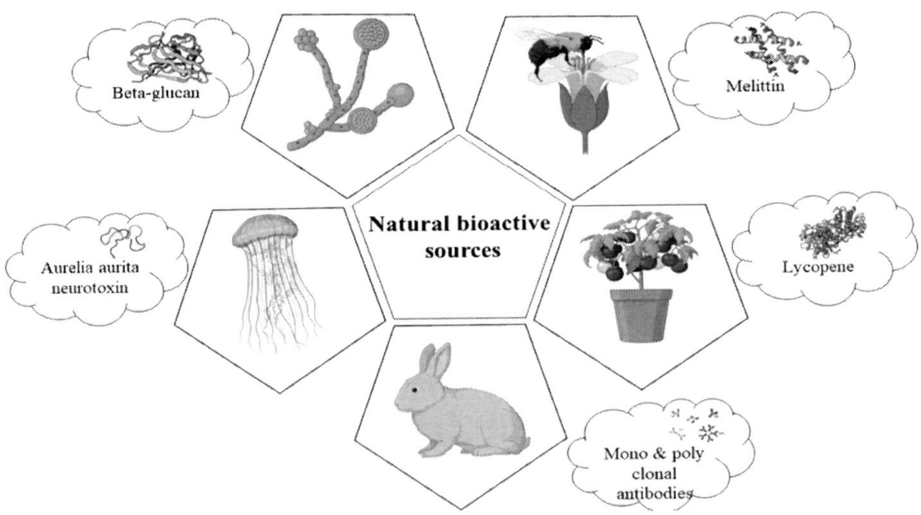

Figure 1. Schematic diagram of different natural bioactive sources.

Nature remains an enormous repository of diverse and biologically active chemical compounds. Natural products have been the most efficacious medications for combating cancer. The majority of clinical drugs, over 70%, are sourced from natural substances, as shown in Figure 1. This is predominantly evident in the cancer chemotherapeutic approach, where natural materials constitute most of the drugs now used for treatment [3]. However, very few of the extracted natural products are directly transformed into medications that are beneficial in clinical applications. These unique composites or molecules also serve as models for developing more effective analogs and pro-drugs by applying chemical approaches. Extensive research is being conducted primarily focused on nature-derived bioactive compounds, which is the discovery and development of novel pharmaceutical medications that can combat cancer. Additionally, the importance ofcollaboration across different fields of study to extract unique molecular properties from natural resources has also been thoroughly examined [4]. Nevertheless, natural product development and advancement stages are costly, laborious, and time-consuming [1].

2. Oceans as a Reservoir for Diverse Marine Bioactive Compounds

Oceans contain an extensive array of marine animals that possess natural medicinal components with significant promise. Research on aquatic life remains mostly unexplored [5]. Marine bioprospecting is a newly emerging practice that involves exploring and exploiting marine resources to generate commercial and public health benefits [6]. Terrestrial life has been the primary source of the majority of pharmacologically potent natural substances up to the present day [7]. However, marine compounds are gaining attention and have profoundly influenced drug discovery and the development of novel pharmaceuticals. The vast diversity of marine ecosystems, ranging from deep-sea trenches to coral reefs, harbors numerous unique organisms, each producing distinct bioactive compounds. These marine-derived substances have provided novel chemical structures and mechanisms of action that are often not found in terrestrial sources. Several marine metabolities have been successfully extracted from extensive oceanic aquatic flora fauna and have been developed into drugs, chiefly in the spheres of oncology, pain management, and other contagious diseases.

3. Marine Bioprospecting and Human Health

Earth's marine ecology is dynamic and sensitive, covering roughly 70.8% of the planet [8]. This vast range of biodiversity includes a diverse collection of microorganisms as well as larger life forms [9]. Marine plants have been extensively employed for their potential therapeutic benefits worldwide. Marine organisms such as bacteria, fungi, actinobacteria, marine seaweeds, and oceanic benthic sponges have been used for remedying cancer [10]. Trabectedin, an anticancer medication, is an example of a therapeutic application of marine natural products. It is obtained from the sea squirt *Ecteinascidia turbinata*. Marine compounds' intricate and distinctive characteristics also present novel opportunities for surmounting drug resistance, a formidable obstacle in contemporary medical therapies. Furthermore, progress in marine biological research and sustainable harvesting techniques improve the capacity to investigate and exploit marine biodiversity for pharmaceutical development [11].

Marine microalgae's role in the ecosystem has been important and utilized as sustenance and therapeutic items since ancient times. Marine microalgae, being eukaryotic plants, play a role in drug development due to their ability to adapt metabolically. This adaptability can stimulate the synthesis of many chemicals potentially used in fighting different diseases, including cancer [12]. These aquatic marine species have become highly known for their biologically active constituents, which provide an exceptional potent for various pharmacological activities. The dissemination of marine genera varies based on the specific type of marine ecosystems, including the pelagic zone, epipelagic zone, bathypelagic layer, mesopelagic layer, benthic environment, and abyssopelagic layer. The various layers of the aquatic environment contain a wide range of diverse marine microorganisms and microalgae. Microflora and, to some extent, microalgae play an important role in driving nutrient fluxes in an underwater environment. The changes in non-living variables of marine environments, such as oxygen, food availability, salinity, temperature, sunlight, water transparency, waves, tides, aerial exposure, and oceanic currents, have significantly impacted the production as well as dissemination of biologically active organic metabolites from marine animals. To endure the challenging abiotic conditions, the marine microflora tends to establish asymbiotic associations with other marine microbes. These symbiotic interactions allow them to survive and produce several rare secondary metabolites/bioactive substances that they would not produce naturally [13]

In the past twenty years, there has been a momentous change in organic compound-based drug research, shifting the focus from terrestrial areas to the oceans. Over the past thirty years, some 15,000 metabolites derived from marine sources have been discovered. These metabolites have shown the ability to kill tumorigenic cells in various *in vitro* models and animal studies. Currently, there are 28 of these metabolites being tested in clinical investigations for their potential as anticancer medicines [1]. There are numerous unexplored structural patterns in the open sea ecosystem, and these substances can be used as powerful novel anticancer drugs.

4. Bioactive Constitution of Marine Organisms

The primary marine bioactive chemicals derived from marine sources with anti-oncogenic properties include alkaloids, polyketides, terpenes, peptides, nucleosides, and carbohydrates [14]. The bioactive constitution of marine organisms is as follows:

4.1. Alkaloids

Alkaloids are a class of organic molecules, both naturally derived and synthesized, that are characterized by their alkaline-like characteristics, containing a minimum N-atom in their structure [15]. Currently, these heterocyclic compounds containing nitrogen atoms can be categorized according to their similarity in carbon structure to metabolic precursors such as different proteins and amino acids ornithine, tyrosine, tryptophan, and lysine; they can have either indole and isoquinoline or pyridine components [16].

Marinoquinoline A is an alkaloid with anticancer properties Catalinimonas alkaloidigena, a marine bacterium, synthesizes. This bacterium also produces 13 additional alkaloid metabolites. *Pantoea citrea* and *Pseudoalteromonas tunicata* are two types of oceanic bacteria that produce hay-colored alkaloids belonging to a category known as tambjamines. These tambjamines have been found to exhibit antitumor properties, as well as antibacterial, antifungal, and antimalarial effects [17].

4.2. Polyketides

Polyketides are a diverse category of compounds produced using a series of modular enzymes acting as biocatalysts. These enzymes build analogs from precursors, including switching carbonyl and methylene groups (-CO-CH2-). *Streptomyces koyangensis* (*S. koyangensis*) is an oceanic bacterium that yields two chemicals called abyssomicins with anticancer properties. However, this topic has limited research [1]. The marine *Nostoc spp.* cyanobacteria (*Nostoc spongiaeforme* and *Nostoc linckia*) produce a polyketide derivative compound termed borophycin A, which has revealed a potent anticancer activity against human colon tumorigenic cell lines [18].

4.3. Terpenes

Terpenes are organic molecules composed of isoprene units containing five carbon atoms. These units combine to form a wide variety of structures, which are then modified by enzymes to add different functional groups and change their oxidation state. These cyclic molecules may be categorized based on the total number of isoprene (CH_2=CCH=CH_2) units these organic marine compounds contain [19]. It is widely recognized that marine microflora is a highly valuable source of terpenes [1].

4.4. Peptides

Peptides are specific oligomers of proteins that have beneficial implications for human health. Peptides are typically inactive during part of the original protein chain and can perform various physiological functions when they undergo proteolysis [20]. Enzymatic hydrolysis has been instrumental in producing peptide molecules in marine organisms [21]. Several studies have been piloted on using marine-based peptides for pharmacological advancements, specifically concerning their anticancer capabilities [1].

Polydiscamide A, a novel peptide, and its derivatives exhibit anti-tumor properties. A specific quantity of marine peptides underwent successful evaluation in clinical research and are now commercially available as manufactured medications under different brand names. The peptides derived from various *Lyngbya spp.* as well as *Nostoc spp.* have been found to exhibit significant anticancer properties by disrupting microfilaments, inhibiting secretory processes, and affecting many intracellular pathways [1].

4.5. Nucleosides

Nucleosides are a class of biological molecules referred to as the nitrogen glycosides of purine bases and pyrimidine bases. Combined with esterified phosphates, known as nucleotides, are essential in all live cells and are involved in several key physiological processes [22]. Marine microorganisms are highly proficient in producing various nucleosides with unique structures and specific biological characteristics [1]. The powerful biological characteristics of nucleosides produced from marine sources have prompted the development of some analogies. Apart from anti-cancerous properties, marine nucleosides have shown numerous bioactive properties, such as muscle relaxant, antiviral, hypertensive, and vasodilator activity [23].

4.6. Carbohydrates

Marine species mostly rely on carbohydrates as their primary food source, with algae being a significant source. These compounds exhibit an eclectic range of molecular structures similar to the sulphated polysaccharides produced by the sulphur and carbon-producing photosynthetic organisms [24]. Saccharides are categorized based on their origin into three categories: plant-based polysaccharides, animal-based polysaccharides, and microbial-based polysaccharides [25]. These groupings consist of saccharides of different sizes, both neutral and negatively charged. Examples of oligosaccharides found in glycoproteins include those related to either nitrogen or oxygen. Nevertheless, thes substances' chemical constitution and conformations are intricate and varied [1].

The carbohydrates produced from marine sources, such as laminarin, algin, glucan, Irish sea moss, chitin, agar, and cellulose, have a variety of significant biologically active qualities, including the ability to inhibit cancer [26]. These carbohydrate-based compounds demonstrated anti-cancer effects on several oncogenic cells by influencing the innate or non-specific immune system. As a result, macrophages (Mφ) and natural killer (NK) cells exhibited chemotactic reactions towards the target location, where they secreted tumoricidal cytokines.

5. Marine Organisms to Anticancer Drugs

There are over ~22000-23000 microbial marine secondary metabolites, with marine actinomycetes producing seventy percent, oceanic fungus producing twenty percent, and oceanic *Bacillus spp.* producing seven percent, as well as other oceanic microbes producing one to two percent, Figure 2 represents the different marine sources [27]. It is important to mention that around 10% of all presently recognized biologically active natural compounds originate from microorganisms. Several marine antineoplastic medicines have progressed to clinical phase testing. Such compounds include dolastatin 10, ET-743, and bryostatin 1.

Bryostatin 1 is currently undergoing a 2nd phase clinical trial for the treatment of melanoma, renal cancer, breast cancer, as well as non-Hodgkin's lymphoma. Bryostatin 1 exerts its biological impact by facilitating the healthy development of bone-marrow predecessor cells. In addition, Ecteinascidian-743, tetrahydroisoquinoline alkaloids obtained from the oceanic tunicate *E. turbinate*, has progressed to the first phase of clinical trials [24]. This compound exerts anti-neoplastic effects by selectively modifying GC residues in the minor groove of the DNA. Similarly, dolastatin 10, a peptide derivative of the mollusc *Dolabella auricularia*, has advanced to the second phase of the clinical trials [28]. Its mechanism of action encompasses inhibiting the microtubule assemblage, leading to a metaphasic halt during cell division.

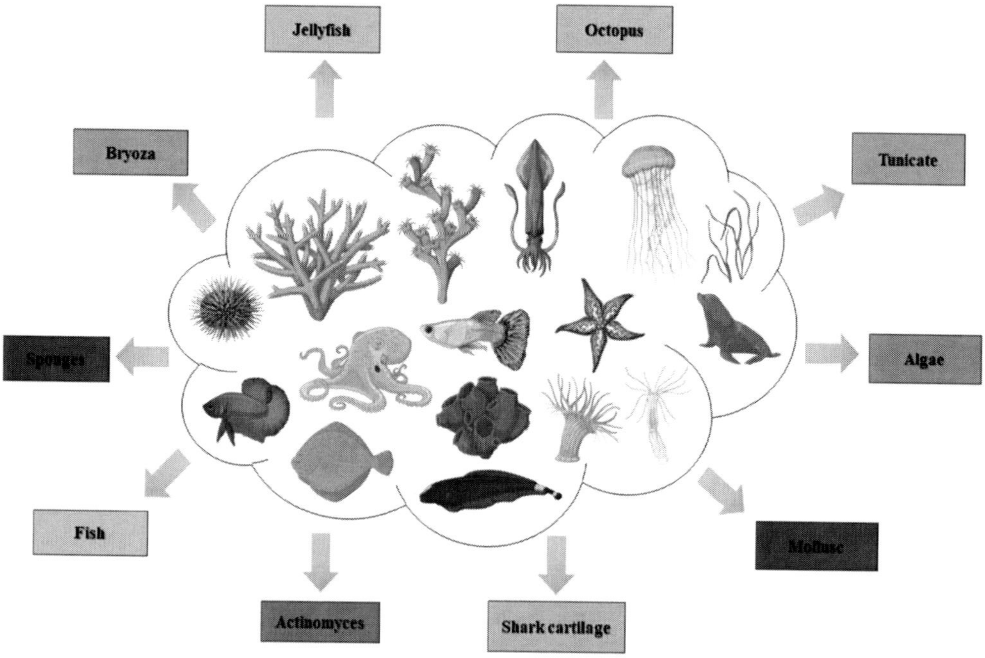

Figure 2. Different marine sources being used as an anticancer agent.

5.1. Marine Bacteria

The biologically active compounds extracted from Oceanic Pseudomonas spp. are varied as well as encompass pyrrolidine-dione, phloroglucinol, phenazine, benzaldehyde, pseudo-peptides, quinoline, phenanthrene, phthalate, quinolone, andrimid, moiramides, bushrin, zafrin, as well as pyrroles [29]. Several of these bioactive chemicals exhibit antibacterial properties, while di-n-butyl phthalate (DNBP), as well as bis-(2-ethylhexyl) phthalate (DOP) have been identified as inhibitors of cathepsin B. Discodermolide, sarcodictyin, eleutherobin, and bryostatins, are highly potent anticancer medicines primarily synthesized by marine microorganisms.

5.2. Marine Actinomycetes

Marine actinomycetes comprise *Dietzia, Rhodococcus, Streptomyces, Salinispora, and Marinispora* [30] genera. Actinomycetes, marine microorganisms, are undeniably the primary suppliers of secondary marine metabolites, exhibiting the highest production rate [31]. Secondary metabolites derived from actinomycetes constitute around 45% (around 10,000 compounds) of all known antimicrobial metabolites. Out of the chemicals derived from actinomycetes, 75% were derived from *Streptomyces*, while the remaining 25% were derived from uncommon actinomycetes [1].

5.3. Marine Fungi

Marine fungus offers a valuable and encouraging supply of new anticancer substances. Biologically active compounds, including leptosphaerolide and its derivative and leptosphaerin, have been obtained from higher-order fungi (basidiomycetes), endophytic-fungi, as well as filamentous-fungi found in oceanic habitats. These compounds were extracted from the oceanic fungus named *Leptosphaeria oraemaris* (*Pleosporaceae*) [32]. Acremonin A from marine fungal *Acremonium* species and a xanthone derivative from a marine fungus named *Wardomyces anomalus* have demonstrated antioxidative properties that counteract the harmful effects of free radical reactions linked to cancer. Oceanic organic products continue to be a valuable source of innovative cum potent drugs for cancer treatment.

5.4. Marine Sponge

These organisms have significantly contributed to the discovery of about thirty percent of all oceanic organic products identified to date [33]. The preliminary findings from oceanic sponges fostered the expectation that genuine medications derived from marine sources would soon be available. An exemplary instance is the successful exploration and recognition of two unusual marine-derived nucleosides: spongo-thymidine and spongo-uridine, derived from the Caribbean sponge *Tethya crypta* (*Tethya crypta*). Antiviral activity was detected in these molecules, and further investigation into synthetic analogs eventually resulted in the creation of nucleoside cytosine containing arabinoside (AraC) as a clinically effective anti-cancer drug [34]. Eribulin, a synthetic derivative of halichondrin B obtained from the oceanic sponge named *Halichondria okadai*, exhibits promising therapeutic effects against metastatic breast cancer cells previously treated.

5.5. Soft Corals

Sarcophyton is a highly prevalent genus of soft corals found in tropical and subtropical oceans. Approximately 30 species belonging to the *sarcophyton* genus have been analyzed to determine the existence of beneficial secondary metabolites, particularly fatty acids such as arachidonic ($C20H_{32}O_2$), eicosapentaenoic ($C_{20}H_{30}O_2$), and docosahexaenoic acids (DHA OR $C_{22}H_{32}O_2$) [35]. Cembranoids, which are found in soft coral, are a crucial element, occurring in significant amounts (up to 5% of the coral's dry weight). Cembranoids wield numerous biological properties, including cytotoxic, anti-inflammatory, and antagonistic action. The results of *in vitro* cell-based toxicity testing demonstrated that furanocembranoids and membrane diterpene-based decaryiol, were extracted from *Nephthea spp.* As well as *Sarcophyton cherbonnieri (S. cherbonnieri)* exhibits significant efficacy against various carcinoma cell lines, including breast, epithelium, and hepatic cells.

6. Marine Bioactive Compounds and Their Mechanism of Action

Nature's bounty has proven to be a valuable reservoir for exploring and developing cancer therapeutics. Roughly 75% of the anti-cancer medications currently in use are derived from natural sources or compounds closely related to natural chemicals [36]. Natural products interact powerfully and selectively with a wide range of biological components, making them useful for therapeutic purposes and as tools to investigate unexplored biological mechanisms in cells [37].

Most marine bioactive compounds induce apoptosis and autophagy through different mechanisms such as regulating intracellular proteins like p53, Bcl-2, Bax, pRB, AKT, Beclin-1, mTOR, PI3K, CDKs, cyclins; inducing non-repairable DNA damage, activating proteasomal degradation and ubiquitination by phosphorylating or dephosphorylating different secondary messenger molecules, altering the structure of the intracellular translocator molecules [38].

Table 1. Compilation of drug derivatives from marine sources exhibiting promising anticancer properties

Compound	Marine based Source	Cancer type	Mode of action	References
Apratoxin-A	*Lyngbya bouillon*, bacteria	Cervical carcinoma	Inhibition of the cell-cycle	[42, 43]
Trabectedin	Sea squirt (*Ecteinascidia turbinata*)	Soft tissue sarcoma, ovarian cancer	Binds to DNA, disrupting cell cycle and inducing apoptosis	[41, 44, 45].
Plitidepsin	Tunicate (*Aplidium albicans*)	Multiple myeloma, leukemia, lymphoma	Targets eEF1A2 induce oxidative stress	[45].
Eribulin Mesylate	Sponge (*Halichondria okadai*)	Breast cancer, liposarcoma	Suppresses the movement of microtubules, resulting in the halting of the cell cycle	[40, 46],
Salinosporamide A (Marizomib)	Bacterium (*Salinispora tropica*)	Multiple myeloma, solid tumors	The proteasome inhibitor induces apoptosis	[45, 47]
Dolastatin	Sea hare (*Dolabella auricularia*)	Cancers	Inhibits tubulin polymerization, leading to cell cycle arrest	[48]
Coibamide A	Cyanobacterium (Leptolyngbya)	Glioblastoma, lung cancer	Disrupts cell cycle induces apoptosis	[49]
Bryostatin 1	Bryozoan (*Bugula neritina*)	Leukemia, melanoma, solid tumors	Activates protein kinase C, modulates apoptosis	[50]
Discodermolide	Sponge (*Discodermia dissoluta*)	Solid tumors	Stabilizes microtubules, causing a halt in the cell cycle	[51]
Sansalvamide A	Marine fungi	Breast, renal, and prostate cancer	Prevents to the development of protein complexes	[52]

Apoptosis, also called programmed cell death, is a crucial mechanism by which cells are genetically designed to be abolished. It is a key process connected with the development of the organism, its physiology, and body homeostasis. Alteration or deregulation of apoptosis and accumulation of mutations can enable cancerous cells to persevere and spread between local body organs [39]. Any disruption in the equilibrium between pro- and anti-apoptotic signals might result in various clinical diseases, such as cancer. Apoptosis, or programmed cell death, leads to the activation of caspases, enzymes that play a role in cell demise. This activation causes the cleavage of various caspases, resulting in the fragmentation of DNA, degradation of peptides as well as macromolecules in the cytoskeleton and nucleus, cross-linking of intracellular macromolecules, development of intracellular apoptotic bodies, spatial and temporal expression of molecules that bind to receptors on phagocytic cells, and ultimately, uptake of the dying cell by phagocytic cells [40].

Autophagy is a catabolic process that is evolutionarily conserved and serves as a cellular breakdown system, eliminating defective or unwanted proteins as well as organelles. The autophagy mechanism is commonly triggered by the buildup of detrimental free radicals, low oxygen levels, lack of nutrients, drug exposure, or stress on the endoplasmic reticulum. The key proteins implicated in the autophagic mechanism are the mammalian target of rapamycin complex-1 (mTORC-1), PIP-3, LC-3B, kinase AKT (AKT), Beclin-1, phosphatidylinositol-3 kinase (PI-3K), p110, p53, PDK-1, as well as p85 [41]. The PI3K/AKT/mTOR signal transduction cascade is the prime target preceding the kinase mTOR. The activation of PI3K, and AKT, leads to the inhibition of autophagy through the activation of mTOR [42]. On the other hand, the suppression of the mTOR would stimulate the autophagy process, which is the primary focus of some marine-derived substances and medications, as shown in Table 1.

6.1. Apratoxins

Apratoxins are cytotoxic marine organic products that are cyclodepsipeptides derived from oceanic cyanobacteria. They possess significant anticancer efficacy against multiple types of carcinoma cells in vitro. Among the various apratoxins, Apratoxin A exhibits the highest level of cytotoxicity. Apratoxin-A induces G-1 cellular cycle arrest as well as disrupts the fibroblast growth factor receptor (FGFR) signaling cascade by reducing STAT3 phosphorylation. Additionally, Apratoxin A has been proposed to stabilize the association between Hsc70/Hsp70 as well as Hsp-90 client macromolecules, thereby preventing their binding to Hsp-90. This stabilization leads to the degradation of EGFR as well as ErbB2 through chaperone-mediated autophagy. Furthermore, Apratoxin-A impedes the secretory process in a cellular-free in vitro system by blocking the co-translational translocation of newly synthesized proteins, potentially targeting Sec61α [42].

Additionally, Apratoxin A reduces interleukin 6 signal transducer (gp130) promptly, suppressing Janus kinase (JAK)/STAT signaling. It also inhibits the N-glycosylation of carcinoma-associated receptor tyrosine kinases, rapidly degrading the proteasome and depleting cancer cells [43]. Exposure to Apratoxin A results in the down-regulation of many membrane-associated receptors, including c-MET, HER-2, PDGFR-β, as well as IGF1R-β.

6.2. Trabectedin

Trabectedin is a tetrahydroisoquinoline alkaloid, a potent alkylator obtained from the Caribbean Sea squirt *E. turbinata*. Notably, it was the 1st marine-based antineoplastic medication sanctioned by the European Union. Trabectedin has unique anticancer mechanisms, including transcription regulation, DNA binding, interference with DNA reparation pathways, and modulation of the tumor's microenvironment. Trabectedin has demonstrated significant clinical benefits and is used as a standalone treatment, especially for sufferers with advanced soft tissue sarcoma who have not responded to anthracycline and ifosfamide therapies or are resistant to these medications. Both trabectedin monotherapy and combination therapy regimens are employed to treat various malignancies, such as soft tissue sarcomas, breast carcinoma, non-small-cell lung carcinoma, as well as ovarian carcinoma [41]. Additionally, it is employed in conjunction with pegylated-liposomal-doxorubicin for those experiencing a recurrence of platinum-sensitive ovarian cancer.

Trabectedin exhibits a distinctive and intricate mode of action that affects transcription regulation and DNA repair mechanisms. It specifically binds to the N-2 position of guanine in the minor groove of the dsDNA, showing a preference for GC-rich sequences. This binding causes the DNA to bend towards the major groove and effectively blocks activated transcription by binding to transcribing RNA polymerase II, leading to its degradation in several biological systems without affecting constitutive transcription. Trabectedin suppresses cellular cycle advancement during the S-phase and induces a cell cycle halt in the G2-phase. It also alters and activates transcription-coupled nucleotide excision repair (TC-NER) as well as homologous recombination repair (HRR). These intriguing mechanisms are key factors in its ability to suppress cell proliferation, as validated in laboratory models [44, 45].

6.3. Plitidepsin

Plitidepsin, a cyclic depsipeptide derived from *Aplidium albicans*, has a structure similar to bis-benzophenone. It is known for its significant anticancer efficacy coupled with low toxicity. Plitidepsin received Food and Drug Administration (FDA) approval in 2004 to cure various myelomas. The therapeutic mechanism of the drug plitidepsin in cancer treatment has been well-documented. It exhibits antineoplastic properties by suppressing carcinoma cell growth as well as inducing programmed cell demise across several carcinoma types, including myelomas, pancreatic carcinoma, prostate as well as ovarian carcinoma. The administration of plitidepsin induces oxidative stress as well as triggers the activation of Rac1 GTPase, leading to sustained activation of mitogen-activated protein kinase (p38/MAPK) as well as c-Jun N-terminal kinase (JNK). This cascade results in caspase activation as well as subsequent apoptosis of carcinoma cells. Recent research has identified that plitidepsin primarily targets the eukaryotic translation elongation factor-1 α-2 (eEF1A2). It binds to the interface of the subdomains 1 and 2 of eEF1A2 in its GTP-bound configuration, effectively disrupting the protein's function and contributing to its antitumor effects [45].

6.4. Eribulin

Eribulin mesylate is a naturally occurring compound in the oceanic sponge *Halichondria okadai*. It is an analog of halichondrin-B. Eribulin, a member of the halichondrin class of antineoplastic medicines, is a suppressor of microtubule dynamics. It has a distinct mode of action, which is the suppression of the polymerization phase of the microtubule while not affecting the depolymerizing phase [40].

Furthermore, eribulin induces the production of ineffective tubulin aggregates, which can potentially hinder the formation of the mitotic spindle by reducing the levels of free tubulin within the cell. The consequence is the impediment of the development of the mitotic spindle, leading to the halt of mitosis as well as the subsequent death of the cell through apoptosis. According to many studies, eribulin causes the tubulin polymerase ch-TOG to be removed from the + ends of microtubules as they expand. This causes cancer cells to migrate and changes the dynamics of microtubules. Additionally, to its antimitotic mechanism, eribulin has shown non-mitotic effects such as vascular remodelling, prevention of carcinoma cell motility, invasion, metastasis, and transformation of the epithelial-to-mesenchymal (EMT) transition [40, 46].

6.5. Salinosporamide A (Marizomib)

Marizomib, or Salinosporamide-A, is a proteasome suppressor that includes a gamma-lactam-beta-lactone fragment as well as is produced from the marine bacteria Salinispora tropica. Marizomib is an effective treatment for several neoplastic carcinomas, such as glioma, multiple myeloma, as well as brain tumors. The treatment of multiple myeloma, was authorized by the FDA in 2013 and the EMA in 2014. Inducing mitochondrial membrane potential disturbance, activating caspases, increasing ROS production, promoting cytochrome C/Smac release, and ultimately killing multiple myeloma cells is what marizomib does. It triggers the death receptor system and the inner mitochondrial apoptotic pathway by inhibiting the proteasome and suppressing the NF-κB pathway. Increased amounts of IκBα are the outcome of this inhibition, which subsequently blocks the NF-κB pathway. Marizomib inhibits the proteasome's caspase-like, chymotrypsin-like and trypsin-like activities; it is quickly transported from the vascular compartment to different organs and tumors.

The production of proteins that block cell death, angiogenesis-promoting factors, and apoptosis is reduced due to this inhibition. Furthermore, Marizomib causes an imbalance and disruption of cell division by suppressing the proteasome, which in turn disturbs the regulation of cyclins, CDK, as well as other macromolecules that control the cell cycle. In addition, by lowering levels of CRBN, IRF4, MCL1, and MYC and triggering programmed cell death by caspase activation, Marizomib increases the effectiveness of the immunosuppressive medication pomalidomide in treating multiple myeloma [45, 47].

6.6. Dolastatin 10

The sea hare, or Indian Ocean mollusk, *Dolabella auricularia*, is primarily found in the Indian Ocean. Its primary pentapeptide, Dolastatin 10 (Dol-10), has three distinct amino acid residues. Dol-10 has garnered a lot of attention recently since it has been developed into a commercial medicine for treating certain particular lymphomas and because it may trigger the death of tumor cells, including lung cancer cells, at nanomolar concentrations. A number of in vitro studies have shown that Dol-10, as well as its analogs, were extremely toxic to typical carcinoma cells, including L1210 leukemia cells, DU-145 cells from human prostate cancer, NCI-H69 cells from small cell lung carcinoma, as well as others. Dol-10 disrupts the equilibrium between tubulin and microtubules; it is an antimitotic medication as well. The protein binds to various sites on spindle microtubules and then interacts with the β-tubulin subunits. This sets off a series of reactions that eventually lead to cell death, such as apoptosis, tubulin-GTP hydrolysis, cross-linking of β-tubulin subunits, and improved microtubule depolymerization. Dol-10 also increases the overexpression of p53 or c-myc, which, as the primary signal of reactivating the apoptosis cascade, would additionally drive apoptosis of carcinoma cells, as well as down-regulate certain oncoproteins or suppress the production of anti-apoptotic Bcl-2 macromolecules. Not only does Dol-10 block microtubule accumulation as well as increase Bcl-2 phosphorylation, but it also causes chromosomal abnormalities and apoptosis-mediated loss of telomere repeat sequence [48].

6.7. Coibamide A

Leptolyngbya spp., a marine cyanobacterium discovered in Panama's Coiba National Park, is the source of the highly methylated cyclodepsipeptide known as coibamide A (CbA). This chemical has a unique mechanism of action that causes cell death by disrupting critical cellular processes; it is thus highly cytotoxic. An important signaling route for regulating cell growth, survival, and proliferation, CbA mainly works by blocking the mTOR pathway. By obstructing this pathway, CbA disturbs vital cellular functions required for survival as well as the fast proliferation of carcinoma cells. The development of numerous cell lines is markedly and potently inhibited by this naturally occurring macrocyclic chemical. There appears to be a distinct mode of action based on the reported selectivity pattern. Autophagosome aggregation and autophagy induction are both induced by CbA through a route that is not dependent on mTOR and ATG5. In order to stop various molecules from being translocated both before and after translation, CbA attaches directly to the alpha-subunit of a Sec-61 translocon (Sec61α) [49].

6.8. Bryostatin 1

Bryostatins represent a noteworthy class of pharmacologically miraculous marine bioactive compounds. The synthesis of this compound is conducted by symbiotic microorganisms that naturally inhabit marine invertebrates, specifically bryozoans. Bryostatin 1, a macrolide compound with a polyacetate backbone that is rich in oxygen, has undergone substantial research. The oceanic organism *Bugula neritina L.* was the original source of the substance. This species typically lives in colonies that resemble moss and adhere themselves to various surfaces, such as rocks, shells, hulls of vessels, and docksides. By attaching to the regulatory unit; bryostatin 1 exerts its regulatory effect by suppressing CDKs and protein kinase-C [50].

According to Clinical studies, bryostatin-1 stimulates PKC activation and triggers maturation, with or without apoptosis, in several leukemic cell lines. The NB4 human promyelocytic leukemia cell line underwent differentiation into specialized cells of the myeloid cell lineage upon exposure to bryostatin-1. Modifications such as alterations in cell shape, elevated levels of CD11b, and reduced expression of highly active myeloperoxidase, as well as c-myc mRNA, were observed. Additionally, bryostatin-1 induces the excessive addition of phosphate groups to the mutant p53 protein, a process known as hyperphosphorylation, through the MAPK pathway. This ultimately results in the breakdown of the mutant p53 protein through the ubiquitin-proteasome pathway [50].

6.9. Discodermolides

The unique deep-sea sponge Discodermia dissolute synthesizes chemical molecules called discodermolides, which occur naturally. In 1990, they were initially discovered in isolation. As an immunosuppressant, it stops many tumor cell types from dividing and triggers cell death. To improve microtubule stability, discodermolides and paclitaxel, another tubulin-binding medication, can competitively bind to tubulin. Even in paclitaxel-resistant cell types, it exerts its effects by forming stronger connections with tubulin. Despite their antagonistic nature, discodermolides and paclitaxel work together to achieve a synergistic effect. Instead of having similar properties, the binding sites may overlap, which could explain the conflicting impact. Though both compounds interact with the taxane binding pocket, discodermolide orients itself towards the H1-S2 loop, whereas taxol binds with the M-loop. Also, microtubules seem to be stabilized by the two substances in a way that promotes itself. In NSLCS cells, Discodermolide treatment causes delayed caspase-3 and -8 activation and PARP cleavage. In addition, cytochrome c is released from the mitochondria after discodermolide treatment. Despite these findings, preventing cell death by overexpressing Bcl-2, using FADD-negative cells, or inhibiting caspases was not successful. Additionally, discodermolide appears to promote cell death mostly through cathepsin b release rather than caspase activation. Reducing apoptosis in NSLCS cells may be achieved by reducing Cathepsin B. Discodermolide had promising results in mice models and was more effective in tumor cells than paclitaxel; however, the Phase I studies were halted by Novartis, the pharmaceutical company, because of cytotoxicity concerns. Nevertheless, discodermolide's fascinating properties make it a potential candidate for combination drug therapy [51].

6.10. Sansalvamide A

The maritime plant Halodule wrightii on Little San Salvador Island, Bahamas, is home to a Fusarium fungus that produces *Sansalvamide A*. A suppressor of topoisomerase I, sansalvamide-A is a pentacyclic depsipeptide. Sansalvamide A blocks the formation of a covalent compound by preventing DNA binding, according to research into the topoisomerase reaction cycle. As it was reported that premade covalent compounds can still reseal a DNA nick unhindered. This presents the potential mechanism for inducing cell death. Sansalvamide A and its analogs exhibit various notable antiproliferative properties. New derivatives of sansalvamide have been synthesized to enhance their effectiveness and decrease their toxicity to cells. These analogs have shown significant impacts on breast, liver, and prostate cancer and melanoma [52, 53].

7. Several Clinical Trials of Different Marine-Based Pharmaceutical Drugs

Few FDA-approved marine anticancer drugs have been granted licenses that are listed in Table 2. Limited marine-based pharmaceutical drugs have been licensed, and some potent anti-cancer medications are undergoing clinical investigation in Phases I, II, and III, as shown in Table 3 [54, 55].

Table 2. Compilation of marine anticancer medicines authorized by the FDA and EMA

Generic Name	Brand Name/s	Nature-based Source	Category Chemical	Clinically treated for
Cytarabine (cytosine arabinoside or Ara-C)	Cytosar-U Aracytin C.- Hospira	Oceanic-Sponge	Nucleosides	Leukemia (Blood carcinoma)
Fludarabine	Fludara	Oceanic-Sponge	Nucleosides	Leukemia (Blood carcinoma)
Nelarabine	Arranon (US) Atriance (EU)	Oceanic-Sponge	Nucleosides	
Trabectedin	Yondelis	Oceanic-Tunicate	Alkaloids	Ovarian carcinoma
Eribulin mesylate	Halaven (*Eribulin mesilate*)	Oceanic-Sponge	Macrolides	Breast carcinoma
Brentuximab vedotin (SGN-35)	Adcetris (BV)	Mollusc/ blue green algae	Antibody-drug conjugate	Lymphomas
Lurbinectedin (PM01183)	Zepzelca (PM1183)	Oceanic-Tunicate	Alkaloids	Ovarian carcinoma
Polatuzumab vedotin (DCDS4501A or RG7596)	Polivy (polatuzumab vedotin-piiq)	Mollusc/ cyanobacterium	Antibody-drug conjugate	Breast cancer
Enfortumab vedotin	Padcev	Mollusc/ cyanobacterium	Antibody-drug conjugate	Urothelial cancer
Belantamab mafodotin	Blenrep	Mollusc/ cyanobacterium	Antibody-drug conjugate	Multiple myeloma

Table 3. List of anti-cancer marine drugs under different phases of clinical investigations

Drug	Natural Source	Chemical Class	Phase
Samrotamab vedotin	Mollusk/cyanobacterium	Antibody-drug conjugate	I
Sirtratumab vedotin (ASG-15ME)	Mollusk/cyanobacterium	Antibody-drug conjugate	I
SGN-CD48A	Mollusk/cyanobacterium	Antibody-drug conjugate	I
ALT-P7	Mollusk/cyanobacterium	Antibody-drug conjugate	I
ARX788	Mollusk/cyanobacterium	Antibody-drug conjugate	I
Upifitamab rilsodotin (XMT-1536)	Mollusk/cyanobacterium	Antibody-drug conjugate	I
AGS62P1	Mollusk/cyanobacterium	Antibody-drug conjugate	I
PF-06804103	Mollusk/cyanobacterium	Antibody-drug conjugate	I
Cofetuzumab pelidotin (ABBV-647)	Mollusk/cyanobacterium	Antibody-drug conjugate	I
ZW-49	Mollusk/cyanobacterium	Antibody-drug conjugate	I
MRG003	Mollusk/cyanobacterium	Antibody-drug conjugate	I
STRO-002	Sponge	Antibody-drug conjugate	I
MORAb-202	Sponge	Antibody-drug conjugate	I
RC-88	Mollusk/cyanobacterium	Antibody-drug conjugate	I
SGN-B6A	Mollusk/cyanobacterium	Antibody-drug conjugate	I
SGN-CD228A	Mollusk/cyanobacterium	Antibody-drug conjugate	I
FOR-46	Mollusk/cyanobacterium	Antibody-drug conjugate	I
A-166	Mollusk/cyanobacterium	Antibody-drug conjugate	I
Plocabulin (MN-029)	Oceanic-Sponge	Polyketide	II
Tisotumab vedotin (TV)	Mollusk/blue-green algae	Antibody-drug conjugate	II
Ladiratuzumab vedotin (XMT-1522)	Mollusk/blue-green algae	Antibody-drug conjugate	II
Telisotuzumab vedotin	Mollusk/cyanobacterium	Antibody-drug conjugate	II
CAB-ROR2	Mollusk/blue-green algae	Antibody-drug conjugate	II
CX-2029	Mollusk/cyanobacterium	Antibody-drug conjugate	II
W0101	Mollusk/cyanobacterium	Antibody-drug conjugate	II
Plinabulin (NPI-2358)	Fungi	Diketopiperazine (DKP)	III
Marizomib	Oceanic-Actinomycetes	γ-lactam-β-lactone (lactacystin)	III
Plitidepsin (Aplidin)	Oceanic-Tunicate	Depsipeptide (frugiperdoxin)	III

7.1. Significance of Virtual Screening of Marine Bioactive Compounds toward Drug Discovery

The expansive and uncharted marine ecosystem possesses significant possibilities for the identification of pharmaceuticals, showcasing a wide array of creatures with distinctive and possibly lucrative bioactive substances. Drug discovery, as well as development, is an extensive and intricate process that requires a significant number of resources, time and carries a high level of risk. The clinical success rate, which refers to the probability of a medicine being approved after entering clinical testing, is roughly 12%, indicating a low possibility of success. The process of transforming a drug idea into a marketable product typically spans 13-15 years and necessitates an average expenditure of $2-3 billion in the United States. Various novel approaches have been devised and implemented in the field of drug research and development to expedite the research process and minimize expenses. Virtual screening and computational approaches are essential for unlocking the promise of expediting and improving the drug discovery process. CADD methods have become a potent instrument in the creation of medically significant minor compounds for more than thirty years. These methods have proven to be more effective than relying solely on experimental high-throughput screening (HTS) methodologies, resulting in greater rates of successful discoveries [56]. The significance of marine bioactive compounds is as follows:

1. Expedited Target Identification and Validation:

Marine organisms synthesize a wide array of novel compounds, offering a vast chemical space for screening. Virtual screening allows researchers to quickly identify promising lead compounds with specific targets from large libraries of marine natural products through HTS, eliminating the need for expensive and time-consuming experimental testing of every candidate and leading to a more focused approach to drug discovery.

2. Enhanced Hit Rate and Lead Optimization

Virtual screening models can predict compounds' binding affinity and activity against specific targets with high accuracy using AutoDock, CBDOCK, and PyRx, increasing the likelihood of identifying true hits. It allows for the quick identification and optimization of lead compounds before expensive and time-consuming laboratory experiments. Virtual screening and CADD are used to identify lead compounds with novel structures and analogs, leading to the development of new drug classes.

3. Addressing Challenges in Marine Drug Discovery

Marine organisms can be difficult to access and cultivate, making it challenging to obtain sufficient quantities of compounds for drug development. Virtual screening can overcome this obstacle by focusing on in silico studies. Many marine natural products have complex structures that can be challenging to synthesize. HTS can help identify synthetically accessible analogs of promising lead compounds. Virtual tools such as ADMET analysis can be used to predict potential toxicity and bioavailability issues early in the drug discovery and optimization process, mitigating risks and increasing the likelihood of developing safe and effective drugs.

4. Sustainable Drug Discovery

Marine bioactive compounds have shown promise in treating various diseases, including cancer, infections, and inflammatory disorders. Virtual screening can accelerate the discovery of novel therapeutic compounds from oceanic sources. It significantly reduces the need for laboratory experiments, contributing to a more sustainable and ethical drug discovery process.

8. Isolation Techniques for Marine Bioactive Compounds

Oceanic organisms are an extensive source of new biologically active compounds with potent medicinal applications, and the isolation of these compounds is crucial for further study, characterization, and potential commercialization [57]. A bioactivity-guided isolated strategy is also being incorporated for combined extraction, purification, and bioactivity assays to isolate the marine bioactive compounds that are liable for a precise biological effect [58]. This tactic is intensively used for pinpointing novel bioactive compounds [59]. Here are some commonly used isolation techniques:

1. Solvent Extraction: The most common technique involves the use of organic solvents to dissolve the compounds from the organism. Different solvents have different polarities, allowing for selective extraction of specific compound classes.
2. Methanol: Often used for initial extraction, as it dissolves a wide range of compounds.
3. Dichloromethane: Favored for extracting lipophilic compounds like lipids and sterols.
4. Ethanol: Used for extracting polar compounds like polysaccharides and phenolic compounds.
5. Supercritical Fluid Extraction (SFE): This isolation technique utilizes a supercritical fluid (usually CO2) at high pressure and temperature to extract compounds. It is environmentally friendly and can be used for sensitive compounds.
6. Microwave-based Extraction: The MBE method uses microwave radiation to warm the sample, increasing extraction efficiency and reducing extraction time.

9. Purification of Isolated Marine Bioactive Compounds

Chromatography is being widely used technique for separating and purifying bioactive compounds based on their different affinities for a stationary phase [60, 61]. This technique has been classified into several types:

1. Column Chromatography: This versatile technique puts the sample onto a column stuffed with a stationary phase. Different solvents are used to elute the compounds based on their polarity.
2. High-Pressure Liquid Chromatography: It uses high pressure and a specialized column to achieve fast and efficient separation of compounds.
3. Thin-layer chromatography: A simpler method used for monitoring the separation process and identifying the presence of specific compounds.

4. Crystallization: A process where a pure compound is isolated by forming crystals from a solution.
5. Precipitation: Using specific reagents to precipitate the desired compound from a solution.
6. Ultrafiltration: Separating compounds based on their size by using membranes with specific pore sizes.
7. Dialysis: Removing small molecules from a solution by diffusion across a semipermeable membrane.
8. Lyophilization (Freeze Drying): A gentle method for removing water from a sample by freezing and applying a vacuum.

10. Key Considerations for Choosing Isolation Techniques

The isolation of marine bioactive compounds is a multi-step process requiring careful selection of techniques based on the nature of the compound, the source material, and available resources. These isolated compounds hold immense potential for diverse applications, driving progress in various fields. There are several key considerations, such as the nature of bioactive compounds, their molecular weight, polarity, and stability of the target compound, that need to be considered when choosing isolation techniques. Appropriate and optimized isolation techniques influence the yield and purification of the compound. Additionally, diversified marine organisms and their abundance affect the extraction effectiveness of the compounds. Extraction of certain marine biologically active compounds from rare exotic oceanic species makes the isolation tedious and adds up costs [62].

11. Challenges in Marine Bioactive Compound Isolation

Several challenges have been associated with marine bioactive compound isolation, which are as follows:

1. Complex Matrix: Marine organisms often contain a wide variety of compounds, making isolation difficult.
2. Low Abundance: Bioactive compounds are often present in low concentrations, making extraction and purification challenging.
3. Stability: Some marine bioactive compounds are unstable in air or light, requiring special handling and storage conditions [63].

12. Characterization of Marine Metabolites

Sampling challenges pose significant obstacles to achieving a comprehensive understanding of the chemistry and composition of marine metabolites. Adequate quantities are necessary for thorough analysis and resolution of the instrumental and bioassay methods employed. The main challenges in gaining a deeper understanding of the chemistry and composition of marine metabolites are the sampling problems associated with low microgram quantities, particularly in the context of modern techniques such as mass spectrometry (MS) and nuclear magnetic resonance (NMR) [64]. Adequate numbers are necessary to conduct a thorough analysis and achieve precise results using instrumental and bioassay methods.

In the field of MS and NMR spectroscopy, a small quantity of less than one milligram or a few micrograms is sufficient to fully determine the structure using advanced analytical techniques. For the earlier 3-4 decades, the oceanic community has employed MS [65]. A separation technique is frequently employed in conjunction with MS to improve the clarity and specificity of analysis. This is done by coupled techniques such as Gas chromatography-mass spectrometry (GC-MS), High-Resolution Mass Spectrometry (HR-MS), Liquid chromatography-mass spectrometry (LC-MS), and Direct temperature-resolved mass spectrometry (DT-MS) approaches [66]. Mass spectrometry (MS) has primarily been utilized in previous research to detect and measure the specific portions or small amounts of substances within marine organisms. The discovery of hierridin B from an oceanic cyanobacterium *Cyanobium spp.* strain further supports the identification procedure, as revealed by MS/MS analysis [10]. Undoubtedly, NMR is the predominant method employed for the confirmational analysis of molecules. For instance, *Acremines P*, obtained from a strain of *Acremonium persicinum* found in the sea, was identified using NMR techniques [67]. However, the reduced sensitivity of NMR in comparison to MS limits the ability to detect and identify metabolites that are present in small quantities or at trace levels. Initially employed for tiny, comparatively uncomplicated chemical molecules, this approach has become widely popular for the identification of oceanic metabolites in organisms [10]. LC-HRMS exhibits exceptional sensitivity and is capable of detecting substances with maximum precision, even at extremely low concentrations. Nevertheless, certain categories of chemicals exist that are not detectable using MS. These compounds may exhibit a lack of ion formation, impeded ion formation, or inability to be eluted from the column for detection. Unlike other methods, NMR does not require a separation step and so provides an instantaneous representation of the sample's metabolome. It has lower sensitivity compared to MS, but it is more durable and consistent, with universally applicable characteristics for detecting metabolites. Because of this, it is possible to compare results from experiments run at various times or on different datasets. Furthermore, integrating metabolomics with other "omics" technologies, such as genomics, could enhance the association between the marine metabolome and its genotype [68]. These analytical techniques are also used to further understand how metabolites work. The metabolic variations in MCF-7 breast carcinoma cells exposed to the alkaloid ascididemin were identified by functional metabolomics.

Conclusion

The marine flora consists of bacteria, microalgae, cyanobacteria, fungi, halophytes, mangroves, and actinobacteria have been discovered to have anticancer properties. These oceanic resources are of great significance. Polysaccharides, peptides, Polyphenols, terpenoids (specifically cembranoids), and alkaloids are examples of metabolites found in marine organisms. These metabolites have been found to have a diverse range of beneficial properties, antioxidant and anti-cancer effects, and the ability to stimulate the immune system. Oceanic natural metabolites have demonstrated *in vitro* anti-cancer efficacy against many tumors cell lines, such as lung, bladder, prostate, breast, osteosarcoma, renal, mammary, and cancer-derived cell lines. Furthermore, both *in vitro* as well as animal studies reveal that the majority of the oceanic compounds hinder tumor growth through the processes of necrosis, apoptosis, as well as lysis of the carcinoma cells. Trabectedin, cytarabine, and brentuximab vedotin are examples of oceanic natural products that have demonstrated anticancer activity in laboratory studies.

Future Perspectives

As the scientific exploration of the ocean's depths progresses, marine compounds are anticipated to have a growing significance in the identification and development of groundbreaking pharmaceuticals, offering fresh possibilities for the treatment of various illnesses. Although these metabolites have demonstrated promising choices for cancer therapy, there are other obstacles linked to the advancement of these medications that require careful consideration. While oceans undeniably harbor a vast array of interesting organisms, researchers are now unable to gain access to these areas. Hence, the issue lies in the limited accessibility of marine research, necessitating the establishment of robust collaborative relationships among researchers and oceanographers. By inserting the genetic code of the target chemical into the cells of the host organism, genetic engineering is being developed to enhance the manufacturing of medicines. To controllably extract lead compounds from aquatic environments, research into the isolation and expression of aquatic genes is essential.

References

[1] Mondal, Arijit, Sankhadip Bose, Sabyasachi Banerjee, Jayanta Kumar Patra, Jai Malik, Sudip Kumar Mandal, Kaitlyn L. Kilpatrick, et al., 2020. "Marine Cyanobacteria and Microalgae Metabolites—A Rich Source of Potential Anticancer Drugs." *Marine Drugs* 18 (9): 476. https://doi.org/10.3390/md18090476.

[2] Newman, David J., and Gordon M. Cragg. 2012. "Natural Products As Sources of New Drugs over the 30 Years from 1981 to 2010." *Journal of Natural Products* 75 (3): 311–35. https://doi.org/10.1021/np200906s.

[3] Sithranga Boopathy, N., and K. Kathiresan. 2011. "Anticancer Drugs from Marine Flora: An Overview." *Journal of Oncology* 2010 (February):e214186. https://doi.org/10.1155/2010/214186.

[4] Banerjee, Sabyasachi, Sankhadip Bose, Subhash C. Mandal, Subhojit Dawn, Ujjwal Sahoo, Mohmmed A. Ramadan, and Sudip Kumar Mandal. 2019. "Pharmacological Property of Pentacyclic Triterpenoids." *Egyptian Journal of Chemistry* 62 (Special Issue (Part 1) Innovation in Chemistry): 13–35. https://doi.org/10.21608/ejchem.2019.16055.1975.

[5] G. Grothaus, Paul, Gordon M. Cragg, and David J. Newman. 2010. "Plant Natural Products in Anticancer Drug Discovery." *Current Organic Chemistry* 14 (16): 1781–91. https://doi.org/10.2174/138527210792927708.

[6] Basmadjian, Christine, Qian Zhao, Embarek Bentouhami, Amel Djehal, Canan G. Nebigil, Roger A. Johnson, Maria Serova, et al., 2014. "Cancer Wars: Natural Products Strike Back." *Frontiers in Chemistry* 2 (May). https://doi.org/10.3389/fchem.2014.00020.

[7] Gohary, Osmo Otto Paivio Hanninen; Mustafa Atalay; B.P. Mansourian; A. Wojtezak; S. M. Mahfouz; Harry Majewski; Elaine Elisabetsky; Nina L. Etkin; Ralph Kirby; T. G. Downing and M.I.El. 2010. *MEDICAL AND HEALTH SCIENCES-Volume XIII.* EOLSS Publications.

[8] Townsend, Michael, Kate Davies, Nicholas Hanley, Judi E. Hewitt, Carolyn J. Lundquist, and Andrew M. Lohrer. 2018. "The Challenge of Implementing the Marine Ecosystem Service Concept." *Frontiers in Marine Science* 5 (October). https://doi.org/10.3389/fmars.2018.00359.

[9] Snelgrove, Paul V.R. 1999. "Getting to the Bottom of Marine Biodiversity: Sedimentary Habitats: Ocean Bottoms Are the Most Widespread Habitat on Earth and Support High Biodiversity and Key Ecosystem Services." *BioScience* 49 (2): 129–38. https://doi.org/10.2307/1313538.

[10] Khalifa, Shaden A. M., Nizar Elias, Mohamed A. Farag, Lei Chen, Aamer Saeed, Mohamed-Elamir F. Hegazy, Moustafa S. Moustafa, et al., 2019. "Marine Natural Products: A Source of Novel Anticancer Drugs." *Marine Drugs* 17 (9): 491. https://doi.org/10.3390/md17090491.

[11] Newman, David J., and Gordon M. Cragg. "Marine-sourced anti-cancer and cancer pain control agents in clinical and late preclinical development." *Marine drugs* 12, no. 1 (2014): 255-278.

[12] Lauritano, Chiara, Jeanette H. Andersen, Espen Hansen, Marte Albrigtsen, Laura Escalera, Francesco Esposito, Kirsti Helland, Kine Ø Hanssen, Giovanna Romano, and Adrianna Ianora. 2016. "Bioactivity Screening of Microalgae for Antioxidant, Anti-Inflammatory, Anticancer, Anti-Diabetes, and Antibacterial Activities." *Frontiers in Marine Science* 3 (May). https://doi.org/10.3389/fmars.2016.00068.

[13] Mahapatra, Gyana Prakash, Surabhi Raman, Suman Nayak, Sushanto Gouda, Gitishree Das, and Jayanta Kumar Patra. 2020. "Metagenomics Approaches in Discovery and Development of New Bioactive Compounds from Marine Actinomycetes." *Current Microbiology* 77 (4): 645–56. https://doi.org/10.1007/s00284-019-01698-5.

[14] Deshmukh, Sunil K., Ved Prakash, and Nihar Ranjan. 2018. "Marine Fungi: A Source of Potential Anticancer Compounds." *Frontiers in Microbiology* 8 (January). https://doi.org/10.3389/fmicb.2017.02536.

[15] Carroll, Anthony R., Brent R. Copp, Rohan A. Davis, Robert A. Keyzers, and Michèle R. Prinsep. 2022. "Marine Natural Products." *Natural Product Reports* 39 (6): 1122–71. https://doi.org/10.1039/D1NP00076D.

[16] Aniszewski, Tadeusz. 2007. *Alkaloids-Secrets of Life: Alkaloid Chemistry, Biological Significance, Applications and Ecological Role*. Elsevier.

[17] CH, Ramesh, N.V . Vinithkumar, and Ramalingam Kirubagaran. 2019. "Marine Pigmented Bacteria: A Prospective Source of Antibacterial Compounds." *Journal of Natural Science, Biology and Medicine* 10 (January):104. https://doi.org/10.4103/jnsbm.JNSBM_201_18.

[18] Davidson, Bradley S. 1995. "New Dimensions in Natural Products Research: Cultured Marine Microorganisms." *Current Opinion in Biotechnology* 6 (3): 284–91. https://doi.org/10.1016/0958-1669(95)80049-2.

[19] Kandi, S., Godishala, V. , Rao, P. , & Ramana, K. (2015). Biomedical Significance of Terpenes: An Insight. Biomedicine and Biotechnology, 3(1), 8-10.

[20] Ghanbari, Raheleh. 2019. "Review on the Bioactive Peptides from Marine Sources: Indication for Health Effects." *International Journal of Peptide Research and Therapeutics* 25 (3): 1187–99. https://doi.org/10.1007/s10989-018-9766-x.

[21] Shukla, Shruti. 2016. "Therapeutic Importance of Peptides from Marine Source: A Mini Review." *IJMS Vol.45(11) [November 2016]*, November. http://nopr.niscpr.res.in/handle/123456789/38626.

[22] Bhakuni, D.S., and D.S. Rawat, eds. 2005. "Bioactive Marine Nucleosides." In *Bioactive Marine Natural Products*, 208–34. Dordrecht: Springer Netherlands. https://doi.org/10.1007/1-4020-3484-9_8.

[23] Huang, Ri-Ming, Yin-Ning Chen, Ziyu Zeng, Cheng-Hai Gao, Xiangdong Su, and Yan Peng. 2014. "Marine Nucleosides: Structure, Bioactivity, Synthesis and Biosynthesis." *Marine Drugs* 12 (12): 5817–38. https://doi.org/10.3390/md12125817.

[24] Helbert, William. 2017. "Marine Polysaccharide Sulfatases." *Frontiers in Marine Science* 4 (January). https://doi.org/10.3389/fmars.2017.00006.

[25] Poli, Annarita, Gianluca Anzelmo, and Barbara Nicolaus. 2010. "Bacterial Exopolysaccharides from Extreme Marine Habitats: Production, Characterization and Biological Activities." *Marine Drugs* 8 (6): 1779–1802. https://doi.org/10.3390/md8061779.

[26] Thomas, Noel Vinay, and Se-Kwon Kim. 2013. "Beneficial Effects of Marine Algal Compounds in Cosmeceuticals." *Marine Drugs* 11 (1): 146–64. https://doi.org/10.3390/md11010146.

[27] John Faulkner, D. 2001. "Marine Natural Products." *Natural Product Reports* 18 (1): 1R-49R. https://doi.org/10.1039/B006897G.

[28] Vázquez, José Antonio, Isabel Rodríguez-Amado, María Ignacia Montemayor, Javier Fraguas, María Del Pilar González, and Miguel Anxo Murado. 2013. "Chondroitin Sulfate, Hyaluronic Acid and Chitin/Chitosan Production Using Marine Waste Sources: Characteristics, Applications and Eco-Friendly Processes: A Review." *Marine Drugs* 11 (3): 747–74. https://doi.org/10.3390/md11030747.

[29] Taori, Kanchan, Valerie J. Paul, and Hendrik Luesch. 2008. "Structure and Activity of Largazole, a Potent Antiproliferative Agent from the Floridian Marine *Cyanobacterium symploca spp.*" *Journal of the American Chemical Society* 130 (6): 1806–7. https://doi.org/10.1021/ja7110064.

[30] Mevers, Emily, Wei-Ting Liu, Niclas Engene, Hosein Mohimani, Tara Byrum, Pavel A. Pevzner, Pieter C. Dorrestein, Carmenza Spadafora, and William H. Gerwick. 2011. "Cytotoxic Veraguamides, Alkynyl Bromide-Containing Cyclic Depsipeptides from the Marine Cyanobacterium Cf. Oscillatoria Margaritifera." *Journal of Natural Products* 74 (5): 928–36. https://doi.org/10.1021/np200077f.

[31] Han, Bingnan, Doug Goeger, Claudia S. Maier, and William H. Gerwick. 2005. "The Wewakpeptins, Cyclic Depsipeptides from a Papua New Guinea Collection of the Marine Cyanobacterium Lyngbya Semiplena." *The Journal of Organic Chemistry* 70 (8): 3133–39. https://doi.org/10.1021/jo0478858.

[32] Weiss, Christine, Eduard Figueras, Adina N. Borbely, and Norbert Sewald. 2017. "Cryptophycins: Cytotoxic Cyclodepsipeptides with Potential for Tumor Targeting." *Journal of Peptide Science* 23 (7–8): 514–31. https://doi.org/10.1002/psc.3015.

[33] Luesch, Hendrik, Wesley Y. Yoshida, George G. Harrigan, James P. Doom, Richard E. Moore, and Valerie J. Paul. 2002. "Lyngbyaloside B, a New Glycoside Macrolide from a Palauan Marine Cyanobacterium, Lyngbya Sp." *Journal of Natural Products* 65 (12): 1945–48. https://doi.org/10. 1021/np0202879.

[34] Matthew, Susan, Lilibeth A. Salvador, Peter J. Schupp, Valerie J. Paul, and Hendrik Luesch. 2010. "Cytotoxic Halogenated Macrolides and Modified Peptides from the Apratoxin-Producing Marine Cyanobacterium Lyngbya Bouillonii from Guam." *Journal of Natural Products* 73 (9): 1544–52. https://doi.org/10.1021/np1004032.

[35] Navarro, Gabriel, Milo E. Cummings, John Lee, Nathan Moss, Evgenia Glukhov, Frederick A. Valeriote, Lena Gerwick, and William H. Gerwick. 2015. "Isolation of Polycavernoside D from a Marine Cyanobacterium." *Environmental Science & Technology Letters* 2 (7): 166–70. https://doi. org/10.1021/acs.estlett.5b00116.

[36] Jimenez-Lopez, C., A. G. Pereira, C. Lourenço-Lopes, P. Garcia-Oliveira, L. Cassani, M. Fraga-Corral, M. A. Prieto, and J. Simal-Gandara. 2021. "Main Bioactive Phenolic Compounds in Marine Algae and Their Mechanisms of Action Supporting Potential Health Benefits." *Food Chemistry* 341 (March):128262. https://doi.org/10.1016/j.foodchem.2020.128262.

[37] Hamed, Imen, Fatih Özogul, Yesim Özogul, and Joe M. Regenstein. 2015. "Marine Bioactive Compounds and Their Health Benefits: A Review." *Comprehensive Reviews in Food Science and Food Safety* 14 (4): 446–65. https://doi.org/10.1111/1541-4337.12136.

[38] Debbab, Abdessamad, Amal H. Aly, Wen H. Lin, and Peter Proksch. 2010. "Bioactive Compounds from Marine Bacteria and Fungi." *Microbial Biotechnology* 3 (5): 544–63. https://doi.org/10.1111/j.1751-7915.2010.00179.x.

[39] Schwartzman, Robert A., and John A. Cidlowski. 1993. "Apoptosis: The Biochemistry and Molecular Biology of Programmed Cell Death*." *Endocrine Reviews* 14 (2): 133–51. https://doi.org/10.1210/edrv-14-2-133.

[40] Swami, Umang, Umang Shah, and Sanjay Goel. 2015. "Eribulin in Cancer Treatment." *Marine Drugs* 13 (8): 5016–58. https://doi.org/10.3390/md13085016.

[41] Wang, Jiali, Pengfei Wang, Zheng Zeng, Caiji Lin, Yiru Lin, Danli Cao, Wenqing Ma, et al., 2022. "Trabectedin in Cancers: Mechanisms and Clinical Applications." *Current Pharmaceutical Design* 28 (24): 1949–65. https://doi.org/10.2174/1381612828666220526125806.

[42] Huang, Kuan-Chun, Zhihong Chen, Yimin Jiang, Sandeep Akare, Donna Kolber-Simonds, Krista Condon, Sergei Agoulnik, et al., 2016. "Apratoxin A Shows Novel Pancreas-Targeting Activity through the Binding of Sec 61." *Molecular Cancer Therapeutics* 15 (6): 1208–16. https://doi.org/10.1158/1535-7163.MCT-15-0648.

[43] Liu, Yanxia, Brian K. Law, and Hendrik Luesch. 2009. "Apratoxin A Reversibly Inhibits the Secretory Pathway by Preventing Cotranslational Translocation." *Molecular Pharmacology* 76 (1): 91–104. https://doi.org/10.1124/mol.109.056085.

[44] Chuk, Meredith K., Frank M. Balis, and Elizabeth Fox. 2009. "Trabectedin." *The Oncologist* 14 (8): 794–99. https://doi.org/10.1634/theoncologist.2009-0104.

[45] Wu, Lichuan, Ke Ye, Sheng Jiang, and Guangbiao Zhou. 2021. "Marine Power on Cancer: Drugs, Lead Compounds, and Mechanisms." *Marine Drugs* 19 (9): 488. https://doi.org/10.3390/md19090488.

[46] O'Shaughnessy, Joyce, Virginia Kaklamani, and Kevin Kalinsky. 2019. "Perspectives on The Mechanism of Action and Clinical Application of Eribulin for Metastatic Breast Cancer." *Future Oncology* 15 (14): 1641–53. https://doi.org/10.2217/fon-2018-0936.

[47] Ma, Long, and Aipo Diao. 2015. "Marizomib, A Potent Second Generation Proteasome Inhibitor from Natural Origin." *Anti-Cancer Agents in Medicinal Chemistry- Anti-Cancer Agents)* 15 (3): 298–306.

[48] Gao, Gang, Yanbing Wang, Huiming Hua, Dahong Li, and Chunlan Tang. 2021. "Marine Antitumor Peptide Dolastatin 10: Biological Activity, Structural Modification and Synthetic Chemistry." *Marine Drugs* 19 (7): 363. https://doi.org/10.3390/md19070363.

[49] Kitamura, Takashi, Rikito Suzuki, Shinsuke Inuki, Hiroaki Ohno, Kerry L. McPhail, and Shinya Oishi. 2022. "Design of Coibamide A Mimetics with Improved Cellular Bioactivity." *ACS Medicinal Chemistry Letters* 13 (1): 105–10. https://doi.org/10.1021/acsmedchemlett.1c00591.

[50] Kortmansky, Jeremy, and Gary K. Schwartz. 2003. "Bryostatin-1: A Novel PKC Inhibitor in Clinical Development." *Cancer Investigation* 21 (6): 924–36. https://doi.org/10.1081/CNV-120025095.

[51] Schwarzenberg, Karin von, and Angelika M. Vollmar. 2013. "Targeting Apoptosis Pathways by Natural Compounds in Cancer: Marine Compounds as Lead Structures and Chemical Tools for Cancer Therapy." *Cancer Letters, Apoptosis Targeting Drugs in Cancer*, 332 (2): 295–303. https://doi.org/10.1016/j.canlet.2010.07.004.

[52] Heiferman, Michael J., Mohammad R. Salabat, Michael B. Ujiki, Matthew J. Strouch, Eric C. Cheon, Richard B. Silverman, and David J. Bentrem. 2010. "Sansalvamide Induces Pancreatic Cancer Growth Arrest through Changes in the Cell Cycle." *Anticancer Research* 30 (1): 73–78.

[53] Hwang, Young, David Rowley, Denise Rhodes, Jeff Gertsch, William Fenical, and Frederic Bushman. 1999. "Mechanism of Inhibition of a Poxvirus Topoisomerase by the Marine Natural Product Sansalvamide A." *Molecular Pharmacology* 55 (6): 1049–53. https://doi.org/10.1124/mol.55.6.1049.

[54] Ruiz-Torres, Verónica, Jose Antonio Encinar, María Herranz-López, Almudena Pérez-Sánchez, Vicente Galiano, Enrique Barrajón-Catalán, and Vicente Micol. 2017. "An Updated Review on Marine Anticancer Compounds: The Use of Virtual Screening for the Discovery of Small-Molecule Cancer Drugs." *Molecules* 22 (7): 1037. https://doi.org/10.3390/molecules22071037.

[55] Cappello, Emiliano, and Paola Nieri. 2021. "From Life in the Sea to the Clinic: The Marine Drugs Approved and under Clinical Trial." *Life* 11 (12): 1390. https://doi.org/10.3390/life11121390.

[56] Pereira, Florbela, and Joao Aires-de-Sousa. 2018. "Computational Methodologies in the Exploration of Marine Natural Product Leads." *Marine Drugs* 16 (7): 236. https://doi.org/10.3390/md16070236.

[57] Riguera, Ricardo "Isolating Bioactive Compounds from Marine Organisms." 1997, 5(4):187-193

[58] Zhang, Lixin, Rong An, Jinping Wang, Nuo Sun, Si Zhang, Jiangchun Hu, and Jun Kuai. 2005. "Exploring Novel Bioactive Compounds from Marine Microbes." *Current Opinion in Microbiology*, Ecology and industrial microbiology/Edited by Sergio Sánchez and Betty Olson ·Techniques/Edited by Peter J Peters and Joel Swanson, 8 (3): 276–81. https://doi.org/10.1016/j.mib.2005.04.008.

[59] Odeleye, Tinu, William Lindsey White, and Jun Lu. 2019. "Extraction Techniques and Potential Health Benefits of Bioactive Compounds from Marine Molluscs: A Review." *Food & Function* 10 (5): 2278–89. https://doi.org/10.1039/C9FO00172G.

[60] Wang, Xueqin, Huahua Yu, Ronge Xing, and Pengcheng Li. 2017. "Characterization, Preparation, and Purification of Marine Bioactive Peptides." *BioMed Research International* 2017 (July): e9746720. https://doi.org/10.1155/2017/9746720.

[61] Ebada, Sherif S., Ru Angelie Edrada, Wenhan Lin, and Peter Proksch. 2008. "Methods for Isolation, Purification and Structural Elucidation of Bioactive Secondary Metabolites from Marine Invertebrates." *Nature Protocols* 3 (12): 1820–31. https://doi.org/10.1038/nprot.2008.182.

[62] Pauli, Guido F., Samuel M. Pro, and J. Brent Friesen. 2008. "Countercurrent Separation of Natural Products." *Journal of Natural Products* 71 (8): 1489–1508. https://doi.org/10.1021/np800144q.

[63] Kiuru, Paula, M. Valeria D'Auria, Christian D. Muller, Päivi Tammela, Heikki Vuorela, and Jari Yli-Kauhaluoma. 2014. "Exploring Marine Resources for Bioactive Compounds." *Planta Medica* 80 (14): 1234–46. https://doi.org/10.1055/s-0034-1383001.

[64] Kim, Woo Jung, Young Ok Kim, Jin Hee Kim, Bo-Hye Nam, Dong-Gyun Kim, Cheul Min An, Jun Sik Lee, et al., 2016. "Liquid Chromatography-Mass Spectrometry-Based Rapid Secondary-Metabolite Profiling of Marine *Pseudoalteromonas spp.* M2." *Marine Drugs* 14 (1): 24. https://doi.org/10.3390/md14010024.

[65] Fries, Elke, Jens H. Dekiff, Jana Willmeyer, Marie-Theres Nuelle, Martin Ebert, and Dominique Remy. 2013. "Identification of Polymer Types and Additives in Marine Microplastic Particles Using Pyrolysis-GC/MS and Scanning Electron Microscopy." *Environmental Science: Processes & Impacts* 15 (10): 1949–56. https://doi.org/10.1039/C3EM00214D.

[66] Panagiotopoulos, Christos, Daniel J. Repeta, Laura Mathieu, Jean-François Rontani, and Richard Sempéré. 2013. "Molecular Level Characterization of Methyl Sugars in Marine High Molecular Weight Dissolved Organic Matter." *Marine Chemistry* 154 (August):34–45. https://doi.org/10.1016/j.marchem.2013.04.003.

[67] Garson, Mary J., Warren Hehre, Gregory K. Pierens, and Suciati. 2017. "Revision of the Structure of Acremine P from a Marine-Derived Strain of *Acremonium persicinum*." *Molecules* 22 (4): 521. https://doi.org/10.3390/molecules22040521.

[68] Viegelmann, Christina, Lekha Menon Margassery, Jonathan Kennedy, Tong Zhang, Ciarán O'Brien, Fergal O'Gara, John P. Morrissey, Alan D. W. Dobson, and RuAngelie Edrada-Ebel. 2014. "Metabolomic Profiling and Genomic Study of a Marine Sponge-Associated Streptomyces *spp.*" *Marine Drugs* 12 (6): 3323–51. https://doi.org/10.3390/md12063323.

Chapter 8

Novel Marine Metabolites in Cancer Treatment: Mechanisms and Therapeutic Strategies

Krishna Gupta*
Kalyani Thombre
and Milind Umekar
Department of Pharmaceutical Chemistry, Smt. Kishoritai Bhoyar College of Pharmacy, Kamptee, Nagpur (MS), India

Abstract

Cancer is the leading cause of death worldwide. There are no effective cancer therapies since some of the medications that are already on the market have negative side effects. The pharmaceutical industry and scientific institutions worldwide have invested much effort in the past several decades to isolate and identify novel marine metabolites. Marine metabolites proved a significant and most unexplored pool of bioactive substances with a potential source of novel compounds with various pharmacological characteristics. They provide a vital opportunity to identify novel anticancer medications since they are chemically distinct, taxonomically diversified, mostly effective, and physiologically active.

Marine flora, including bacteria, actinobacteria, cyanobacteria, fungi, microalgae, seaweeds, mangroves, and other halophytes, accounts for over 90% of the ocean's biomass. These marine floras are invaluable metabolite resources.

Marine metabolites obtained from different marine flora are a diverse group of chemical classes, including polyphenols, sulfated polysaccharides, peptides, polyketides, terpenoids, glycosides, phenols, and alkaloids. Marine metabolites exhibit remarkable anticancer effects through targeting various hallmarks of cancer via different mechanisms, like proliferation, activation of macrophages, angiogenesis inhibition, metastasis, apoptosis evasion, and prevention of DNA oxidative damage. Their mechanisms of action and medicinal applications are highlighted in this chapter, which investigates the potential of marine metabolites in cancer management.

Further evidence supporting the possibility of marine metabolites as substitutes or supplements to traditional chemotherapy comes from their effectiveness against multidrug-resistant cancer cells. Numerous chemicals originating from marine sources have already

* Corresponding Author's Email: krg1903@gmail.com

In: Healing from the Depths Marine Metabolites in Disease Management
Editors: Pranav Kumar Prabhakar and Tarun Kumar Upadhyay
ISBN: 979-8-89530-832-5

progressed into preclinical and clinical stages to treat diverse forms of cancer. The importance of recently recognized compounds from marine sources discussed in this chapter, along with their information on their origin, organism, structure, and mode of action.

1. Introduction

Cancer is described as "a group of diseases characterized by the uncontrolled growth and spread of abnormal cells" and is one of the deadliest diseases globally [1]. Cancer currently ranks as the top cause of death globally, killing more than 19.3 million people in 2020, or around one in every six, despite several substantial breakthroughs achieved recently. By 2040, it is projected that there will be 28 million new cancer diagnoses worldwide, with a death rate exceeding 16 million [2].

In terms of incidence and death, breast and lung cancers are the most frequent, accounting for almost 2 million cases in the year 2020. Men are more likely to develop lung, prostate, and colorectal cancers [3]. Lung cancer is the leading cause of mortality (almost 1 million), followed by colorectal cancer. Breast cancer is the most common among women, followed by colorectal and lung cancers. It has the major mortality rate among women, accounting for roughly 600,000 fatalities [4].

Cancer is a devastating illness that is becoming more prevalent due to changes in lifestyle, diet, and global warming. High incidences of cancer have been linked to aging and unhealthy lifestyles. Cancer risk factors include external variables such as tobacco, chemicals, radiation, and infectious organisms, as well as internal factors including genetics and immunological disorders [5, 6].

Recent advancements have been made in surgical techniques, radiation, and chemotherapeutic medicines, including combining chemotherapy and hormone treatment with immunotherapy. Still, cancer therapy remains a constantly evolving field for improving efficacy and selectivity. Cancer therapies lack potency and might lead to unwanted effects [7, 8]. To address critical challenges in achieving more effective cancer therapy for tumor heterogeneity, resistance to anticancer medicines, and reducing adverse effects. Drug development for cancer therapy is currently being explored extensively. Innovative drug discovery methodologies and compounds are needed to address the issue from a scientific perspective [9, 10].

Thirty years ago, the pharmaceutical industry relied heavily on combinatorial chemistry to develop a variety of chemicals in a straightforward procedure. In recent years, drug research has shifted towards natural compounds [11]. Natural products play a significant role in drug discovery and have great potential in the healthcare domain. Natural products (NP, sometimes termed secondary or specialized metabolites) are found in cells/tissues of microbes, fungi, plants, or animals from aquatic and terrestrial circumstances. They have been utilised for medicinal purposes since ancient times and are key pillars of modern pharmacy [12]. The WHO reports mention that plant-based medications are used by 80% of the world's population, particularly in developing nations, for healthcare. Natural products and derivatives account for about 50% of clinically used medications globally [13, 14]. Higher plants account for at least 25% of the total. Natural medicines account for about 60% of cancer therapy approvals. Fruits and vegetables have high levels of vitamins C, B, and E, carotenoids, and fiber, which may aid in cancer protection [15, 16]. Consuming more natural antioxidants has been linked to lower rates of coronary heart disease, cancer mortality, and a higher life expectancy [17]. Herbal medication formulations for cancer prevention and therapy have emerged in the last 30 years, and researchers continue to explore natural sources as possible chemotherapeutic agents [18].

Surprisingly, marine microbes remain the most unexplored, a significant source of several bioactive compounds [19]. In recent years, marine sources have played an imperative role in determining the origin of lead compounds for diverse pharmaceutical applications [22]. The research findings suggest that a vast taxonomic diversity is linked to a diverse range of natural products. The ocean houses over 250,000 species, making it a rich source of biodiversity. Marine flora has traditionally been used for therapeutic purposes, with a higher prevalence in oriental nations. Marine-derived compounds have been employed to prevent and cure malignancies such as leukemia, metastatic breast cancer, soft tissue sarcoma, and ovarian cancer. Due to the diversity of creatures in the marine environment, researchers have been looking for novel marine natural compounds to turn into therapies [23, 24].

Despite the success of terrestrial sources in manufacturing novel medications, our assessment focuses on those of marine origin due to their promising future.

Thus, the present chapter highlights marine metabolites isolated from various marine resources, their unique attributes, chemical classes, molecular targets, and mechanisms of anticancer marine metabolites. Their activity level, chemical structure, and the potential for therapeutic leads will be explored.

2. Unique Characteristics of Marine Metabolites

The biological and chemical variability of marine ecosystems is incalculable, serving as a unique resource for the finding of new bioactive compounds, pharmaceuticals, toxin pigments, and enzymes, as well as biopolymers, bio-adhesives, elastomers, and hierarchically structured bio-composites [25]. Oceans account for 95% of the Earth's ecosystem and more than 70% of the planet's surface [26]. Despite having various habitats and containing bulk of the world's biodiversity, they remain mostly unexplored. Just over five percent of the deep sea has been studied, with an even smaller fraction (0.01%) of the deep-sea floor sampled and analysed [27].

Microorganisms occupy a sustainable length from the shallow water in the shoreline to the deep seaward zones that cover 70% of the biosphere. The variations in pressure, temperature, and illumination in the marine system compared to the terrestrial environment may facilitate the production of novel secondary marine metabolites by particular marine organisms to endure severe circumstances including temperature, salinity, pressure, and predators [20, 21].

Marine metabolites are the outcome of enzymatic activities, and they possess structural three-dimensional characteristics that aid in binding to active sites, resulting in high selectivity and unique bioactivity [28]. Compared to synthetic pharmaceuticals, they have more sp3-hybridized carbons, chiral centres, condensed aliphatic rings, fewer aromatic rings, more carbon, hydrogen, and oxygen atoms, and greater nitrogen content [29]. Furthermore, the presence of halogen atoms in their structures, primarily chlorine and bromine, is caused by their abundance in the surrounding saltwater and specific halogenating enzymes found in organisms [30]. Natural products are often classified into structural types such as nucleosides, polyphenols, polysaccharides, peptides, alkaloids, polyketides (including macrolides), diketopiperazines, terpenoids (including steroids), and mixed biogenesis metabolites [31].

Microorganisms (bacteria, actinobacteria, cyanobacteria, autotrophic flagellates, diatoms, microalgae and fungi, etc.), macroalgae (seaweeds), invertebrates (sponges, corals, worms, molluscs, echinoderms, etc.), and flowering plants (seagrasses, mangroves, and other halophytes, etc.) thrive in the marine environment, with nearly 250,000 previously described. Marine microflora, with their 3.5 billion years of expertise in biosynthesis, are the most reliable source of compounds in nature. Marine species generate unique compounds to adapt to harsh environmental conditions, including pressure, salt, and temperature. These molecules are diverse in structure and function [32, 33].

The endeavour to extract pharmaceuticals from the water began in the late 1960s. However, the methodical examination started in the mid-1970s. From 1977 to 1987, around 2500 novel metabolites were discovered in various marine species [34]. These studies indicate that the marine environment is a rich source of new composites, not found in terrestrial sources. Over 10,000 chemicals have been extracted from marine species, and hundreds more are being discovered currently. Each year, discoveries are made. Between 1969 and 1999, around 300 patents were issued for bioactive marine natural compounds. Marine organisms serve as potent sources of pharmaceuticals. The search focuses on marine faunal species, whereas floral species are mainly disregarded. Marine species have chemicals with antioxidant and anticancer effects, which have not been completely investigated. Marine flora has been utilized for therapeutic purposes in India, China, the Near East, and Europe since ancient times. The geographical distribution of marine chemicals reported since 2008 reveals that over a quarter of them are from Australia (24%), with a significant rise in metabolites detected from the South China Sea (18%) and the Pacific Ocean (17%) [35-40].

The prodigious marine biodiversity is due to the variety of distinct environments under which species exist. Temperatures range from −1.5°C in Antarctic seas to 350°C in hydrothermal systems. In the ocean, pressure increases by around one atmosphere for every 10 meters, thus it might be 1000 times higher than the pressure at sea level. Nutrient levels vary from oligotrophic (low nutrient content) to eutrophic (nutrient-rich zones). Light intensity decreases fast as depth increases, and there is no substantial brightness beyond 200 m [41]. The near-surface photic zones get enough sunshine to support photosynthesis in marine plants and plankton. The deepest and most widespread non-photic zones are insufficiently lighted to undertake photosynthesis, thus various physical, chemical, and biological processes, such as bioluminescence, have evolved to compensate for the absence of light. Similarly, the quantity of free oxygen dissolved in the water ranges from 7 mg/L at the surface to less than 0.1 mg/L in oxygen-deprived zones based on temperature, latitude, salinity, and depth. Typically, pH is about 8, and the salinity is usually 35g/kg. Because of the tremendous diversity of circumstances in marine environments, animals have evolved and adapted processes that terrestrial creatures do not require. As a result, marine-derived molecules often have distinct structural frameworks, and biological modes of action, and serve as vital hit or lead compounds in drug development [42, 43].

Seaweed dietary fiber exhibits several activities, including antioxidant, antimutagenic, anticoagulant, and anticancer properties. Consuming more calcium, potassium, and salt can lower systolic pressure and reduce the risk of hypertension. Seaweeds have a high potassium level, comparable to our normal plasma levels, Seaweed extract is similar to human blood plasma. Two Japanese surgeons successfully substituted seaweed components with water for whole blood transfusions in over 100 surgeries. Seaweed polymer extract is widely used in pharmacy, medicine, and biochemistry, though its usage in medicine is less common now.

Clinical studies are underway to eliminate the need for insulin injections for diabetes patients by a "jelly capsule" manufactured from seaweed-derived alginic acid. This capsule protects white blood cells and the immune system in patients. Seaweed gums, such as carrageenan (from red seaweed) and algin (from brown seaweed), contain high levels of soluble fibre [44-46].

3. Resources of Marine Metabolite

3.1. Actinobacteria

Actinobacteria, belonging to the Actinomycetaceae family, are a kind of Gram-positive bacteria recognized for producing a diverse spectrum of bioactive chemicals. For over 50 years, soil-derived actinomycetes of terrestrial origin have been a valuable resource for discovering bioactive secondary metabolites [47]. Mycelium-forming actinobacteria were in marine sediments some decades ago, despite previously being isolated from terrestrial sources. Marine actinobacteria have recently been identified as potential sources of new anticancer drugs due to their unique structures and properties. These bacteria can produce anticancer drugs through different mechanisms, and their potential for commercial use has been proven in many studies [48, 49].

Thiocoraline, a naturally occurring depsipeptide isolated from the mycelium cake of the marine actinomycete strain L-13-ACM2-092. The bioactive chemical exhibits pronounced affinity targeting lung and colon cancer cell lines, alongside melanoma. Interestingly, the drug has selective antiproliferative activity in colon cancer cell lines with defective p53 systems. In vitro studies reveal that thiocoraline arrests the cell cycle in the G1 phase and slows the passage of the S phase to the G2/M phase [50, 51]. Lucentamycin's A-D were identified as four novel 3-methyl-4-ethylideneproline-containing peptides from the fermentation broth of a marine-derived actinomycete, Nocardiopsis Lucentensis. Lucentamycin's A and B exhibited significant in vitro cytotoxicity against HCT-116 human colon cancer [52-54]. Tartrolon D, a macrodiolide, was discovered by Perez and coworkers from the fermentation broths of Streptomyces sp. MDG-04-17-069. Tartrolons are a family of chemicals whose biological features have piqued the curiosity of many researchers. The isolated tartrolon in this study demonstrated high cytotoxic action against three human tumor cell lines: lung (A549), colon (HT29), and breast (MDA-MB-231) [55]. Another study examined the secondary metabolites of a marine Saccharomonospora sp. obtained from the La Jolla Submarine Canyon. Lodopyridone, a novel alkaloid, was discovered to be cytotoxic (IC50 = 3.6μM) to HCT-116 human colon cancer cells [56, 57].

Salinosporamide A (Marizomib), a potent proteasome inhibitor, is a promising therapeutic candidate obtained from these marine actinomycetes. It is biosynthesized by Salinispora tropica, discovered by Fenical and Jensen. Currently undergoing clinical studies for the treatment of multiple myeloma and other malignancies. This marine natural product has a highly functionalized γ-lactam-β-lactone pharmacophore that irreversibly binds to the β subunit of the 20S proteasome. Three years after discovery, Salinosporamide A progressed into phase I clinical trials for multiple myeloma management. The potent pharmacological properties and complex framework of this compound have prompted intense research in academia and industry in recent years, resulting in the development of over ten syntheses, a better understanding of its biosynthetic route, and the establishment of promising structure-activity relationships and oncological data [58-62]. Staurosporine was discovered from the bacterium Streptomyces staurosporeus in 1977, its specific structure was revealed using crystal structure in 1994. Staurosporine is an indolocarbazole containing a bi-indole structural motif. Indolocarbazoles are now under chemical investigation for their anticancer efficacy. Staurosporine, a common indolocarbazole, inhibits several kinases. It is a competitive inhibitor that blocks ATP from binding to the specific kinases. The USFDA authorized in 2017 Staurosporine derivative midostaurin for acute myeloid leukemia disease. Despite exhibiting significant toxicity, its compounds, such as UCN-01, are being studied in clinical trials for cancer treatments [63-67]. Gilvocarcin HE was isolated from Streptomyces sp. QD01-2 has demonstrated cytotoxic activity against the MCF-7, K562, and P388 cell lines, with IC50 values of 36, 39, and 45μM, suggesting that the vinyl side chain enhanced the cytotoxic activity [68, 69].

Lomaiviticin A and B are anticancer chemicals obtained by marine actinobacteria, especially the species Salinispora pacifica. Preclinical studies reveal that it has high cytotoxicity against numerous cancer cell lines at low doses. Their ability to induce DSBs makes them particularly effective against rapidly dividing cells, characteristic of cancer. Their capacity to cause DNA double-strand breaks makes them interesting candidates for developing novel cancer therapies [70, 71].

3.2. Algae

Marine algae, particularly macroalgae (seaweeds) and microalgae, are significant reservoirs of bioactive chemicals. Microalgae are photosynthetic, autotrophic microorganisms crucial for the aquatic food chain. Algae contain unique metabolites such as polysaccharides, polyphenols, and terpenoids, which have shown promise in cancer treatment due to their antioxidant, anti-inflammatory, and antiproliferative properties [72, 73].

Fucoidans isolated from Dictyota Ciliolata, Padina sanctaecrucis, and Sargassum fluitans brown algae have been demonstrated to exhibit considerable antioxidant activity and can protect HepG2 cells from oxidative stress. Furthermore, at a dosage of 2 mg/mL, none of the fucoidan extracts demonstrated cytotoxic effects [74]. Carotenoids derived from microalgae have been identified as a promising treatment for inflammatory diseases and cancers. It has been researched in vitro, in vivo, and in humans for their anti-inflammatory, anti-tumor, and anticancer properties [75]. Specifically, β-carotene from D. salina, C. reinhardtii, T. suecica, and I. galbana has anticancer properties against a spectrum of malignancies, including neuroblastoma, non-Hodgkin lymphoma, prostate, breast, liver, pancreatic, colorectal, and gastric cancers [76]. D. salina is the primary source of β-carotene, which has been linked to prostate cancer cell death by apoptosis [77].

The antitumor efficacy of chrysolaminarin isolated from S. acus was tested on human colon cancer cell lines HTC-116 and DLD-1 [78]. Nigjeh et al., investigated the effect of an ethanolic extract derived from the planktonic diatom Chaetoceros calcitrans on breast cancer (MCF-7), breast epithelial (MCF-10A), and peripheral blood mononuclear cells (PMBC). The comparison to tamoxifen demonstrates that extract from C. calcitrans exhibits promising anti-cancer properties [79]. Miralto et al., identified three polyunsaturated aldehydes (PUAs) from the marine diatoms Thalassiosira rotula, S. costatum, and P. delicatissima. Their findings indicated that 2-trans-4-cis-7-cis-decatrienal, 2-trans-4-trans-7-cis-decatrienal, and 2-trans-4-trans-decadienal inhibited the proliferation of the human colon adenocarcinoma cell line (Caco-2) [80]. Meroditerpenoids are thought to be found predominantly in brown algae. Metabolites, such as fallahydroquinone, fallaquinone, fallachromenoic acid, sargaquinone, sargaquinoic acid, sargahydroquinonic acid, and sargachromenol, have been isolated from the Sargassum fallax brown algae and reported to possess lower to moderate antitumor activity against a P388 murine leukemia cell line [81]. Alginate, laminarin, fucoidan, and many other seaweed polysaccharides have been shown to have anticancer properties. Seaweeds, including Ulva, Ascophyllum, Porphyra, and Palmaria, contain around 65% polysaccharides in total dry weight [82].

However, further research is still needed to fully understand the anticancer action mechanisms of algae-derived chemicals and profit from their application as functional additives in the pharmaceutical sciences.

3.3. Bacteria

Antagonism is nature's inherent response to survival and existence. Bacteria synthesize secondary metabolites to defend themselves against other microbes, and such secondary metabolites are a source of bioactive substances used in human therapeutics. Marine bacteria are prolific makers of secondary metabolites because they flourish in hostile marine environments [83, 84].

Pseudomonas is a Gram-negative bacterium that inhabits both the lithosphere and marine environments. The investigation into the marine environment continues in search of new, unique Pseudomonas strains that are potential sources of bioactive chemicals. The chemical structure of these marine Pseudomonas-derived bioactive compounds includes pyrroles, pseudo-peptide pyrrolidinedione, phloroglucinol, phenazine, benzaldehyde, quinoline, quinolone, phenanthrene, phthalate, andrimid, moiramides, zafrin, and bushrin [85]. Pseudoalteromonas is another recently identified species of marine bacteria that produces bioactive substances. Tambjamines have been isolated from marine invertebrates and have previously demonstrated antibacterial, antitumorigenic, immunosuppressive, anti-proliferative, and ichthyodeterrent properties [86]. Discodermolide, bryostatins, sarcodictyin, and eleutherobin are some of the most potent anticancer medications derived generally from marine bacteria [87]. In vivo, Lactobacilli and Noctiluca scintillans have shown chemopreventive benefits against colon and melanoma cancer [88]. Lactobacilli and Noctilucascintillans both have shown chemopreventive effects against colon and melanoma cancers in vivo. Lactobacilli can inhibit the enzymes azoreductase, nitroreductase, and β-glucuronidase, leading to lower enzyme levels in the intestinal tract. These findings imply that Lactobacilli may lower the risk of colon cancer. Probiotic bacteria like Lactobacillus and Bifidobacterium create anticancer compounds [89, 90]. The isolation of cytotoxic hydroxyphenyl pyrrole dicarboxylic acids, such as 3-(4-hydroxyphenyl)-4-phenylpyrrole-2,5-dicarboxylic acid (HPPD-1), 3,4-di-(4-hydroxyphenyl) pyrrole-2,5-dicarboxylic acid (HPPD-2) and the indole derivatives 3-(hydroxyacetyl)-indole, indole-3-carboxylic acid, indole-3-carboxaldehyde, and indole-3-acetic acid, have also been observed in a marine Halomonas sp. [91]. The two most active extracts were derived from isolates of Sulfitobacter pontiacus (P1-17B (1E)) and Halomonas axialensis (P5-16B (5E)), which suppressed the growth of HeLa and DU145 cells by 50-70% [92].

Novel culture techniques are needed to successfully isolate slow-growing marine bacteria, as laboratory isolation is limited. There is an urgent requirement for a microbe capable of synthesizing new natural compounds.

3.4. Bryozoa

Bryozoa (also known as Ectoprocta, Polyzoa, sea mats, or moss creatures) represent a phylum of aquatic, filter-feeding invertebrates that are widespread, speciose, ubiquitous, and significant components of many benthic ecosystems from the intertidal to deep-sea levels in a range of marine environments [93]. It is composed of little modules known as zooids. The feeding zooids usually comprise a calcified body wall and a soft-bodied component known as the polypide, with a ciliated tentacle crown (lophophore), a stomach, and related muscles and nerves. Around 6000 living species and new taxa are constantly reported, especially in previously inaccessible regions [94]. Bryozoa, or moss creatures, are marine invertebrates that create a wide range of bioactive chemicals. Bryostatins, a kind of macrolide lactone generated from bryozoans, have been demonstrated to exhibit potent anticancer properties by modifying protein kinase C and influencing cell differentiation and death. They are ecologically significant suspension feeders that live in freshwater, brackish, and marine environments [95]. Bryozoans are a massive source of pharmacologically intriguing compounds, including alkaloids and polyketides with various biological functions. Despite their diversity (about 6000 extant species), bryozoans remain understudied [96].

3.5. Cyanobacteria

Cyanobacteria, often known as blue-green algae, are considered among the oldest species on Earth, having inhabited since several billion years. Cyanobacteria are photoautotrophic, Gram-negative bacteria that may live in various habitats, including freshwater, marine, and soil [97]. Over 1600 cyanobacteria molecules have been identified and classified into 260 families of metabolites, with 148 (53%) found in marine atmospheres [98]. Marine cyanobacteria are classified into two major ecological groups: planktonic and benthic. Planktonic cyanobacteria float freely in water columns, and manage depth through buoyancy management and creating aerotopes carrying gas vesicles [99]. The cyanobacterial taxa Prochlorococcus, Synechococcus, Synechocytis, and Cynobium are prevalent in marine planktonic ecosystems. Benthic cyanobacteria have distinct gliding motility, enabling interaction with solid surfaces such as sediments, boulders, stones, algae, and aquatic plants. Lyngbya, a marine cyanobacterial genus, is prevalent in benthic habitats worldwide, creating dense and broad cyanobacterial mats [100]. More than half of the marine cyanobacteria create bioactive chemicals and generate a variety of secondary metabolites with selective bioactivity against vertebrates, animals, plants, microalgae, fungi, bacteria, viruses, and cell lines. The bioactive chemicals generated by marine cyanobacteria are a rich source of molecules with potential anticancer properties, capable of killing cancer cells by influencing cell signaling via activation of protein kinase-C family members or triggering apoptosis [101].

Cai et al., investigated the anticancer efficacy of two cyclic lipopeptides, laxaphycins B4 and A2, obtained from the marine cyanobacterium Hormothamnion enteromorphoides, using an MTT test against human colon cancer HCT116 cells. They found laxaphycin B4 had anticancer activity in HCT116 cells with an IC50 value of 1.7 μM, but laxaphycin A2 had less activity with an IC50 value of 2 μM [102]. Phycocyanin, a highly water-soluble phycobiliprotein derived from the marine cyanobacterium Spirulina platensis, was evaluated against the Vero and Hep-G2 (liver cancer) cell lines using the MTT assay. The highest anticancer activity was at 500μg/mL, whereas the lowest was at 100μg/mL against Hep-G2 cells [103]. Shishido et al., also studied the anticancer effects of 62 Brazilian cyanobacteria isolates against acute myeloid leukemia (AML) cancer cell lines and discovered that extracts from 39 strains cause cell death in AML cancer cell lines [104]. Lyngbyabellins are a peptolide occurring in several marine cyanobacteria species, including Moorea and Okeanea. Lyngbyabellins are found to have anticancer effects against certain cancer cell lines. Ten lyngbyabellin compounds (lyngbyabellin G, lyngbyabellin O, lyngbyabellin P, lyngbyabellin H, 27-deoxylyngbyabellin A, R=H, homohydroxydolabellin, and lyngbyabellin A, R=OH) were isolated from Malaysian Moorea bouillonii and Red Sea Okeania Sp. and tested for their cytotoxic activity against human breast cancer cells (MCF7) [105]. Caldoramide, the formal pentapeptide metabolite of the marine cyanobacterium Caldora penicillata, was tested on the human colon adenocarcinoma cell line HT-29, the human colon carcinoma cell line HCT116, and the human breast adenocarcinoma cell line MCF-7 using an MTT assay [106]. Fayyad et al., also reported that the hot methanolic extract of the cyanobacterium S. Platensis exhibited anticancer action against the L20B and MCF-7 human cancer cell lines. According to the scientists, the methanolic extract of S. platensis was found to have antiproliferative effects against the L20B and MCF-7 cell lines after 24 hours of incubation [107]. In addition, Elkomy et al., found that Oscillatoria simplicissima has an anticancer effect against the A-549 cancer cell line, with 49.465% cell viability, verified the data using an SBR test [108]. Furthermore, an MTT assay was used to assess the methanol extract of the marine cyanobacterium Nostoc sp. N42 and Fischerella Sp. S29 towards a human liver cancer cell line (HepG2) as well as a human non-small-cell lung carcinoma (A-549). Fischerella Sp. demonstrated anticancer properties, with an IC50 of 254.51μg mL^{-1} against HepG2 and 171.74μg mL^{-1} towards A-549. Nostoc Sp. N42 demonstrated anticancer activity with an IC50 of 583.1μg mL^{-1} against HepG2 and 792.17μg mL^{-1} against A-549 cancer cells. Compounds displayed various levels of cytotoxic activity depending on their structure [109]. Lyngbya sp., a marine cyanobacterium, possesses naturally occurring chlorinated leucine compounds, sharing structural similarities with polychlorinated leucine, known to have a variety of bioactivities, including anti-cancer action [110]. The aplysia toxin derivatives neo-debromoaplysia toxin I and neo-debromoaplysia toxin J were isolated from the marine cyanobacterium Lyngbya sp. The cytotoxicity of these compounds was tested on the SW480 human colorectal carcinoma cell line, SGC7901 human gastric cancer cells, LoVo human colorectal carcinoma cells, and PC-9 non-small-cell lung cancer cells. According to the collected findings, neo-debromoaplysiatoxin I displayed modest cytotoxicity, whereas neo-debromoaplysiatoxin J demonstrated substantial cytotoxicity against the tested cancer cell lines, with less than 20% cell viability [111]. Caldorin is a cytotoxic polyketide discovered in the marine cyanobacteria Caldora penicillata. Caldorin was discovered to have moderate to mild cytotoxic action against the HeLa and HL 60 cell lines, which are extensively utilized in cancer research [112].

Based on the findings of the preceding investigations, marine cyanobacteria may be a potential source of natural anticancer compounds. Marine cyanobacteria extracts and compounds have substantial anticancer activity against various cancer cell lines, indicating a possibility for creating new anticancer treatments with fewer adverse effects than standard chemotherapy medications. However, further study is needed to find and extract additional bioactive chemicals from marine cyanobacteria and understand their mode of action. More research is required to fully understand the potential of marine cyanobacteria as a source of anticancer drugs.

3.6. Fungi

Marine fungi produce hydrolytic and/or oxidative enzymes, such as alginate lyase, amylase, cellulase, chitinase, glucosidase, inulinase, keratinase, ligninase, lipase, nuclease, phytase, protease, and xylanase. The capacity of marine mushrooms to adapt to high saline environments and severe pH constitutes a key biological advantage over terrestrial fungi, enhancing their biotechnological applications and possible resistance to ocean acidification [113,114]. Fungi have a variety of medicinal uses due to their capacity to produce numerous enzymes and antibacterial compounds. They generate more antibacterial and anticancer compounds than bacteria. Higher fungi (basidiomycetes), endophytic fungi, and filamentous fungi from marine habitats produce physiologically active principal compounds, such as leptosphaerin, leptosphaerolide, and its O-dihydroquinone derivative, and leptosphaerrodione from the lignicolous fungus Leptosphaeria oraemaris (Pleosporaceae) [115].

Aspergiloid A, produced using the marine filamentous fungus A. glaucus, showed cytotoxicity in A-549, HL-60, BEL-7402, and P388 cell lines. Marine fungi can generate a variety of bioactive compounds. Their biotechnological promise is still undeniable. In 2016, 1277 novel marine natural products were identified, and marine fungi accounted for 36% of the compounds. Cephalosporins, a kind of β-lactam antibiotic discovered by Giuseppe Brotzu in 1945 from Acremonium chrysogenum (formerly Cephalosporium), were the first bioactive chemicals from marine fungi. Marine fungi are a possible source of additional pharmacologically active metabolites (Imhoff, 2016) with unique anticancer, antibacterial, antiviral, antiplasmodial, anti-inflammatory, and antifouling properties. They are also beneficial in generating biosurfactants, enzymes, and bioremediation [116, 117].

3.7. Sponges

Sponges belong to Marine invertebrates that live solely in water. They are either sessile or sluggish-moving, have soft bodies, and lack morphological defensive features like shells or thorns. Consequently, these creatures evolved chemical defence mechanisms against predation and overgrowth of fouling microbes. The large variety of secondary metabolites generated for chemical defence allows researchers to investigate chemical entities with potential therapeutic uses [118, 119].

Marine sponges are the primary research focus for discovering physiologically active secondary marine metabolites. Sponges are excellent filter feeders, and microbes in the surrounding environment are actively swirling by sponge-driven currents. While some germs are immediately digested, others remain within the sponge's body. According to reports, related microorganisms can account for up to 60% of the fresh weight of marine sponges and are more varied than present technology can determine [120].

Several statistical studies on marine natural products have shown that sponges contain the maximum number of physiologically active chemicals that have been identified. The scientists discovered that nearly 75% of the chemicals were obtained from marine invertebrates, with sponges containing the highest number of bioactive compounds [121].

A variety of chemicals with antitumor activity have been found in marine sponges. Among all Marine organisms, sponges, which belong to the phylum Porifera, are recognized as the most flexible source of marine metabolites, accounting for 30%, with many biological uses. So far, particular nucleosides, sterols, alkaloids, peroxides, terpenes, fatty acids, amino acid derivatives, and cyclic peptides have been discovered as active compounds, with at least 60 of them having chemopreventive and anticancer potential [122]. Sponges are an abundant source of numerous natural chemicals that provide chemical scaffolding for novel medications.

Renieramycin M is a tetrahydroisoquinoline extracted from the blue sponge Xestospongia Sp., the first marine anticancer agent approved by the European Union and the Food and Drug Administration (FDA) for the treatment of advanced soft tissue sarcoma [123, 124]. At nanomolar concentrations, it is cytotoxic for human colon, lung, prostate, brain, and breast cancer cells [125]. Aaptamines are alkaloids isolated from several species of Aaptos sp. that have significant anticancer action. Dyshlovoy and colleagues focused on three chemicals (aaptamine, demethyl(oxy)aaptamine, and isoaaptamine) and established their potential to induce apoptosis in a monocytic leukemia cell line (THP-1). Among the three compounds, demethyl(oxy)aaptamine and isoaaptamine showed the most activity [126, 127]. Psammaplysene A (PsA), a bromotyrosine derivative from the marine sponge Psammaplysilla sp., was investigated in two endometrial cancer cell lines, Ishikawa and ECC-1. In both cell lines, PsA triggered apoptosis and elevated the expression of nuclear Forkhead box O1 (FOXO1) [128]. Stellettin B, a triterpene derived from the marine sponge Jaspis stellifera, shows anticancer action in a variety of cell lines, including human glioblastoma (SF295), non-small cell lung cancer (A549), and human chronic myeloid leukemia cells (K562 and KU812) [129, 130]. The dibromo tyrosine derivative (1R,5′S,6′S)-2-(3′,5′-dibromo-1′,6′-dihydroxy-4′-oxocyclohex-2′-enyl) acetonitrile (DT), isolated from the sponge Pseudoceratina sp., was shown to exert proapoptotic action on leukemia K562 cells [131, 132].

Heteronemin is a sesterterpene derived from the sponge Hyrtios sp., affecting the transcriptional level of various cell survival pathways, including cell cycle, apoptosis, and inflammation. Heteronemin activates internal and extrinsic apoptotic pathways in chronic myeloid leukemia K562 and human renal cell carcinoma A498 cells [133]. Dideoxypetrosynol A, a polyacetylene derived from Petrosia sp., is recognized for its proapoptotic effect on human skin melanoma SK-MEL-2 and human leukemia U937 cells via the mitochondrial route [134, 135]. Aragusterol A, a steroid obtained from Xestospongia sp., exhibits antiproliferative action against numerous human cancer cell lines (breast, ovarian, cervical cancer, lung, oral epidermoid, gastric, colon cancer, and leukemia) [136]. Chondrosia corticata produces (19Z)-halichondramide, a macrolide, inhibiting the proliferation of non-small cell lung cancer cells (A549) by blocking the G2/M phase [137, 138].

The information presented in this study emphasizes the crucial importance of marine sponge derivatives as a potential source of anticancer chemicals. The marine chemicals reported here show their potential anticancer action through several pathways.

3.8. Halophytes

Halophytes are emerging as a novel type of marine vegetable with unique organoleptic and functional qualities. Halophytes are salt-tolerant plants that thrive and reproduce in an environment with salt concentrations over 200 mM sodium chloride. The halophytic flora is thus abundant in marshy areas, coastal locations, sandy beaches, and large swaths of deserts with low rainfall, including the mountains across the Arabian desert, plains, as well as prairies, where the salt content of the soil is comparatively very high compared to tropical and subtropical soils. Halophytes adapt to harsh environments, withstanding the abiotic oxidative stress of high salinity on a more complex level through multiple antioxidant enzymes and related secondary metabolites produced by the activity of the plant's intrinsic enzyme systems. Superoxide dismutase, catalases, ascorbate peroxidases, and glutathione reductase are components of the enzymatic defence system [139, 140]. These plants have complex defence mechanisms, and the enzymatic system contributes to them through various physiological processes and biochemical pathways [141, 142]. Additionally, some halophytes treat cancer symptoms and prevent cancer. There are few published reports on the ethno-medical applications of halophytes in cancer treatment. Nevertheless, Salicornia herbacea (glasswort), Rubia cordifolia (manjistha, Indian madder), and Plantago species (fleaworts, plantains) are well-known medicinal halophytes used against cancer [143]. Chinese traditional medicine has utilized several species from the halophytic genus Salsola, including Salsola tragus (Russian-thistle), Salsola foetida (Zri-che, Ecchi, Ressal, Aghacel), Salsola baryosma, and Salsola richteri (Kata-kara, Cherkez), as a treatment for cancer. Algerian patients with breast cancer also use the seeds of Atriplex halimus, also known as saltbush. Newly discovered and previously unidentified cytotoxic substances from various halophytic plants, including alkaloids, phenolics, flavonoids, coumarins, and saponins, are on the list. Numerous novel compounds have demonstrated highly encouraging anticancer properties [144].

3.9. Mangroves

Mangroves are shrubs and trees that use their prop roots, which resemble stilts, to perch on the water. On the coastal edges where land meets water, mangroves are found. They are specially adapted to brackish waters flooded with salt water during high tides and low-oxygen soils, which would kill other trees. Recent studies show that different species can treat fungal infections, constipation, toothaches, and sore throats. Sponges, soft corals, and tunicates can effectively treat debilitating diseases. Moreover, mangroves supply toxic substances with antibacterial, anti-cancer, fungal, and pesticidal qualities [145-148]. Rhizophora mucronata, a mangrove plant species, contains a bioactive component called prenylated terpenoids, having antioxidant and anticancer activity. Mangrove flower species include antioxidants and bioactive chemicals, which have been utilized to treat cancer disorders [149].

3.10. Mollusc

Mollusca is a phylum with a wide range of morphological, ecological, and chemical diversity. It is divided into two subphyla: Aculifera and Conchifera. Aculifera are of two sub-types:

Aplacophora (worm-like) and Polyplacophora (dorsal shell with eight valves); they do not have a complete shell. Marine molluscs span from tropical and temperate oceans to the Arctic, with diverse morphologies and biological niches. Molluscs can store food metabolites in specific anatomical structures. The chemical study of marine molluscs has identified a wide range of bioactive metabolites that originated in marine species as advantageous adaptations for survival in various conditions. Chemical analysis of this phylum has yielded several chemicals of interest as possible anticancer medicines, which are especially significant given the expanding literature on cancer biology and chemotherapy [150-151].

3.11. Tunicates

Tunicates are widely distributed worldwide and are recognized as abundant marine bioresources with many potential applications. Tunicates, often called sea squirts or ascidians, are sessile marine invertebrates with sac-like structures that filter feed. They belong to the subphylum Tunicate, categorized under the chordate group. There are roughly 2000 marsupial species worldwide, with tunicates accounting for the majority [152, 153]. Tunicates are sedentary dwellers of the marine benthos, occupying habitats ranging from the intertidal zone to a maximum depth of 6000 meters. A chemical composition study is required for bioresource utilization and guides particular applications. Anatomically, all tunicates are separated into two parts: the outer shell and the interior organs. The outer shell, a tunic, is an exterior supporting tissue that retains the body form and helps animals filter seawater. In addition, the tunic prevents predators from attacking, is comprised of ~60% cellulose, with an additional ~27% dry weight from nitrogen-containing organic components [155]. More than 1200 active molecules have been discovered from tunicates and tunicate-associated microbial species, indicating their abundance in diverse physiologically active compounds. Screening of bioactive compounds from tunicates is predominantly employed in anticancer and anti-inflammatory drugs, with antimalarial and other treatments ranking second and third, respectively. Peptides and alkaloids isolated from tunicates demonstrate anticancer effects. Aplidine (dehydrodidemnin B, DDB, Aplidin), a cyclic depsipeptide found in the Mediterranean tunicate Aplidium albicans, has been shown to inhibit the growth of breast, melanoma, and non-small-cell lung cancer cells. The activity increased oxidative stress, activating JNK and p38 MAPK phosphorylation and promoting apoptosis [156].

Several molecules or their derivatives have been found in clinical or pharmaceutical settings. These include the antitumor drugs plitidepsin (used in multiple myeloma), lurbinectedin (used in small cell lung cancer), midostaurin (used in acute myeloid leukemia), and trabetidine (ET-743/yondelis, used in soft tissue sarcomas and ovarian tumors) [157].

3.12. Others

Viruses are commonly regarded as pathogens that cause illness in animals and plants. In recent years, however, it has become increasingly understandable that they play vital roles in the world's seas. Very little is known about the bioactive potential of marine viruses. Researchers are increasingly focusing on discovering additional marine viruses, their physiological processes, and secondary metabolites generated [158-161].

Shark cartilage was developed as a cancer-fighting diet because it was thought to possess an ingredient that might inhibit tumor development. Sharks are becoming endangered due to overfishing, making it hard to gather natural components from shark cartilage for medicinal development research. Using recombinant technology, a shark-derived angiogenesis inhibition factor (SAIF) with 33 amino acid residues was isolated from shark cartilage [162, 163].

4. Chemical Classes of Marine Metabolites

Marine floras are high in biologically active and medicinally beneficial compounds. However, marine sources remain mostly unexplored for anticancer medicinal compounds. Regardless of how various classes have been built up, thinking about their compound structures, the most outstanding marine metabolites incorporate different engineered classes of alkaloids, glycosides, phenols, terpenes, peptides, and polyketides with varied structural diversity.

4.1. Alkaloids

Alkaloids are a varied class of widely dispersed chemicals. "Alkaloids" refers to "cyclic organic compounds containing nitrogen in a negative oxidation state having limited distribution among living organisms." Alkaloids are classified based on their chemical structure, biological function, biosynthesis mechanism, and heterocyclic or nonheterocyclic composition. Alkaloids are categorized into seven subclasses: pyridoacrine, indole, pyrrole, isoquinoline, guanidine, aminoimidazole, and sterol alkaloids. Alkaloids have been identified from marine species, including sponges, tunicates, anemones, and molluscs. These organisms are known for their vibrant colors and patterns, commonly associated with alkaloids. Alkaloids have several biological functions, including anticancer properties [164]. Marine alkaloids are compounds that may be isolated from marine organisms. They feature complex and unique chemical structures exhibiting diverse biological functions [165].

Agelamide D (1) aids in curing cancer with radiation in mice with xenograft tumors (hepatocellular carcinoma). According to the researchers, treatment with this alkaloid increased radiation therapy's effectiveness, with a possible mechanism involving a rise in stress in the endoplasmic reticulum (an increase in levels of activating transcription factor 4) and eventual apoptosis during radiotherapy [166]. Scientists investigated the effects of Ascomylactams A (2) on human lung cancer (A549, NCI-H460, and NCI-H1975). The prominent outcomes of this alkaloid therapy were a considerable reduction in the volume and weight of NCI-H460 tumors and a decrease in the volume of A549 tumors in mice [167]. Polycarpines (3) are sulphur-containing alkaloids extracted from the ascidian Polycarpa aurata. These alkaloids were cytotoxic against varied human cancer cells, including central nervous system, colon, melanoma, and leukemia. Polycarpine and its synthetic derivative, dimethylpolycarpine, were reported to trigger apoptosis in JB6 cells in a p53- and caspase 3-dependent manner [168]. Lissoclinidine B, isolissoclinotoxin B, and diplamine B are three alkaloids extracted from Lissoclinum badium. However, only Lissoclinidine B (4) was discovered to prevent p53 degradation and cause cell death in p53 wild-type cancer cells [169].

The pentacyclic aromatic alkaloid ascididemin (ASC) (5), discovered from the Mediterranean ascidian Cystodytes dellechiajei, is a potent cytotoxic anti-cancer drug [170]. Penicillium janthinellum produces three indole alkaloids, shearinine A, D, and E (6), which have been found to trigger apoptosis in human promyelocytic leukemia cells HL-60 [171]. Two indolocarbazole alkaloids (K252C (7) and arcyriaflavin (8)) have been extracted from Actinomycete Z2039-2 using bioassay-guided fractionation, triggering apoptosis in the myelogenous leukemia cell line (K562) [172]. Granulatimide (9) and isogranulatimide (10) were identified and extracted from marine species, demonstrating G2/M checkpoint inhibition in the cell cycle. Monanchoxymycalin C (11) (Mom C) is a novel marine pentacyclic guanidine alkaloid recently identified from the marine sponge Monanchora pulchra. The chemical was discovered to be cytotoxic to human cervical cancer HeLa cells at low micromolar concentrations, produced S-phase cell cycle arrest, and was synergistic in conjunction with cisplatin [173]. Structures of all alkaloids listed above 1-11 are shown in Figure 1.

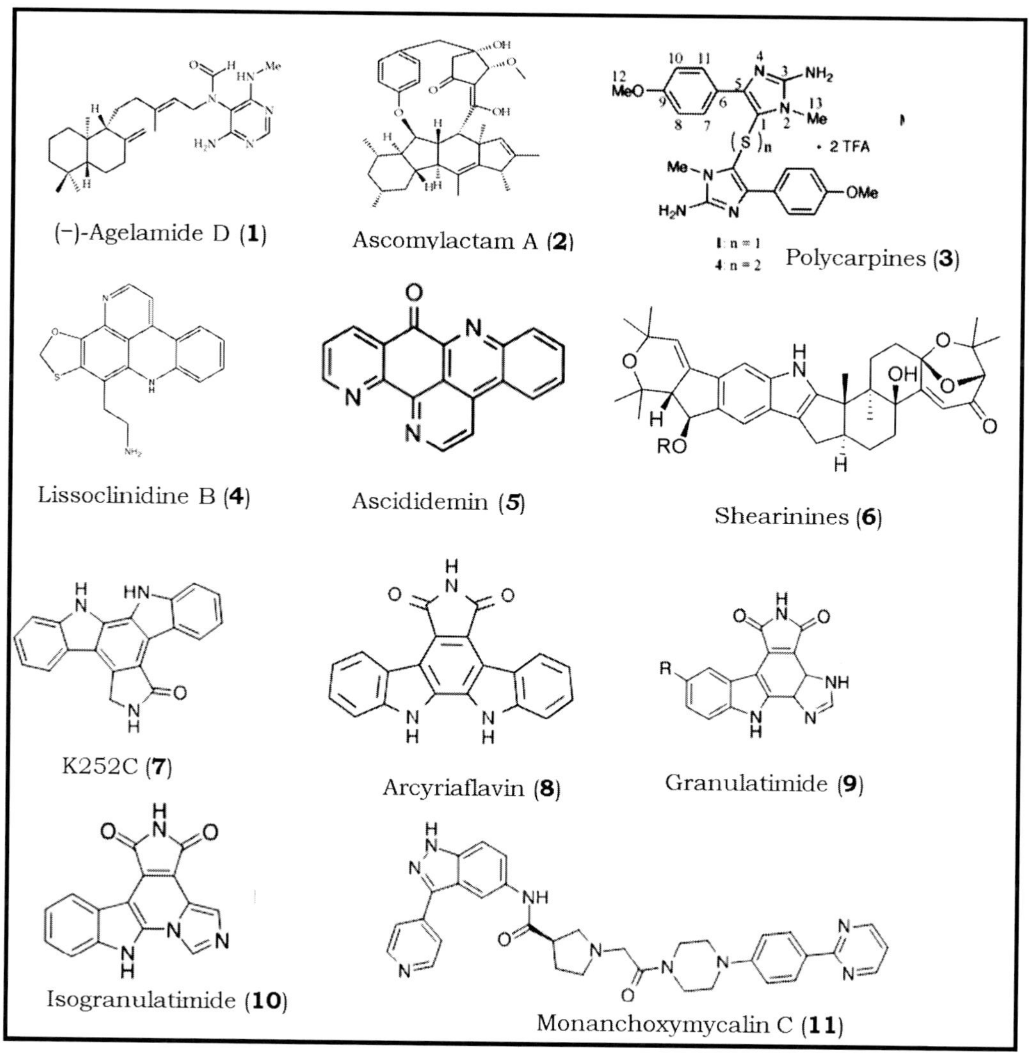

Figure 1. Marine Alkaloids as anticancer agents.

4.2. Phenols

Marine polyphenols are a class of bioactive substances found in diverse marine creatures, including algae, fish, and crustaceans. These substances are distinguished based on numerous hydroxyl groups (-OH) in their molecular structures, which provide various biological activities, including antioxidant, anti-inflammatory, and anticancer effects. These compounds have a complex chemical structure and are classified as flavonoids, phenolic acids, tannins, lignans, and stilbenes [174, 175].

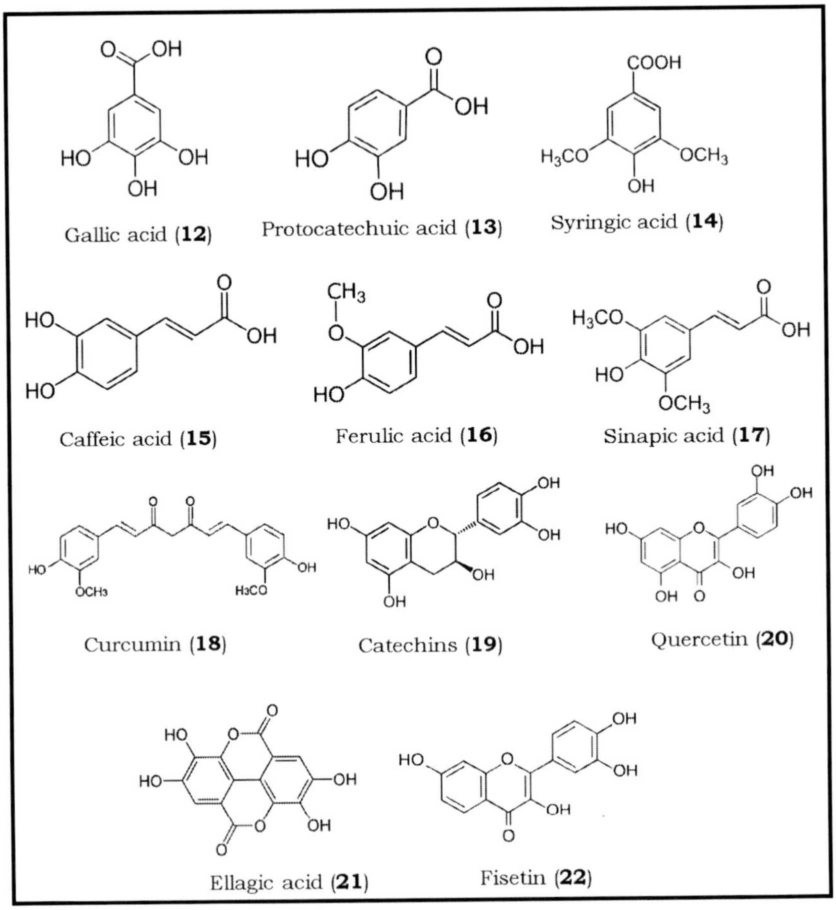

Figure 2. Marine Polyphenols as anticancer agents.

Terrestrial and marine polyphenols share few characteristics but differ fundamentally in their chemical structures. Researchers are interested in marine polyphenols because of their numerous health. They have also been explored as potential treatments for various disorders such as cardiovascular disease, diabetes, and cancer. Polyphenols are polymers composed of flavonoids or gallic acids. Flavonoids include catechins, quercetin, and rutin, commonly found in algae and fish. There are two types of Phenolic acids: hydroxybenzoic acids (HBAs) and hydroxycinnamic acids (HCAs). HBAs include compounds like gallic acid (12), protocatechuic acid (13), and syringic acid (14), whereas HCAs include caffeic acid (15), ferulic acid (16), and sinapic acid (17) obtained from marine organisms found to have anticancer properties.

Although most research has focused on the anticancer characteristics of marine polyphenols, newer studies have highlighted the necessity of exploring the compounds' modes of action, their bioavailability, and metabolism in humans [176, 177]. Some fish, including Tambaqui (Colossoma macropomum), contain curcumin (18), a natural polyphenol. It has been studied extensively due to its antioxidant, anti-inflammatory, and anticancer effects. Catechins (19) are a type of polyphenol with anticancer effects found in fish like tuna and salmon. Catechins are recognized for their capacity to neutralize free radicals, which are unstable chemicals that the body produces in reaction to stress, pollution, and other reasons [178].

Quercetin (20) is a flavonol, a subclass of flavonoid found in various plant foods such as fruits, vegetables, and herbs. Furthermore, quercetin may be found in several fish, including salmon and trout. Quercetin has also been investigated for its ability to prevent and cure cancer. In vivo and in vitro studies have demonstrated that quercetin can help limit cancer cell proliferation and block the creation of new blood vessels that feed tumors. Furthermore, quercetin may contribute to the success of other cancer therapies, such as chemotherapy [179]. Ellagic acid (21) is a form of phenolic acid found in many foods, including fruits, vegetables, and some seafood [180]. Fisetin (22) is a natural flavonoid in several fish, especially salmon. Research has demonstrated that fisetin offers potent antioxidant, anti-inflammatory, and anticancer capabilities that can assist in defending the body from different illnesses and health issues [181, 182]. Structures of all phenols listed above 12-11 are shown in Figure 2.

4.3. Peptides

Peptides range in size from 2 to 20 amino acid residues and are a nitrogen source with diverse physiological roles. Bioactive peptides are protein fragments that gain functionality after being separated from their parent protein [183]. Peptides were formerly thought to have neurotoxic properties linked to cardiotonic, antiviral, anticancer, cardiotoxic, and antibacterial activities [184]. Peptides provide considerable therapeutic potential because of their good binding capabilities, low off-target toxicity, and high stability. Peptides account for around 60% of described natural compounds [185]. Peptides are commonly found in sponges but may also be produced by commensal bacteria or fungi. Many marine species contain peptides, and a wealth of research has revealed that they are most frequently found in sponges, akin to polyketides. Peptides can also be produced by commensal or symbiotic fungi or bacteria [186].

AE-941 (Neovastat) (23), derived from shark cartilage, possesses antiangiogenic qualities. Phase II clinical trial for multiple myeloma is still ongoing, and it has reached Phase III clinical trial evaluation for the treatment of solid tumors (non-small cell lung cancer and renal cell carcinoma) [187]. Tasidotin (24), or synthadotin (ILX-651), is a water-soluble dolastatin pentapeptide derivative with microtubule targeting efficacy in advanced solid tumors. It is now in phase II studies [188]. Soblidotin (TZT-1027) (25), currently in phase II studies, is a microtubule-active medication developed from antimitotic dolastatin-10 obtained from the sea hare Dolabella auricularia. It targets tumor vasculatures and has a significant antivascular and excellent cytotoxic impact [189]. Ziconotide (26), a natural marine peptide, and brentuximab vedotin (SGN-35) (27), a peptide derivative, are the most well-known pharmaceutical drugs made from marine peptides [190].

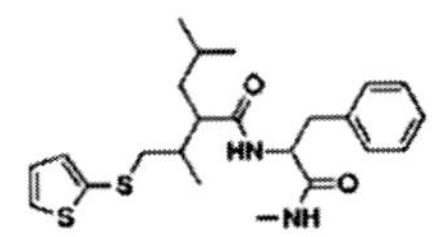

AE-941 (Neovastat) (**23**)

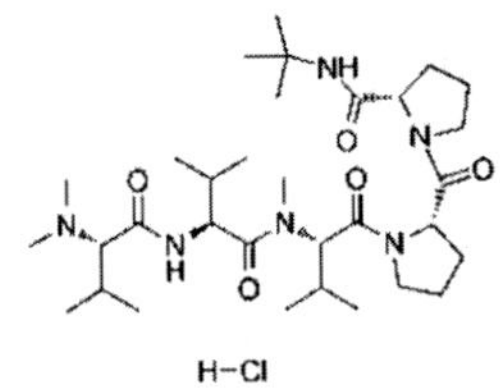

Tasidotin (**24**)

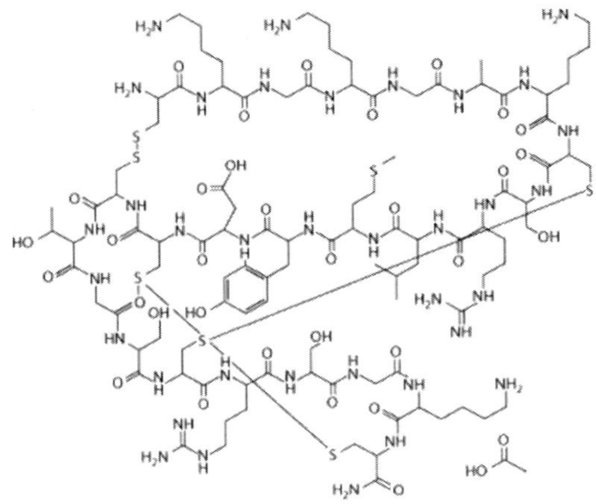

Ziconotide (**26**)

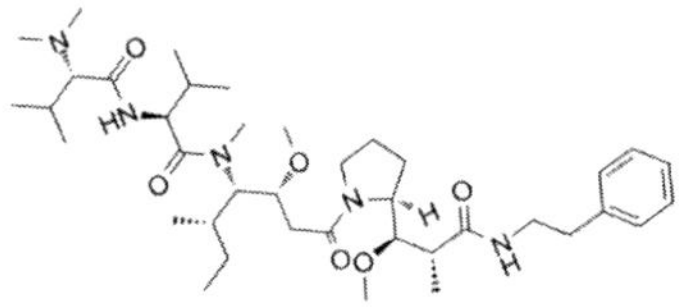

Soblidotin (**25**)

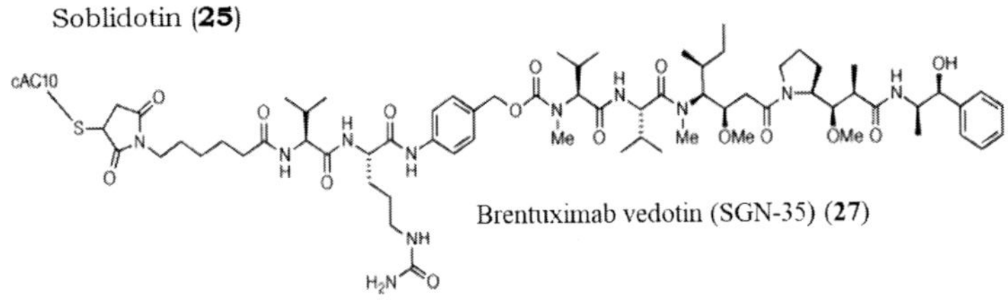

Brentuximab vedotin (SGN-35) (**27**)

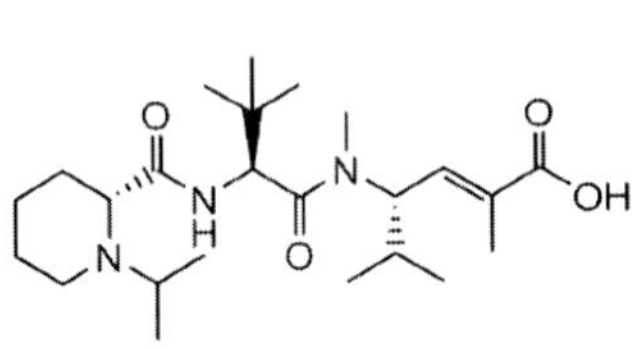

Hemiasterlin (E7974) (**27**)

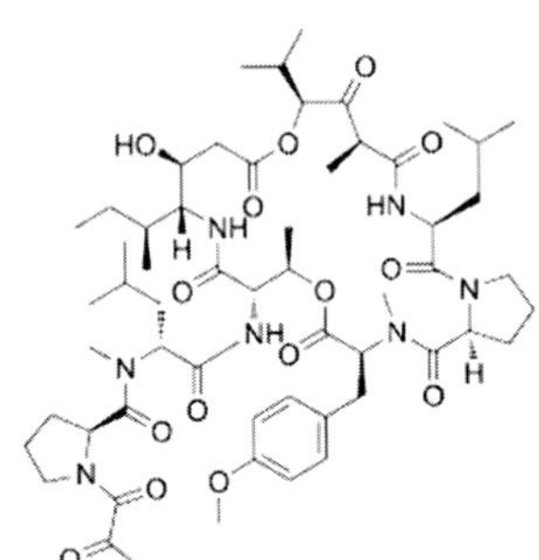

Plitidepsin (**28**)

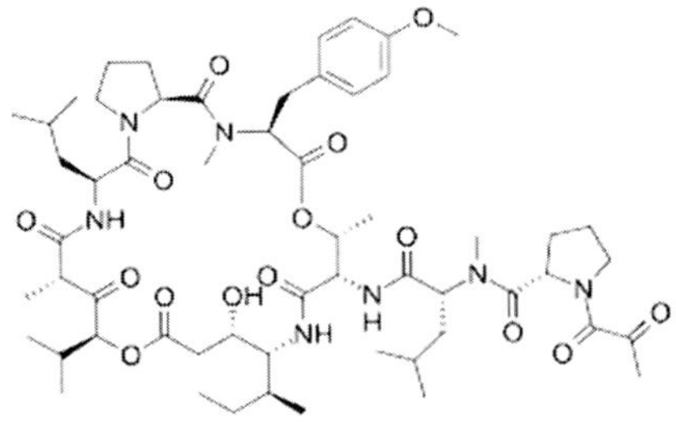

Aplidin (**29**)

Figure 3. Marine Peptides as anticancer agents.

Brentuximab vedotin is an authorized peptide derivative produced by chemical synthesis. This chemical is an antibody-drug combination that has a high antimitotic activity and targets the cell membrane protein CD30. The FDA authorized it in 2011 as a therapy for Hodgkin and systemic anaplastic large-cell lymphomas [191]. Pinatuzumab vedotin and Tisotumab vedotin, obtained from mollusks and cyanobacteria, are now in phase I studies. Pinatuzumab vedotin has efficacy against non-Hodgkin lymphoma and chronic lymphocytic leukemia. Tisotumab vedotin (HuMax®-TF-ADC), an antibody-drug conjugate, has suppressed tumor development in preclinical studies [192]. Hemiasterlin (E7974) (27) is a cytotoxic tripeptide found in marine sponges Hemiasterella minor, Cymbastela sp., and Auletta sp. It is now in phase I trials [193]. Elisidepsin (PM02734, Irvalec®) and glembatumumab vedotin are currently in phase II clinical studies. Elisidepsin, a synthetic cyclic depsipeptide from Elysia rufescens belonging to the Kahalalide family of chemicals, produced from a sacoglossal sea slug species. [194]. Glembatumumab vedotin, a phase II peptide derived from mollusks and cyanobacteria, is being developed to treat breast cancer and melanoma [195]. Plitidepsin, a peptide-based drug, is in phase III clinical studies for breast, melanoma, and non-small cell lung malignancies. Plitidepsin (28) causes dose-dependent cell cycle arrest and apoptosis in solid tumor cells via activating c-Jun N-terminal Kinase (JNK) [196]. Aplidin (29), a marine cyclic depsipeptide (a polymeric molecule having peptide and ester bonds), was isolated from the Mediterranean tunicate Aplidium albicans and is chemically synthesizable [197]. Structures of all Peptides listed above 23-29 are shown in Figure 3.

4.4. Nucleosides

Marine nucleosides are glycosylamines made from a nitrogen-containing heterocyclic nucleobase (purine or pyrimidine) and a modified 5-carbon sugar (arabinose instead of ribose or deoxyribose) joined by a β-glycosidic bond. The compounds inspired by the sponge Tectitethya crypta are known as ara-nucleosides to underline the presence of the arabinose moiety. Their discovery resulted in a new therapeutic principle: nucleosides might be utilized as chemotherapeutics. Arabinonucleosides ara-A and ara-G inspired the anticancer medicines fludarabine phosphate and nelarabine, which were authorized by the FDA in 2005 [198, 199].

Cytarabine (30) is a pyrimidine nucleoside that includes a cytosine base. It is synthesized from 1-β-d-arabinofuranosyluracil by acylation of its hydroxyl groups and conversion of the carbonyl group at position 4 of the pyrimidine ring to a thiocarbonyl group. The final pyrimidine structure, known as 1-arabinofuranosylcytosine, is formed by replacing the mercaptan group with an amino group and hydrolysing the acetyl groups. Cytarabine is a prodrug that is intracellularly converted to the corresponding triphosphate-activated metabolite (arabinosylcytosine triphosphate, ara-CTP), which competes with the physiological building block of nucleic acids, 2′-deoxycytidine, for incorporation into DNA [200]. Nelarabine (31) is a water-soluble prodrug of the cytotoxic deoxyguanosine analog 9-β-d-arabinofuranosyl guanine (ara-G) that shares a similar mode of action with cytarabine. In 2005, the FDA approved for relapsed or refractory T-cell lymphoblastic leukemia and T-cell lymphoblastic lymphoma after at least two previous chemotherapy regimens [201]. Gemcitabine (32) is a nucleoside derivative (a fluorinated derivative of cytarabine) in phase II and III clinical research, used in non-small cell lung cancer, pancreatic cancer, bladder cancer, and breast cancer [202]. Fludarabine phosphate (33) is the phosphate salt of a purine arabinose-nucleoside with 2-fluoro adenine as the nucleobase. Its clinical applications include the treatment of adult patients with chronic lymphocytic leukemia who have not responded to at least one conventional alkylating-agent-containing regimen, and since 2010, with cyclophosphamide and rituximab [203]. Structures of all nucleosides listed above 30-33 are shown in Figure 4.

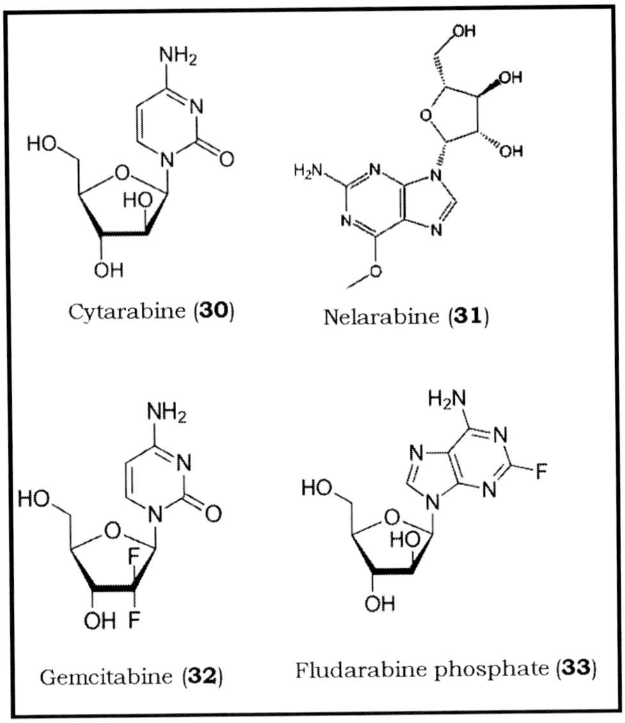

Figure 4. Marine Nucleosides as Anticancer Agents.

4.5. Macrolides

A significant class of marine-derived natural compounds for therapeutic application includes macrolides bearing a macrocyclic lactone ring. Furthermore, glycosidic linkages can bind one or more sugars to the macrocycle. The complex structure of macrolides frequently indicates tough structural elucidation, especially absolute configuration at each chiral center. The chemical synthesis of their complicated chemical characteristics is a demanding field of research in pharmaceutical chemistry. Several marine macrolides have been identified from Marine resources, each with unique and fascinating biological features, however, they are often generated by their associated microbiota [204, 205]. Latrunculins A (34) are marine-derived macrolides from the Red Sea sponge Negombata magnifica, damage cell microfilaments, and have an anti-migratory effect against highly metastatic human prostate cancer PC-3M-CT+ and mouse brain-metastatic melanoma B16B15b cells [206]. Laulimalide (35), a macrolide derived from the marine sponge Cacospongia mycofijiensis, inhibits cell growth by anti-tubulin action. This chemical showed a cytotoxic effect at low concentrations, inhibiting blood vessel formation and VEGF-induced endothelial cell migration [207]. Structures of macrolides listed above 23-29 are shown in Figure 5.

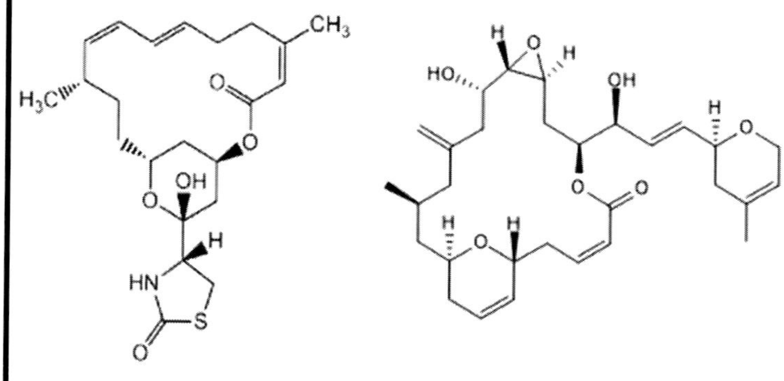

Latrunculins A (**34**) Laulimalide (**35**)

Figure 5. Marine Macrolides as Anticancer Agents.

4.6. Terpenes

Terpenes are the end products of biosynthesis with a five-carbon isoprene molecule. Terpenes are categorized as monoterpenes, sesquiterpenes, diterpenes, sesterterpenes, triterpenes (steroids), and tetraterpenes (carotenoids) based on their unit count [208]. Terpenes produced by marine organisms include biological actions, including cytotoxicity, antiproliferative, antifouling, antifungal, and antibacterial properties. According to several studies, terpenoids may be molecules that aid in preventing the growth of different types of cancer, providing new therapeutic options for the disease. Trials show terpenoids decrease cancer development, perhaps leading to novel therapy options. Many terpenoids have promising anticancer properties, but further research is needed [209, 210].

Two sulfated triterpenoids (Holothurin A (36) and 24-dehydroechinoside A (37)) significantly reduced adhesion, invasion, and migration in HepG2 cells in a dose-dependent manner [211]. Frondoside A (38), a triterpene molecule found in the sea cucumber Cucumaria frondosa, inhibited migration and invasion in the human estrogen receptor-negative breast cancer cell line MDA-MB-231 in wound-healing and Matrigel invasion assays at concentrations ranging from 0.1 to 1 μM [212]. Heteronemin (39), a natural sesterterpene found in the sponge Heteronema erecta, inhibited tumor cell intravasation via the lymph-endothelial barrier in a 3D cell culture model using MCF-7 breast cancer spheroids [213]. Pachycladins A (40), derived from the Red Sea soft coral Cladiella pachyclados, had significant anti-migratory efficacy against prostate cancer PC-3 cells in a wound-healing experiment [214]. Sipholanes, triterpenes having conjugated rings of perhydrobenzoxepine (rings A and B) and bicyclodecane (rings C and D), were identified from the Red Sea sponge Callyspongia siphonella. During a wound-healing experiment, three compounds from this class, sipholenol A, sipholenol E, and sipholenone A, showed significant antimigratory action in MDA-MB-231 breast cancer cells [215]. Sarcophines (41) are marine-derived cembranoids (diterpenes) witha 14-membered macrocyclic ring identified from soft coral. Kashman and collaborators identified Sarcophytum glaucum in the 1970s [216]. Globostellatic acid X methyl esters (42) are isomarabarican-type triterpenes derived from the marine sponge Rhabdastrella globostellata. These chemicals significantly inhibited migration and tube development in human umbilical vein endothelial cells [217]. Smenospongine (43), a sesquiterpene aminoquinone derived from the marine sponge Dactylospongia elegans, inhibits human endothelial cell proliferation, migration, and tube formation, indicating potential anti-angiogenic action [218]. Structures of all terpenes listed above 36-43 are shown in Figure 6.

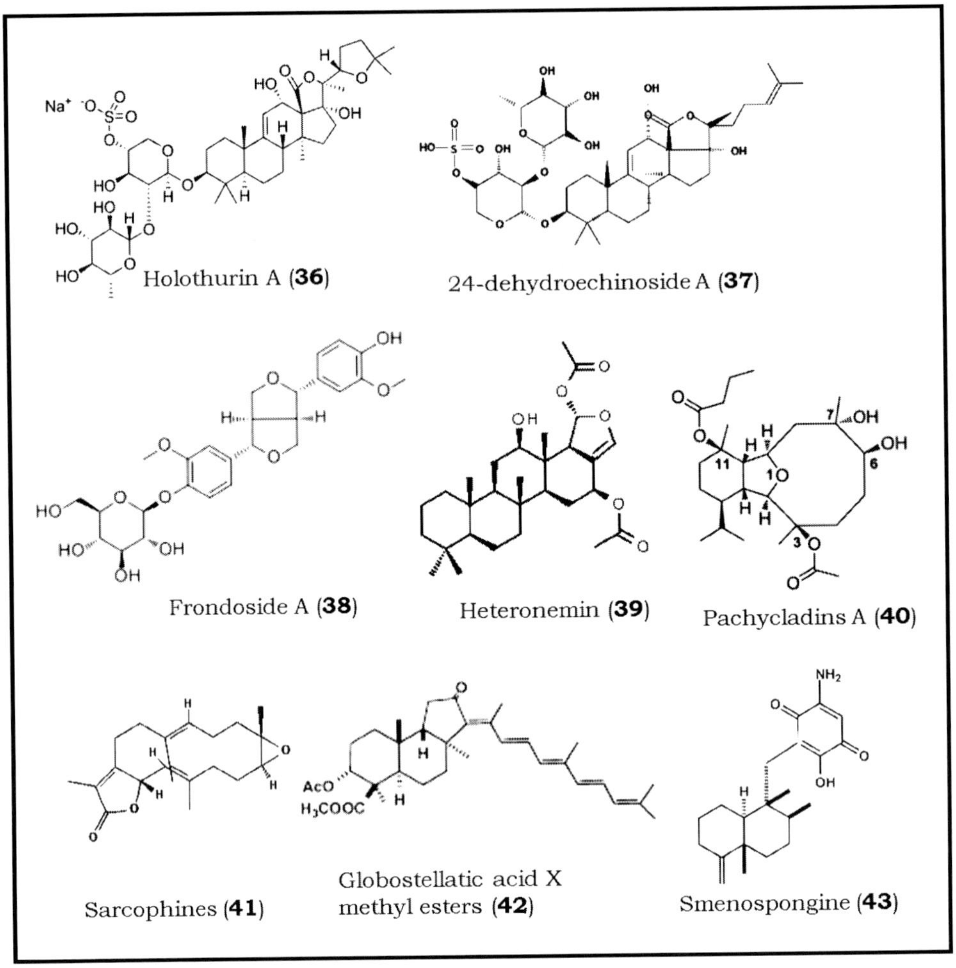

Figure 6. Marine terpenes as Anticancer Agents.

4.7. Polyketides

Polyketides are naturally occurring metabolites with a wide range of chemical configurations. This class contains macrolides, polyethers, polyols, and aromatic chemicals. This class is often highly oxygenated and has several β-hydroxyketone or β-hydroxyaldehyde functional groups. Polyketides are similar to fatty acids in that they are synthesized through the decarboxylation of malonyl-CoA and other acyl-CoAs. However, unlike fatty acids, polyketides can use multiple monomers to create different-sized aromatic groups or reduced chains. Polyketides and fatty acids play crucial roles in cell activity, although polyketides have more complicated metabolic pathways. These metabolites found in sponges, ascidians, soft corals, and bryozoans may be generated by commensal or symbiotic bacteria. Polyketides have many biological actions, including antibiotic, anticancer, antifungal, antiparasitic, and neurotoxic properties [219-222]. Eribulin mesylate (E7389) (44), derived from the marine sponge Halichondria okadai, is a therapeutically licensed polyketide that operates as a non-taxane inhibitor of microtubule dynamics, similar to halichondrin B [223]. The FDA authorized E7389 for the treatment of liposarcoma and breast cancer in 2010 [224]. Discodermolide (45) is a polyketide found in the marine sponge Discodermia dissoluta. It functions as a microtubule interfering agent, attaching to microtubule bundles, breaking mitotic spindles, and causing cell cycle arrest at the G2/M phase [225]. Bryostatin 1 (46), a polyketide found in sponges and tunicates, was isolated from the bryozoan Bugula neritina. This chemical is now in phase I studies for Alzheimer's disease and cancer therapy, being effective against various carcinomas [226]. Structures of polyketides listed above 44-46 are shown in Figure 7.

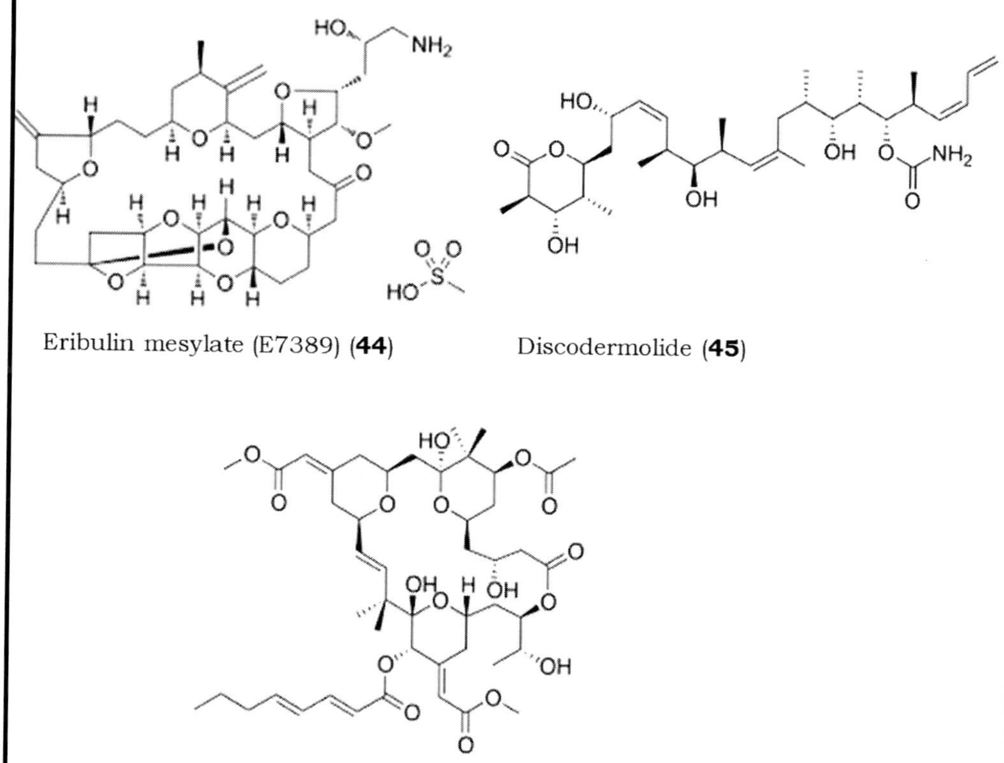

Eribulin mesylate (E7389) (**44**)

Discodermolide (**45**)

Bryostatin 1 (**46**)

Figure 7. Marine Polyketides as anticancer agents.

4.8. Polysaccharides

Polysaccharides are the most often separated carbohydrates from marine creatures; however low molecular weight glycosylated oligosaccharides may also be extracted from sponges and tunicates. Polysaccharides are macromolecules made up of monosaccharides connected with glycosidic residues, having different structural properties. Polysaccharides, also known as glycans, are a class of important chemical substances that include the most prevalent monosaccharide, D-glucose, but also D-fructose, D-galactose, L-galactose, D-mannose, L-arabinose, and D-xylose. These flexible molecules have several pharmaceutical applications, including gel synthesis, drug delivery systems, wound healing, tissue engineering, and blood dialysis membranes. Additionally, they have shown antimutagenic, anticancer, hypocholesterolemic, and anticoagulant properties. Marine oligosaccharides are formed from the hydrolysis of marine polysaccharides. Oligosaccharides typically include 10-12 monosaccharide units, although some, like rhamnogalacturonans, can have 30 or more. Oligosaccharides have complex and varied chemical structures linked to their diverse biological activities. Polysaccharides contain monosaccharide derivatives such as amino sugars (D-glucosamine and D-galactosamine) and their derivatives (N-acetylneuraminic acid and N-acetylmuramic acid), as well as simple sugar acids (glucuronic and iduronic acids) [227, 228].

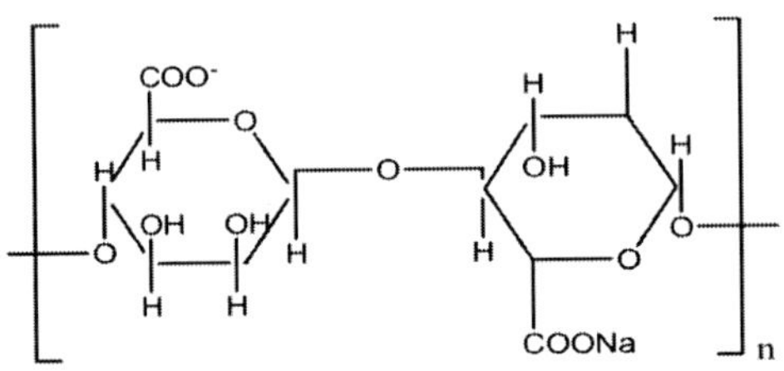

Alginates (**46**)

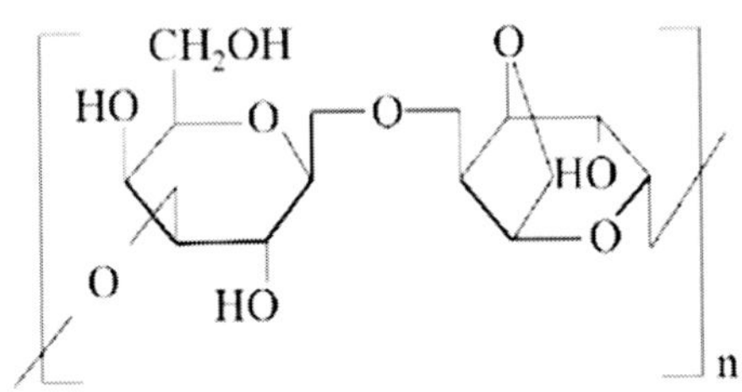

Agar (**47**)

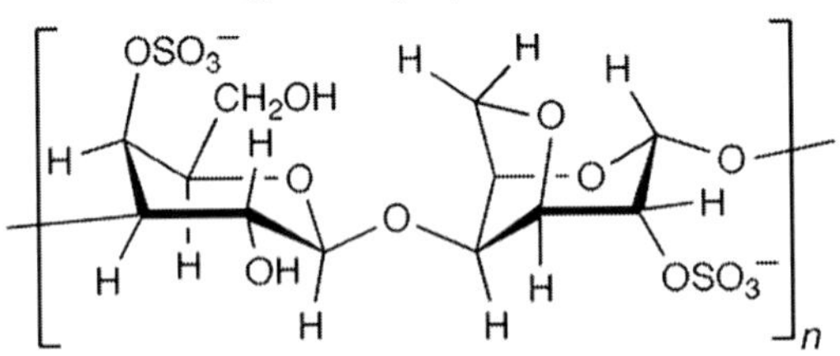

Carrageenans (**48**)

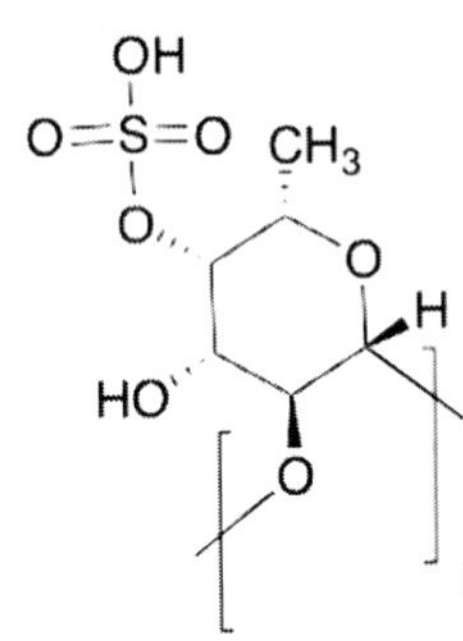

Fucoidan (**49**)

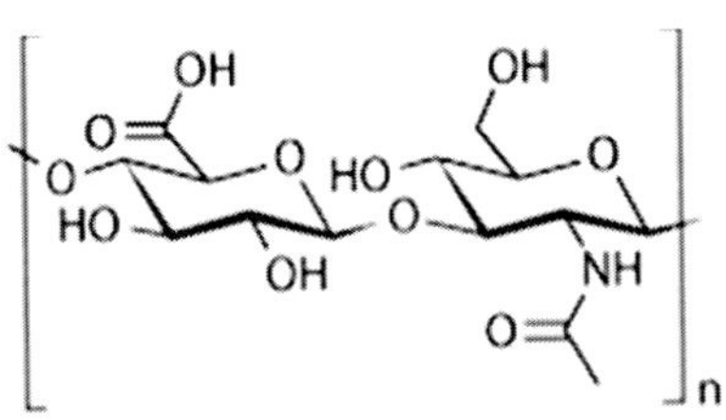

Glycosaminoglycans (**50**)

Figure 8. Marine Polysaccharides as Anticancer agents.

Alginates (46), agar (47), and carrageenans (48) are all polysaccharides derived from algae. Agar is an unbranched polysaccharide found in the cell membranes of red algae, particularly those of the genera Gelidium and Gracilaria, and serves as the principal structural support for algal cell walls. Its chemical structure is made up of galactose sugar molecules. Carrageenans are galactan polysaccharides with alternating 1,3- and 1,4-linked galactose residues that fill gaps in seaweed's cellulose plant structure [229]. The active components in algal polysaccharides are mostly sulfated. Most research suggests that sulfated polysaccharides can improve the innate immune response by enhancing the tumoricidal activities of macrophages and natural killer cells [230]. Fucoidan (49) is a typical sulfated polysaccharide (sulfated L-fucose) produced from the cell walls of brown algae. Fucoidan induces apoptosis in human lymphoma. HS-Sultan cell lines are associated with caspase-3 activation and down-regulation of the extracellular signal-regulated kinase pathway [231]. Glycosaminoglycans (50) have a carbohydrate structure that is both linear and complicated. These molecules are found on all mammalian cell surfaces and exhibit significant sequence variety, resulting in highly specialized interactions with other macromolecules. Despite their biological importance, no research has been conducted on the anticancer activity of marine-derived carbohydrates. [232]. Structures of polysaccharides listed above 46-50 are shown in Figure 8.

4.9. Glycosides

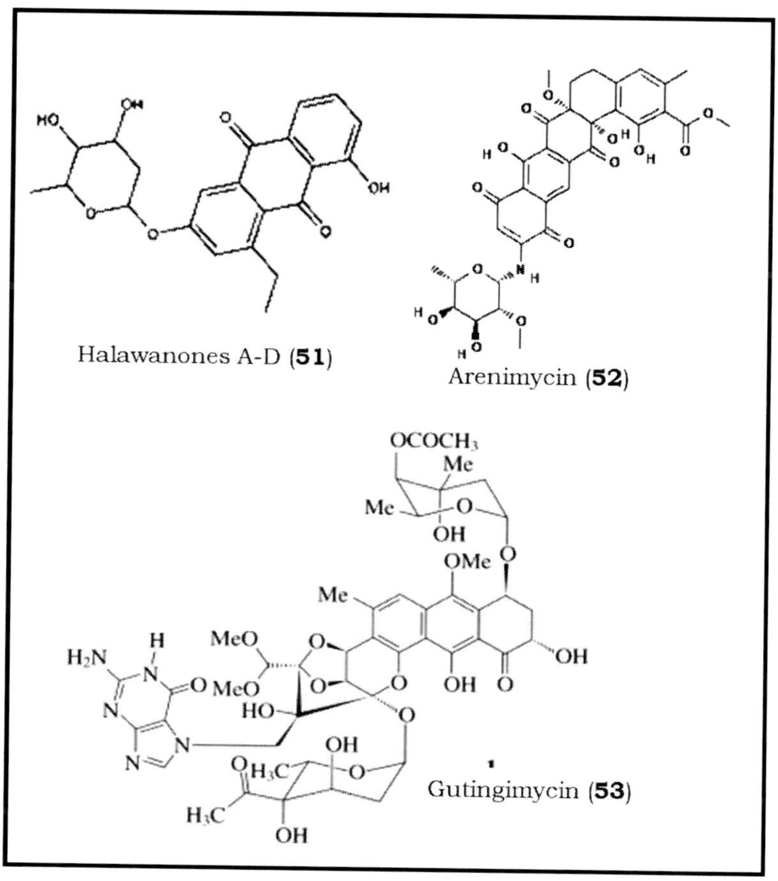

Figure 9. Marine Glycosides as anticancer agents.

Glycosides are a class of carbohydrates. These compounds feature a glycosidic link between a sugar and another functional group. They consist of two parts: sugar and an aglycone chemical group, which may be a terpene, flavonoid, coumarin, or another natural molecule

[233]. Halawanones A-D (51), four quinone-containing metabolites identified from Streptomycete sp. BD-18T was obtained from shallow water sediment [234]. Two antitumor pyranone glycosides, PM050511 and PM006043, and their aglycones, PM050463 and PM060054, were isolated from the marine Streptomyces albus POR-04-15-053 [235]. Arenimycin (52) produced a strain of Salinispora arenicola CNR-647 linked to ascidian Ecteinascidia turbinate, which showed substantial action against HCT-116 cells (IC50 = 1.16 μg/ml) [236]. Gutingimycin (53) is a natural substance with a trioxacarcin skeleton from Streptomyces B86 52, isolated from sediment. The identical Streptomyces species produced trioxacarcins D-F [237]. Structures of glycosides listed above 51-53 are shown in Figure 9.

5. Molecular Targets and Mechanism of Anticancer Marine Metabolites

5.1. Anti-Tubulin Agents

Anticancer medicines that target microtubules are effective for treating breast, ovarian, and lung cancer. Microtubules are cytoskeletal fibers composed of tubulin subunits [238]. These structures are the primary components of the cytoskeleton required in all eukaryotic cells such as cell shape maintenance, vesicle and cell signaling molecule transit, mitochondria homeostasis, motility, and organelle distribution. Furthermore, this microtubule network plays a vital part in mitotic and cell division, which makes it a significant target for anticancer medicines. Microtubules of heteropolymers of α-, y-, and β-tubulin are arranged head-to-tail to form dimers [239]. These dimers polarize, forming 13 parallel protofilaments that create a hard hollow body. Because both α- and β-tubulin dimers attach to GTP before organizing into microtubules, the hydrolysis of the GTP nucleotide governs the balance among the dimeric and polymeric states. GTP-bound heterodimers are more likely to polymerize, while those coupled to GDP move the equilibrium towards the free species. GTP is required for microtubule stability and the balance between polymerization and depolymerization [240]. Microtubules and free tubulin dimers are in a highly dynamic equilibrium that is extremely susceptible to external stimuli. Antitumor medications can disturb the microtubule equilibrium, the mitotic spindle and serving as antimitotic agents, inducing mitotic arrest. Two types of antimitotic medications target microtubules: those that destabilize them and those that stabilize them as shown in Figure 10 [241].

Microtubule-targeting medicines are effective anticancer drugs for breast, ovarian, and lung cancer treatments. The marine metabolites related to this category are given in Tables 1 and 2.

5.2. Apoptosis Promoter

Apoptosis is a complicated process that includes several signaling channels and causes numerous alterations in dying cells. A change in the balance of pro- and anti-apoptotic proteins activates the apoptotic machinery. Apoptosis can be induced in a cell using either the caspase-mediated extrinsic or intrinsic mechanisms as shown in Figure 11. Reduced apoptosis can be achieved by upregulating antiapoptotic proteins, downregulating proapoptotic proteins, and reducing caspase expression. Evading apoptosis can lead to cancer growth by inhibiting differentiation, boosting angiogenesis, and enhancing cell motility, invasion, and metastasis. Dysregulation of apoptotic signaling can cause cancer because of inadequate apoptosis [242-244].

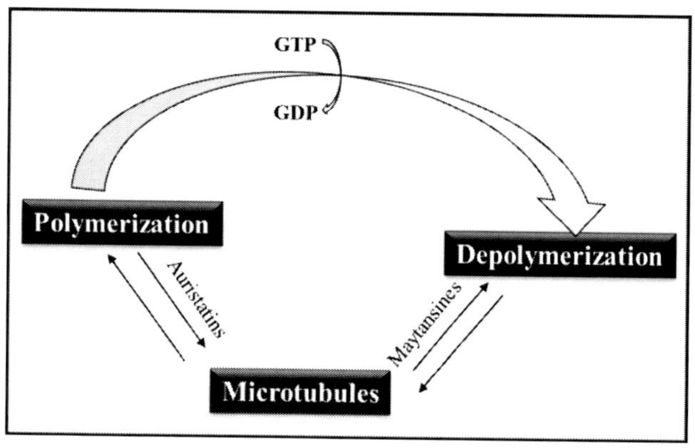

Figure 10. Tubulin polymerization inhibitors (microtubule destabilizers) and polymerization promoters (microtubule stabilizers) are the payloads of tubulin inhibitors that drive their mechanism of action. Two representative medications that affect the formation of microtubules are shown in the figure: several marine metabolites change the way microtubules form by attaching to the β-subunit of α-β tubulin dimers, which causes unchecked microtubule growth. On the other hand, it prevents the formation of tubulin dimers, which hinders the development of mature microtubules.

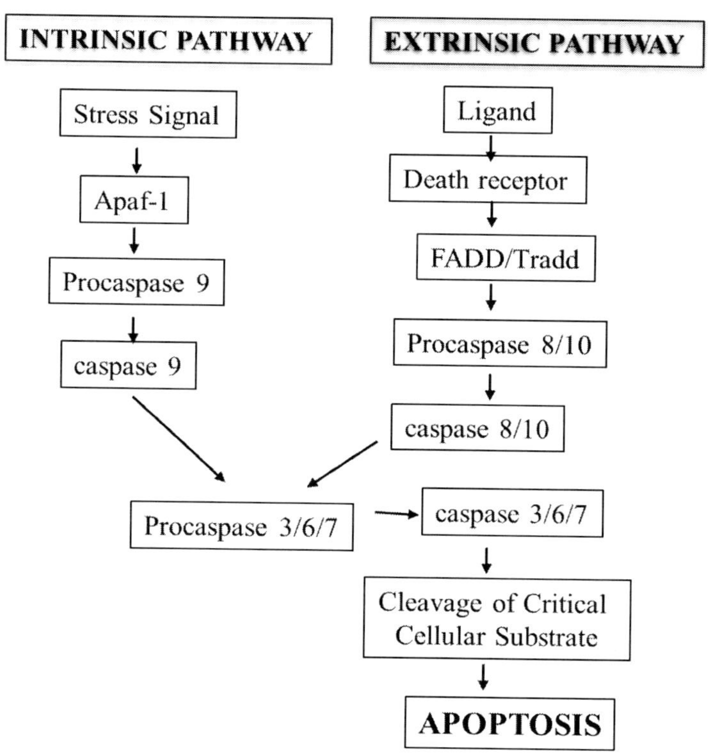

Figure 11. The Intrinsic and Extrinsic cell death pathways.

Inducing apoptosis is an active strategy for preventing cancer cell growth. The most important targets of both apoptosis pathways include death receptors, adaptor proteins, FLIP, NF-κB, pro- and anti-apoptotic proteins from the B cell lymphoma (Bcl)-2 family, intracellular ROS and Ca2+ levels, mitochondrial membrane permeability, p53, and initiating and effector caspases. Cancer can be treated with radiation and pharmacological treatments such as tamoxifen, which induce apoptosis. Chemo-preventive medicines can reduce cancer by triggering apoptosis. Plant extracts may induce apoptosis via increasing immune surveillance, macrophage activity, and death-inducing signaling. Curcumin and resveratrol are natural dietary ingredients that have been shown to trigger apoptosis in malignant cells in vitro [245, 246]. Marine metabolites can activate macrophages, leading to apoptosis as mentioned in Tables 1 and 2.

5.3. Angiogenesis Inhibitors

Angiogenesis is a natural physiological process typified by proliferation of endothelial cells. This process happens during organogenesis or in injured organs and plays a role in wound healing, menstruation, and placenta development during pregnancy. In cancer, angiogenesis facilitates tumor development and distributes tumor cells via circulation to other organs, causing metastases. The intricate interaction between angiogenic stimulants and inhibitors regulates angiogenesis. Angiogenesis includes creating new tubes, encouraging endothelial cell migration and proliferation, and destabilizing the current vessel [247, 248].

The angiogenic process primarily involves vascular endothelial growth factor (VEGF) and its receptor, VEGFR-2 (Flk-1/KDR). Angiogenesis is also linked with extracellular factors, including tumor necrosis factor (TNF), fibroblast growth factor-2 (FGF-2), interleukin-8 (IL-8), platelet-derived endothelial cell growth factor (PD-ECGF), platelet-derived growth factor (PDGF), angiopoietin-1, transforming growth factor alpha (TGF-α), transforming growth factor beta-1 (TGF-β1) and epidermal growth factor (EGF). Blocking the VEGF-VEGFR-2 pathway and subsequent signals is a different approach for inhibiting tumor development. Several cancerous cells excessively express VEGF, which initiates angiogenesis, growth, invasion, migration, and tube formation in endothelial cells [249, 250]. Metastasis is the most dangerous aspect of cancer, determining clinical stages and prognosis. Metastatic tumor cells undergo extensive changes, although several molecular targets have been discovered. Cancer research focuses on substances that inhibit angiogenic and metastatic processes, including proteases, endothelial cell migration and proliferation, angiogenic growth factors, and matrix proteins like integrins [251]. Compounds that disrupt or normalize angiogenesis and/or metastasis have potential as cancer treatments. The marine metabolites that exhibit these features are given in Tables 1 and 2.

5.4. MAPKs Inhibitors

Mitogen-activated protein kinases (MAPKs) phosphorylate the amino acids serine, threonine, and tyrosine, which play a role in cancer cell division, proliferation, development, process, cell death, movement, spread, energy metabolism, and angiogenesis. This family of kinases is activated by external stimuli, which then phosphorylates and activates targets to relay intracellular signals to the nucleus, causing cellular responses. MAPKs act as cellular amplifiers, converting graded inputs into on/off outputs and filtering out low-level noise. MAPKs are often overexpressed or amplified in cancer cells. MAPKs identified in mammals include MEK/ERK, PKC, PI3K/AKT, JNK, and p38.

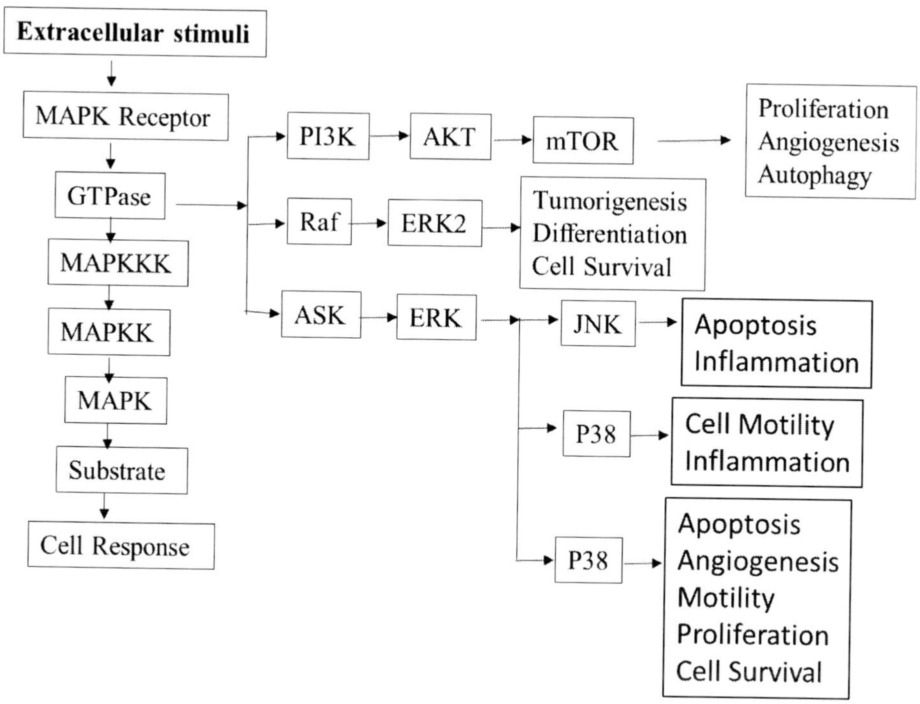

Figure 12. Illustrates the relationship between mammalian MAPK cascades and other kinases that can trigger cell responses in cancer. MAPK-controlled signal transduction pathways involve two steps once extracellular stimuli are transduced by a membrane receptor. Kinases (MAPKKK and MAPKK) lead to unique MAPK responses (seen in the left image). ERK contributes to cell survival, proliferation, angiogenesis, and autophagy. MAPKKK (ASK) activates the JNK and p38 pathways, carcinogenesis, cell motility, osmoregulation, inflammation, and death. ERK 5 is a MAPK that promotes cell proliferation, survival, motility, and angiogenesis. The PI3K/AKT/mTOR axis is a serine-threonine protein kinase pathway activated by the Ras GTPase. It plays a crucial role in cellular quiescence and proliferation.

EGF regulates cell growth and division by activating receptor tyrosine kinase activities, autophosphorylation, and the Ras/Raf/MEK/ERK signaling pathway. Protein kinase C (PKC) is a group of phospholipid-dependent serine/threonine kinases that regulate protein synthesis, gene expression, cell proliferation, differentiation, and cancer. Many PKC enzymes are overexpressed in malignancies. Phosphoinositide-3-kinase (PI3K)/AKT is a serine-threonine-specific protein kinase that regulates oncogenesis and apoptosis in cancer cells, triggered by mitogens. This system, triggered by membrane receptors including VEGFR and PDGFR, is a key target in cancer research. JNK is a member of the MAPK family that affects several biological processes, including cancer and neurodegenerative diseases. This family is susceptible to several stressors, including cytokines, UV and γ-irradiation, cytotoxic medicines, DNA-damaging chemicals, and reactive oxygen species. The p38 mitogen-activated protein kinase (MAPK) regulates cellular responses to external stimuli, including stress, irritation, UV irradiation, osmotic shock, ischemia, and DNA damage. Finally, MEK5 phosphorylates and activates ERK5 the most recently recognized member of the mitogen-activated protein kinase (MAPK) family. This kinase is crucial for cardiovascular development, vascular integrity, pathological conditions like cancer, and angiogenesis, as shown in Figure 12.

6. On-Going Preclinical and Clinical Trials of Anticancer Marine Metabolites

Some marine metabolites have shown encouraging outcomes in preclinical phases, leading to clinical trials and regulatory approval. This section contains updated information on medications in clinical and preclinical trials (summarized in Table 1). The compounds are organized by their chemical makeup and provide information on their origins and probable action methods.

Table 1. Marine metabolites under clinical and preclinical trials as anticancer agents

Sr. No.	Name of compound	Source of marine organisms	Chemical Class	Mechanism of action	Function	References
1.	Agelamide D	Sponge	Alkaloid	Apoptosis	Hepatocellular carcinoma	166
2.	Ascomylactam A	Fungi	Alkaloid	Unknown	Human Lung Cancer (A549, NCI-H460, And NCI-H1975)	167
3.	Polycarpines	Tunicate	Alkaloid	Apoptosis	Colon, Melanoma, and Leukemia	168
4.	Lissoclinidine B	Tunicate	Alkaloid	Cell death	Unknown	258
5.	Ascididemin (ASC)	Tunicate	Alkaloid	Cell cycle arrest	Leukemia	259
6.	Shearinine A, D, and E	Fungi	Alkaloid	Apoptosis	HL-60 Cells	260
7.	K252C	Actinomycetes	Alkaloid	Trigger apoptosis	Leukemia	261
8.	Arcyriaflavin	Actinomycetes	Alkaloid	Trigger apoptosis	Leukemia	262
9.	Granulatimide	Tunicate	Alkaloid	G2/M checkpoint inhibition	Solid tumour	263
10.	Isogranulatimide	Tunicate	Alkaloid	G2/M checkpoint inhibition	Solid tumour	263
11.	Monanchoxymycalin C	Sponge	Alkaloid	Trigger apoptosis	Cervical cancer	173
12.	Catechins	Seaweed	Phenol	Antiangiogenic	Solid tumour	178
13.	Quercetin	Seaweed	Phenol	Antiangiogenic	Solid tumour	179
14.	Ellagic acid	Seaweed	Phenol	Antiangiogenic	Solid tumour	180
15.	Fisetin	Seaweed	Phenol	Antiangiogenic	Solid tumour	181, 182
16.	AE-941 (Neovastat)	Shark cartilage	Peptide	Antiangiogenic	Multiple myeloma	187
17.	Tasidotin	Cyanobacteria	Peptide	Anti tubulin agent	Solid tumors	188
18.	Soblidotin (TZT-1027)	Synthesis derivative from marin	Peptide	Antiangiogenic	Solid tumors	189
19.	Ziconotide	Snail	Peptide	Unknown	Solid tumors	264
20.	Pinatuzumab vedotin	Mollusks	Peptide	Tubulin inhibitor	Non-Hodgkin lymphoma, leukemia	265

Sr. No.	Name of compound	Source of marine organisms	Chemical Class	Mechanism of action	Function	References
21.	Tisotumab vedotin	Cyanobacteria	Peptide	Antibody-drug conjugate	Gastrointestinal Cancer	266
22.	Hemiasterlin (E7974)	Sponge	Peptide	Antitubulin agent	Solid tumors	267
23.	Elisidepsin	Mollusk	Peptide	Cell cycle arrest Apoptosis	NSCLC	268
24.	Glembatumumab vedotin	Mollusks and cyanobacteria,	Peptide	Antibody-drug conjugate	Breast cancer and melanoma	195
25.	Plitidepsin	Tunicate	Peptide	Cell cycle arrest, Apoptosis	Breast, melanoma, and non-small cell lung malignancies	196, 197
26.	MicroviridinToxin BE-4,	Microcystis aeruginosa	Polysaccharide	unknown	Antibiotic, anticancer	269
27.	Apratoxins	Cyanobacteria	Peptide	unknown	Inhibition of a variety of cancer cell lines	270
28.	Borophycin	Cyanobacteria	Polysaccharide	Cell cycle arrest, Apoptosis	Cytotoxicity against human epidermoid carcinoma (lovo) and colorectal adenocarcinoma	271
29.	Aurilide B	L. Majuscula	Peptide	Apoptosis	H-460 lung tumor	272
30.	Belamide A	Symploca sp.	Peptide	Anti tubulin agents	HCT-116 colon cancer	273
31.	Bisebromoamide	Cyanobacteria	Peptide	Unknown	Renal cancer	274
32.	Biselyngbyasides	Cyanobacteria.	Glycoside Macrolides	Apoptosis inhibitor	Hela S3 epithelial carcinoma, SNB-78 central nervous system cancer, and NCI– H522 lung cancer	275
33.	Conidiogenone C	Penicillium chrysogenum QEN-24S	Diterpenoids	Unknown	Antitumor, cytotoxic	276
34.	Halichondrin B	Marine sponge Halichondria okadai	Macrolides	Anti tubulin agents	Antitumor, cytotoxic	277
35.	Thiocoraline	Actinobacteria	Depsipeptide	Arrests the cell cycle in the G1 phase	Lung and colon cancer cell lines	278, 279
36.	Lucentamycins A-D	Actinobacteria	Peptides	Unknown	HCT-116 human colon cancer	280-282
37.	Tartrolon D	Actinobacteria	Macrodiolide	Cell cycle arrest, Apoptosis	Lung (A549), colon (HT29), and breast (MDA-MB-231)	283
38.	Lodopyridone	Actinobacteria	Alkaloid	Cell cycle arrest, Apoptosis	HCT-116 human colon cancer	284, 285
39.	Salinosporamide A	Actinomycetes	Lactone	Proteasome inhibitor	Multiple myeloma	286-290
40.	Staurosporine	Actinobacteria	Alkloids	Apoptosis inducer	Solid Tumors	291-293
41.	Gilvocarcin HE	Actinobacteria	Polyketide glycoside	Unknown	MCF-7, K562, and P388 cell lines	294, 295
42.	Lomaiviticin A and B	Actinobacteria	Glycosides	Growth inhibitors	Breast cancer	296-298
43.	Marineosin A	Actinomycete	Macrolide	Unknown	Leukemia and melanoma	299

Table 1. (Continued)

Sr. No.	Name of compound	Source of marine organisms	Chemical Class	Mechanism of action	Function	References
44.	Anhydrofusarubin	Fungii	Naphthoquinones	Unknown	Cytotoxic Activity against certain human cancer cell	300
45.	Austrocortirubin	Fungi	Quinone	Apoptosis	HCT116 cells	301
46.	Zygosporamide	Fungi	Peptide	Anti tubulin agents	CNS cancer	302
47.	Streptopyrrolidine	Actinobacteria	Benzyl pyrrolidine	Angiogenesis inhibitor	Breast cancer	303
48.	Ammosamide A and B	Actinomycetes	Alkaloids	Cell cycle arrest	Solid tumors	304
49.	Violaxanthin	Microalgae	Xanthophyll	MAPK	Colon cancer	305
50.	Fucoxanthin	Algae	Carotenoid	Cell Cycle Arrest, Apoptosis Induction, Angiogenesis Inhibition	Triple-Negative Breast Cancer	306
51.	Calothrixins A and B	Cyanobacteria	Alkaloids	Inhibits MAPK/ERK pathway Induce apoptosis	Leukemia	307

7. Approved Anticancer Marine Metabolites

This section contains updated information on approved medication as an anticancer agent (summarized in Table 2).

Table 2. List of approved marine metabolites as anticancer agents

Sr. No.	Name of compound	Source of marine organism	Chemical class	Function	Mechanism of action	References
1.	Cytarabine (Ara-C)	Sponge	Nucleoside	Cell cycle arrest	DNA polymerase inhibition	308
2.	Trabectedin (E7389)	Tunicate	Alkloid	Apoptosis	Block tumor cell proliferation and influence tumour microenvironment	309
3.	Brentuximab vedotin (SGN-35)	Mollusk/ Cyanobacteria	Peptide/ Antibody drug conjugate	Lymphomas	CD-30- directed antibody-cytotoxic drug conjugate Microtubule inhibitor	310, 311
4.	Lurbinectedin	Tunicate	Alkaloid	Solid tumors, Lipposarcoma, NSCLC, anti-inflammatory and immunomodulatory	Cell cycle arrest and apoptosis	312, 313
5.	Trabectedin	Tunicate E. Turbinate	Alkaloid	Liposarcoma, anti-inflammatory and immunomodulatory	Cell cycle arrest and apoptosis	312, 314
6.	AE-941 (Neovastat)	Shark cartilage	Peptide	NSCLC, renal cell carcinoma	Angiogenesis inhibitor, Matrix metalloproteinase inhibitors	315

Sr. No.	Name of compound	Source of marine organism	Chemical class	Function	Mechanism of action	References
7.	Nelarabine	Sponge	Nucleoside	Leukemias, lymphomas	Induce apoptosis, cell cycle arrest	316
8.	Fludarabine phosphate	Sponge	Nucleoside	Leukemias, lymphomas	Cell cycle arrest, apoptosis	317
9.	Eribulin mesylate	Sponge	Macrolide polyketide	Liposarcoma Breast cancer	Cell cycle arrest, apoptosis, Microtubule interfering agent	318
10.	Polatuzumab vedotin	Mollusk/ Cyanobacteria	Peptide/ Antibody drug conjugate	Lymphomas	G2/M arrest and apoptosis	319
11.	Enfortumab vedotin	Mollusk/ Cyanobacteria	Peptide/ Antibody drug conjugate	Solid tumors	Growth inhibitor	320
12.	Midostaurin	Tunicate/ Actinobacteria	Indolocarbazole	Leukemias	Inhibit cell proliferation	321
13.	Plitidepsin	Tunicate	Peptide	Multiple myeloma	Cell cycle arrest, apoptosis,	322
14.	Belantamab	Mandolin Mandolin/ Cyanobacteria	ADC/Peptide	Multiple myeloma	G2/M phase arrest and induces tumor cell apoptosis	323

8. Future Trends

Cancer incidence and mortality are still rising despite recent research efforts. The World Health Organization (WHO) reports that cancer is one of the top causes of death globally, with nearly 10 million deaths anticipated in 2020. As a result, cancer poses a serious threat to public health because it takes a significant toll on both human life and financial resources [324, 325]. In response to this public health issue, numerous experimental strategies have recently been put forth to address the difficulties related to the onset and spread of cancer. Regretfully, these efforts have not only resulted in minimal progress, but it is widely recognized that the medications presently employed for anticancer treatments cause noteworthy adverse reactions in these individuals, thereby compromising their standard of living. This environment has prompted researchers to look for novel biologically active substances that indicate viable approaches for creating medicinal agents that will be used to treat this illness [326].

Marine natural products and their derivatives are widely regarded as essential sources of innovative chemical structures with potential as medicinal agents due to their structural diversity and unique modes of action. One of the reasons for the vast spectrum of biological activities is the variety of secondary metabolites found in different organisms. Secondary metabolites function as defence or signaling molecules, playing crucial roles in an organism's interactions with other organisms and its environment. These metabolites, however, may also provide chemical methods that show promise for treating or even eliminating diseases [327].

Even though many compounds have been reported in the chapter, there are still many significant obstacles in the search for naturally occurring marine products with pharmacological qualities. These include the difficulty of accessing different oceanic regions, the scarcity of isolated compounds, and the lack of information on the genomes of several producer organisms. Innovative technologies that can help for obtaining, isolating, assessing, and identifying compounds and their producer organisms are likely to overcome the challenges. Extreme environment sampling, hybrid methods for nanomole structural determination, nanomole NMR techniques, database and computational chemistry, genome mining, molecular networking, and other technologies are some examples of these technologies. These methods will make it possible to find new bioactive marine compounds more quickly and will also help identify new molecules, which will enhance the effectiveness of currently available antitumor treatments and/or serve as a valuable source of novel chemical structures crucial to the creation of antitumor drugs. An attempt has been made to present marine organisms that produce cytotoxic, antineoplastic, and antiproliferative compounds in this chapter, along with the documented mechanisms of action that underlie their pharmacological actions [328, 329].

Moreover, cancer cells exhibit a wide range of genetic and epigenetic alterations, as well as their ability to control specific protein levels to enhance cell survival, proliferation, and/or inhibit natural mechanisms that cause cell death. Due to their ability to escape apoptosis and trigger metabolic stress, SERCA isoform expression levels have recently been the focus of intense cancer research, indicating that controlling these levels could be a viable target for cancer treatments [330]. Modern developments in cancer treatment using molecularly targeted therapies offer patients entirely new therapy options. The extraction of bioactive antineoplastic compounds from marine organisms provides a potential avenue for anticancer medications considered "undruggable." Even though developing and clinically testing these new molecular agents requires extensive research, the potential of these targeted therapies could represent a revolutionary leap for cancer patients with few treatment alternatives. One of the primary objectives of cancer research is to develop customized or combination treatments; therefore, more investigation is needed to clarify the cellular mechanisms or death pathways associated with the antineoplastic molecules originating from marine organisms [331].

Ultimately, there are still more significant obstacles to overcome in this field, such as (i) finding and characterizing novel organisms; (ii) gaining access to these organisms for scientific investigations; and (iii) setting up large-scale culture systems to produce sufficient biomass amounts for the extraction, purification, characterization, and assessment of novel molecules.

Conclusion

Global warming, hunger, and other environmental factors contribute to rising cancer rates. The American Cancer Society predicts that the worldwide burden of cancer will increase by 27 million new cases and 17.5 million deaths by 2050, driven by population growth and aging. Marine resources offer significant biochemical variety and lead to the development of novel anticancer medicines, accounting for around 18% of all marine natural products identified. We must evaluate many variables unique to marine-sourced medication research and development. The rise in publications on marine metabolites in recent years reflects this trend. This chapter discusses recently identified, derived, or manufactured compounds from marine metabolites that have shown potential as cancer therapeutics over the previous two decades. Marine metabolites, including alkaloids, polyketides, terpenes, peptides, and carbohydrates, are chemically diverse and complicated. Marine compounds have been shown to have anticancer properties, with varied mechanisms including reduction of cell growth, antimitotic activity (anti-tubulin effects), apoptosis/autophagy induction, and prevention of migration, invasion, and metastases.

Acknowledgments

The authors are thankful to Smt Kishortai Bhoyar College of Pharmacy, kamptee, for providing the infrastructural facility to carry out the review work.

References

[1] Yu, P., Xu, R., Yang, Z., Ye, T., Liu, Y., Li, S., Abramson, M. J., Kimlin, M., Guo, Y. Cancer and Ongoing Climate Change: Who Are the Most Affected? *ACS Environ.* Au 2023, *3* (1), 5–11. https://doi.org/10.1021/acsenvironau.2c00012.

[2] Sung, H., Ferlay, J., Siegel, R. L., Laversanne, M., Soerjomataram, I., Jemal, A., Bray, F. Global Cancer Statistics 2020: GLOBOCAN Estimates of Incidence and Mortality Worldwide for 36 Cancers in 185 Countries. *CA A Cancer J Clinicians* 2021, 71 (3), 209–249. https://doi.org/10.3322/caac.21660.

[3] Bashar, Mda., Begam, N. Breast Cancer Surpasses Lung Cancer as the Most Commonly Diagnosed Cancer Worldwide. *Indian J Cancer* 2022, 59 (3), 438. https://doi.org/10.4103/ ijc.IJC_83_21.

[4] Chaitanya Thandra, K., Barsouk, A., Saginala, K., Sukumar Aluru, J., Barsouk, A. *Epidemiology of Lung Cancer.* wo 2021, 25 (1), 45–52. https://doi.org/10.5114/wo.2021. 103829.

[5] Marino, P., Mininni, M., Deiana, G., Marino, G., Divella, R., Bochicchio, I., Giuliano, A., Lapadula, S., Lettini, A. R., Sanseverino, F. Healthy Lifestyle and Cancer Risk: Modifiable Risk Factors to Prevent Cancer. *Nutrients* 2024, 16 (6), 800. https://doi.org/10.3390/ nu16060800.

[6] Anand, U., Dey, A., Chandel, A. K. S., Sanyal, R., Mishra, A., Pandey, D. K., De Falco, V., Upadhyay, A., Kandimalla, R., Chaudhary, A., Dhanjal, J. K., Dewanjee, S., Vallamkondu, J., Pérez De La Lastra, J. M. Cancer Chemotherapy and beyond: Current Status, Drug Candidates, Associated Risks and Progress in Targeted Therapeutics. *Genes & Diseases* 2023, 10 (4), 1367–1401. https://doi.org/ 10.1016/j.gendis.2022.02.007.

[7] Xia, Y., Sun, M., Huang, H., Jin, W.-L. Drug Repurposing for Cancer Therapy. *Sig Transduct Target Ther* 2024, 9 (1), 92. https://doi.org/10.1038/s41392-024-01808-1.

[8] Sordo-Bahamonde, C., Lorenzo-Herrero, S., Gonzalez-Rodriguez, A. P., Martínez-Pérez, A., Rodrigo, J. P., García-Pedrero, J. M., Gonzalez, S. Chemo-Immunotherapy: A New Trend in Cancer Treatment. *Cancers* 2023, 15 (11), 2912. https://doi.org/10.3390/cancers 15112912.

[9] Talib, W. H., Alsayed, A. R., Barakat, M., Abu-Taha, M. I., Mahmod, A. I. Targeting Drug Chemo-Resistance in Cancer Using Natural Products. *Biomedicines* 2021, 9 (10), 1353. https://doi.org/ 10.3390/biomedicines9101353.

[10] Debela, D. T., Muzazu, S. G., Heraro, K. D., Ndalama, M. T., Mesele, B. W., Haile, D. C., Kitui, S. K., Manyazewal, T. New Approaches and Procedures for Cancer Treatment: Current Perspectives. SAGE Open Medicine 2021, 9, 205031212110343. https://doi.org /10. 1177/ 20503121211034366.

[11] Greenwell, M., Rahman, P. K. S. M. Medicinal Plants: Their Use in Anticancer Treatment. *Int J Pharm Sci Res* 2015, 6 (10), 4103–4112. https://doi.org/10.13040/IJPSR.0975-8232.6(10).4103-12.

[12] Nadaf, R. *Medicinal Plants & Health Care Systems*, 2024. Intergirty Media-The Publisher, London-New Delhi.

[13] The International Natural Product Sciences Taskforce, Atanasov, A. G., Zotchev, S. B., Dirsch, V. M., Supuran, C. T. Natural Products in Drug Discovery: Advances and Opportunities. *Nat Rev Drug Discov* 2021, 20 (3), 200–216. https://doi.org/10.1038/ s41573-020-00114-z.

[14] Assidi, M., Buhmeida, A., Budowle, B. Medicine and Health of 21st Century: Not Just a High Biotech-Driven Solution. *npj Genom. Med.* 2022, 7 (1), 67. https://doi.org/10.1038/ s41525-022-00336-7.

[15] Tanaka, T., Shnimizu, M., Moriwaki, H. Cancer Chemoprevention by Carotenoids. *Molecules* 2012, 17 (3), 3202–3242. https://doi.org/10.3390/molecules17033202.

[16] Crupi, P., Faienza, M. F., Naeem, M. Y., Corbo, F., Clodoveo, M. L., Muraglia, M. Overview of the Potential Beneficial Effects of Carotenoids on Consumer Health and Well-Being. *Antioxidants* 2023, 12 (5), 1069. https://doi.org/10.3390/antiox12051069.

[17] Marino, P., Pepe, G., Basilicata, M. G., Vestuto, V., Marzocco, S., Autore, G., Procino, A., Gomez-Monterrey, I. M., Manfra, M., Campiglia, P. Potential Role of Natural Antioxidant Products in Oncological Diseases. *Antioxidants* 2023, 12 (3), 704. https://doi. org/10.3390/antiox12030704.
[18] Ekor, M. The Growing Use of Herbal Medicines: Issues Relating to Adverse Reactions and Challenges in Monitoring Safety. *Front. Pharmacol*. 2014, 4. https://doi.org/10.3389/fphar. 2013.00177.
[19] Karthikeyan, A., Joseph, A., Nair, B. G. Promising Bioactive Compounds from the Marine Environment and Their Potential Effects on Various Diseases. *Journal of Genetic Engineering and Biotechnology* 2022, 20 (1), 14. https://doi.org/10.1186/s43141-021-00290-4.
[20] Ghosh, S., Sarkar, T., Pati, S., Kari, Z. A., Edinur, H. A., Chakraborty, R. Novel Bioactive Compounds From Marine Sources as a Tool for Functional Food Development. *Front. Mar. Sci*. 2022, 9, 832957. https://doi.org/10.3389/fmars.2022.832957.
[21] Tangerina, M. M. P., Correa, H., Haltli, B., Vilegas, W., Kerr, R. G. Bioprospecting from Cultivable Bacterial Communities of Marine Sediment and Invertebrates from the Underexplored Ubatuba Region of Brazil. *Arch Microbiol* 2017, 199 (1), 155–169. https://doi.org/10.1007/s00203-016-1290-9.
[22] Pinnaka, A. K., Tanuku, N. R. S. Marine Microbial Diversity for Sustainable Development. In *Microbial Diversity in Ecosystem Sustainability and Biotechnological Applications,* Satyanarayana, T., Johri, B. N., Das, S. K., Eds., Springer Singapore: Singapore, 2019, pp 117–158. https://doi.org/10. 1007/978-981-13-8315-1_5.
[23] Boeuf, G. Marine Biodiversity Characteristics. Comptes Rendus. *Biologies* 2011, 334 (5–6), 435–440. https://doi.org/10.1016/j.crvi.2011.02.009.
[24] Wafar, M., Venkataraman, K., Ingole, B., Ajmal Khan, S., LokaBharathi, P. State of Knowledge of Coastal and Marine Biodiversity of Indian Ocean Countries. *PLoS ONE* 2011, 6 (1), e14613. https://doi.org/10.1371/journal.pone.0014613.
[25] Haefner, B. Drugs from the Deep: Marine Natural Products as Drug Candidates. *Drug Discovery Today* 2003, 8 (12), 536–544. https://doi.org/10.1016/S1359-6446(03)02713-2.
[26] Costello, M. J., Chaudhary, C. Marine Biodiversity, Biogeography, Deep-Sea Gradients, and Conservation. *Current Biology* 2017, 27 (11), R511–R527. https://doi.org/10.1016/ j.cub.2017.04.060.
[27] Busch, K., Slaby, B. M., Bach, W., Boetius, A., Clefsen, I., Colaço, A., Creemers, M., Cristobo, J., Federwisch, L., Franke, A., Gavriilidou, A., Hethke, A., Kenchington, E., Mienis, F., Mills, S., Riesgo, A., Ríos, P., Roberts, E. M., Sipkema, D., Pita, L., Schupp, P. J., Xavier, J., Rapp, H. T., Hentschel, U. Biodiversity, Environmental Drivers, and Sustainability of the Global Deep-Sea Sponge Microbiome. *Nat Commun* 2022, 13 (1), 5160. https://doi.org/10.1038/s41467-022-32684-4.
[28] Jimenez, P. C., Wilke, D. V., Branco, P. C., Bauermeister, A., Rezende-Teixeira, P., Gaudêncio, S. P., Costa-Lotufo, L. V. Enriching Cancer Pharmacology with Drugs of Marine Origin. *British J Pharmacology* 2020, 177 (1), 3–27. https://doi.org/10.1111/bph. 14876.
[29] Wright, G. D. Unlocking the Potential of Natural Products in Drug Discovery. *Microbial Biotechnology* 2019, 12 (1), 55–57. https://doi.org/10.1111/1751-7915.13351.
[30] Al-Adilah, H., Feiters, M. C., Carpenter, L. J., Kumari, P., Carrano, C. J., Al-Bader, D., Küpper, F. C. Halogens in Seaweeds: Biological and Environmental Significance. *Phycology* 2022, 2 (1), 132–171. https://doi.org/10.3390/phycology2010009.
[31] Barreca, M., Spanò, V., Montalbano, A., Cueto, M., Díaz Marrero, A. R., Deniz, I., Erdoğan, A., Lukić Bilela, L., Moulin, C., Taffin-de-Givenchy, E., Spriano, F., Perale, G., Mehiri, M., Rotter, A., P. Thomas, O., Barraja, P., Gaudêncio, S. P., Bertoni, F. Marine Anticancer Agents: An Overview with a Particular Focus on Their Chemical Classes. *Marine Drugs* 2020, 18 (12), 619. https://doi.org/10. 3390/md18120619.
[32] Georgii, A. D. N. P., Teixeira, V. L. Dictyota and Canistrocarpus Brazilian Brown Algae and Their Bioactive Diterpenes—A Review. *Marine Drugs* 2023, 21 (9), 484. https://doi.org/10. 3390/md21090484.
[33] Eze, O. C., Berebon, D. P., Emencheta, S. C., Evurani, S. A., Okorie, C. N., Balcão, V. M., Vila, M. M. D. C. Therapeutic Potential of Marine Probiotics: A Survey on the Anticancer and Antibacterial Effects of Pseudoalteromonas Spp. *Pharmaceuticals* 2023, 16 (8), 1091. https://doi.org/10.3390/ ph16081091.

[34] Malve, H. Exploring the Ocean for New Drug Developments: Marine Pharmacology. *J Pharm Bioall Sci* 2016, 8 (2), 83. https://doi.org/10.4103/0975-7406.171700.

[35] Cario, A., Oliver, G. C., Rogers, K. L. Exploring the Deep Marine Biosphere: Challenges, Innovations, and Opportunities. *Front. Earth Sci.* 2019, 7, 225. https://doi.org/10.3389/ feart.2019.00225.

[36] Gaw, S., Thomas, K. V., Hutchinson, T. H. Sources, Impacts and Trends of Pharmaceuticals in the Marine and Coastal Environment. *Phil. Trans. R. Soc. B* 2014, 369 (1656), 20130572. https://doi.org/10.1098/rstb.2013.0572.

[37] Khan, A. H. A., Barros, R. Pharmaceuticals in Water: Risks to Aquatic Life and Remediation Strategies. *Hydrobiology* 2023, 2 (2), 395–409. https://doi.org/10.3390/ hydrobiology2020026.

[38] Krishna, R., Tejaswini, P., Milind, J. Marine Plants: Extraction and Their Medicinal Importance. *International Journal of Pharmaceutical Chemistry and Analysis* 2023, 2023, 19925. https://doi.org/10.18231/j.ijpca.2023.031.

[39] Pereira, A., Silva, L., Laranjeiro, C., Lino, C., Pena, A. Selected Pharmaceuticals in Different Aquatic Compartments: Part I—Source, Fate and Occurrence. *Molecules* 2020, 25 (5), 1026. https://doi.org/10.3390/molecules25051026.

[40] Boyd, C. E., D'Abramo, L. R., Glencross, B. D., Huyben, D. C., Juarez, L. M., Lockwood, G. S., McNevin, A. A., Tacon, A. G. J., Teletchea, F., Tomasso, J. R., Tucker, C. S., Valenti, W. C. Achieving Sustainable Aquaculture: Historical and Current Perspectives and Future Needs and Challenges. *J World Aquaculture Soc* 2020, 51 (3), 578–633. https://doi.org/10.1111/jwas.12714.

[41] Townsend, M., Davies, K., Hanley, N., Hewitt, J. E., Lundquist, C. J., Lohrer, A. M. The Challenge of Implementing the Marine Ecosystem Service Concept. *Front. Mar. Sci.* 2018, 5, 359. https://doi.org/10.3389/fmars.2018.00359.

[42] Lauritano, C., Andersen, J. H., Hansen, E., Albrigtsen, M., Escalera, L., Esposito, F., Helland, K., Hanssen, K. Ø., Romano, G., Ianora, A. Bioactivity Screening of Microalgae for Antioxidant, Anti-Inflammatory, Anticancer, Anti-Diabetes, and Antibacterial Activities. *Front. Mar. Sci.* 2016, 3. https://doi.org/10.3389/fmars.2016.00068.

[43] Mondal, A., Bose, S., Banerjee, S., Patra, J. K., Malik, J., Mandal, S. K., Kilpatrick, K. L., Das, G., Kerry, R. G., Fimognari, C., Bishayee, A. Marine Cyanobacteria and Microalgae Metabolites—A Rich Source of Potential Anticancer Drugs. *Marine Drugs* 2020, 18 (9), 476. https://doi.org/10.3390/md18090476.

[44] Matos, J., Cardoso, C., Serralheiro, M. L., Bandarra, N. M., Afonso, C. Seaweed Bioactives Potential as Nutraceuticals and Functional Ingredients: A Review. *Journal of Food Composition and Analysis* 2024, 133, 106453. https://doi.org/10.1016/j.jfca.2024.106453.

[45] Choudhary, B., Chauhan, O. P., Mishra, A. Edible Seaweeds: A Potential Novel Source of Bioactive Metabolites and Nutraceuticals With Human Health Benefits. *Front. Mar. Sci.* 2021, 8, 740054. https://doi.org/10.3389/fmars.2021.740054.

[46] Peñalver, R., Lorenzo, J. M., Ros, G., Amarowicz, R., Pateiro, M., Nieto, G. Seaweeds as a Functional Ingredient for a Healthy Diet. *Marine Drugs* 2020, 18 (6), 301. https://doi.org/10.3390/md18060301.

[47] Ranjani, A., Dhanasekaran, D., Gopinath, P. M. An Introduction to Actinobacteria. In *Actinobacteria - Basics and Biotechnological Applications*, Dhanasekaran, D., Jiang, Y., Eds., InTech, 2016. https://doi.org/10.5772/62329.

[48] Selim, M. S. M., Abdelhamid, S. A., Mohamed, S. S. Secondary Metabolites and Biodiversity of Actinomycetes. *Journal of Genetic Engineering and Biotechnology* 2021, 19 (1), 72. https://doi.org/10.1186/s43141-021-00156-9.

[49] Hazarika, S. N., Thakur, D. Actinobacteria. In Ben*eficial Microbes in Agro-Ecology*, Elsevier, 2020, pp 443–476. https://doi.org/10.1016/B978-0-12-823414-3.00021-6.

[50] Tiwari, K., Gupta, R. K. Bioactive Metabolites from Rare Actinomycetes. In *Studies in Natural Products Chemistry*, Elsevier, 2014, Vol. 41, pp 419–512. https://doi.org/10.1016/ B978-0-444-63294-4.

[51] Yan, S., Zeng, M., Wang, H., Zhang, H. Micromonospora : A Prolific Source of Bioactive Secondary Metabolites with Therapeutic Potential. *J. Med. Chem.* 2022, 65 (13), 8735–8771. https://doi.org/10.1021/acs.jmedchem.2c00626.

[52] Cho, J. Y., Williams, P. G., Kwon, H. C., Jensen, P. R., Fenical, W. Lucentamycins A–D, Cytotoxic Peptides from the Marine-Derived Actinomycete Nocardiopsis Lucentensis. *J. Nat. Prod.* 2007, 70 (8), 1321–1328. https://doi.org/10.1021/np070101b.

[53] Mueller, A.-L., Brockmueller, A., Fahimi, N., Ghotbi, T., Hashemi, S., Sadri, S., Khorshidi, N., Kunnumakkara, A. B., Shakibaei, M. Bacteria-Mediated Modulatory Strategies for Colorectal Cancer Treatment. *Biomedicines* 2022, 10 (4), 832. https://doi.org/10.3390/ biomedicines10040832.

[54] Cha, J. W., Park, J.-S., Sim, T., Nam, S.-J., Kwon, H. C., Del Valle, J. R., Fenical, W. Structure Assignment of Lucentamycin E and Revision of the Olefin Geometries of the Marine-Derived Lucentamycins. *J. Nat. Prod.* 2012, 75 (9), 1648–1651. https://doi.org/ 10.1021/np3003854.

[55] Pérez, M., Crespo, C., Schleissner, C., Rodríguez, P., Zúñiga, P., Reyes, F. Tartrolon D, a Cytotoxic Macrodiolide from the Marine-Derived Actinomycete Streptomyces Sp. MDG-04-17-069. *J. Nat. Prod.* 2009, 72 (12), 2192–2194. https://doi.org/10.1021/np9006603.

[56] Hillman, P. F., Lee, J. Y., Lee, J., Lee, J., Oh, D.-C., Fenical, W., Nam, S.-J. Lodopyridones D-G from a Marine-Derived Bacterium Saccharomonospora Sp. *Natural Product Research* 2023, 1–6. https://doi.org/10.1080/14786419.2023.2173192.

[57] Maloney, K. N., MacMillan, J. B., Kauffman, C. A., Jensen, P. R., DiPasquale, A. G., Rheingold, A. L., Fenical, W. Lodopyridone, a Structurally Unprecedented Alkaloid from a Marine Actinomycete. *Org. Lett.* 2009, 11 (23), 5422–5424. https://doi.org/ 10.1021/ol901997k.

[58] Serrano-Aparicio, N., Moliner, V., Świderek, K. On the Origin of the Different Reversible Characters of Salinosporamide A and Homosalinosporamide A in the Covalent Inhibition of the Human 20S Proteasome. *ACS Catal.* 2021, 11 (18), 11806–11819. https://doi.org/10.1021/acscatal.1c02614.

[59] Pokharkar, O., Zyryanov, G. V., Tsurkan, M. V. Natural Products from Marine Actinomycete Genus Salinispora Might Inhibit 3CLpro and PLpro Proteins of SARS-CoV-2: An In Silico Evidence. *Microbiology Research* 2023, 14 (4), 1907–1941. https://doi.org/10.3390/microbiolres14040130.

[60] Chhun, A., Sousoni, D., Del Mar Aguiló-Ferretjans, M., Song, L., Corre, C., Christie-Oleza, J. A. *Phytoplankton Trigger the Production of Cryptic Metabolites in the Marine Actinobacteria Salinispora Tropica.* May 20, 2020. https://doi.org/10.1101/2020. 05.18.103358.

[61] Fenical, W., Jensen, P. R., Palladino, M. A., Lam, K. S., Lloyd, G. K., Potts, B. C. Discovery and Development of the Anticancer Agent Salinosporamide A (NPI-0052). *Bioorganic & Medicinal Chemistry* 2009, 17 (6), 2175–2180. https://doi.org/10.1016/ j.bmc.2008.10.075.

[62] Gulder, T. A. M., Moore, B. S. Salinosporamide Natural Products: Potent 20 S Proteasome Inhibitors as Promising Cancer Chemotherapeutics. *Angew Chem Int Ed* 2010, 49 (49), 9346–9367. https://doi.org/10.1002/anie.201000728.

[63] Serrano-Aparicio, N., Moliner, V., Świderek, K. On the Origin of the Different Reversible Characters of Salinosporamide A and Homosalinosporamide A in the Covalent Inhibition of the Human 20S Proteasome. *ACS Catal.* 2021, 11 (18), 11806–11819. https://doi.org/ 10.1021/acscatal.1c02614.

[64] Yang, Q., Gao, W., Li, X., Li, X., Zhou, X., Li, W., Zhou, C., Luo, A., Liu, Z. Targeting ABCA1 via Extracellular Vesicle-Encapsulated Staurosporine as a Therapeutic Strategy to Enhance Radiosensitivity. *Adv Healthcare Materials* 2024, 13 (16), 2400381. https://doi.org/10.1002/adhm.202400381.

[65] Zambrano, J. N., Williams, C. J., Williams, C. B., Hedgepeth, L., Burger, P., Dilday, T., Eblen, S. T., Armeson, K., Hill, E. G., Yeh, E. S. Staurosporine, an Inhibitor of Hormonally up-Regulated Neu-Associated Kinase. *Oncotarget* 2018, 9 (89), 35962–35973. https://doi.org/10.18632/oncotarget.26311.

[66] Ho, T. T., Tran, Q. T., Chai, C. L. The Polypharmacology of Natural Products. *Future Med. Chem.* 2018, 10 (11), 1361–1368. https://doi.org/10.4155/fmc-2017-0294.

[67] Hou, J., Liu, P., Qu, H., Fu, P., Wang, Y., Wang, Z., Li, Y., Teng, X., Zhu, W. Gilvocarcin HE: A New Polyketide Glycoside from Streptomyces Sp. *J Antibiot* 2012, 65 (10), 523–526. https://doi.org/ 10.1038/ja.2012.61.

[68] Karthikeyan, A., Joseph, A., Nair, B. G. Promising Bioactive Compounds from the Marine Environment and Their Potential Effects on Various Diseases. *Journal of Genetic Engineering and Biotechnology* 2022, 20 (1), 14. https://doi.org/10.1186/s43141-021-00290-4.

[69] Li, K., Cai, J., Su, Z., Yang, B., Liu, Y., Zhou, X., Huang, J., Tao, H. Glycosylated Natural Products From Marine Microbes. *Front. Chem.* 2020, 7, 879. https://doi.org/10.3389/ fchem.2019.00879.
[70] Herzon, S. B. The Mechanism of Action of (−)-Lomaiviticin A. *Acc. Chem. Res.* 2017, 50 (10), 2577–2588. https://doi.org/10.1021/acs.accounts.7b00347.
[71] He, H., Ding, W.-D., Bernan, V. S., Richardson, A. D., Ireland, C. M., Greenstein, M., Ellestad, G. A., Carter, G. T. Lomaiviticins A and B, Potent Antitumor Antibiotics from Micromonospora l Omaivitiensis. *J. Am. Chem. Soc.* 2001, 123 (22), 5362–5363. https://doi.org/10.1021/ja010129o.
[72] Biris-Dorhoi, E.-S., Michiu, D., Pop, C. R., Rotar, A. M., Tofana, M., Pop, O. L., Socaci, S. A., Farcas, A. C. Macroalgae—A Sustainable Source of Chemical Compounds with Biological Activities. *Nutrients* 2020, 12 (10), 3085. https://doi.org/10.3390/nu12103085.
[73] Menaa, F., Wijesinghe, U., Thiripuranathar, G., Althobaiti, N. A., Albalawi, A. E., Khan, B. A., Menaa, B. Marine Algae-Derived Bioactive Compounds: A New Wave of Nanodrugs? *Marine Drugs* 2021, 19 (9), 484. https://doi.org/10.3390/md19090484.
[74] Chale-Dzul, J., Freile-Pelegrín, Y., Robledo, D., Moo-Puc, R. Protective Effect of Fucoidans from Tropical Seaweeds against Oxidative Stress in HepG2 Cells. *J Appl Phycol* 2017, 29 (5), 2229–2238. https://doi.org/10.1007/s10811-017-1194-3.
[75] Ferdous, U. T., Yusof, Z. N. B. Medicinal Prospects of Antioxidants From Algal Sources in Cancer Therapy. *Front. Pharmacol.* 2021, 12, 593116. https://doi.org/10.3389/fphar. 2021.593116.
[76] Ávila-Román, J., García-Gil, S., Rodríguez-Luna, A., Motilva, V., Talero, E. Anti-Inflammatory and Anticancer Effects of Microalgal Carotenoids. *Marine Drugs* 2021, 19 (10), 531. https://doi.org/ 10.3390/md19100531.
[77] Jayappriyan, K. R., Rajkumar, R., Venkatakrishnan, V., Nagaraj, S., Rengasamy, R. In Vitro Anticancer Activity of Natural β-Carotene from Dunaliella Salina EU5891199 in PC-3 Cells. *Biomedicine & Preventive Nutrition* 2013, 3 (2), 99–105. https://doi.org/10.1016/j.bionut. 2012.08.003.
[78] Kusaikin, M. I., Ermakova, S. P., Shevchenko, N. M., Isakov, V. V., Gorshkov, A. G., Vereshchagin, A. L., Grachev, M. A., Zvyagintseva, T. N. Structural Characteristics and Antitumor Activity of a New Chrysolaminaran from the Diatom Alga Synedra Acus. *Chem Nat Compd* 2010, 46 (1), 1–4. https://doi.org/10.1007/s10600-010-9510-z.
[79] Ebrahimi Nigjeh, S., Yusoff, F. M., Mohamed Alitheen, N. B., Rasoli, M., Keong, Y. S., Omar, A. R. B. Cytotoxic Effect of Ethanol Extract of Microalga, Chaetoceros Calcitrans , and Its Mechanisms in Inducing Apoptosis in Human Breast Cancer Cell Line. *BioMed Research International* 2013, 2013, 1–8. https://doi.org/10.1155/2013/783690 .
[80] Martínez Andrade, K., Lauritano, C., Romano, G., Ianora, A. Marine Microalgae with Anti-Cancer Properties. *Marine Drugs* 2018, 16 (5), 165. https://doi.org/10.3390/ md16050165.
[81] Reddy, P., Urban, S. Meroditerpenoids from the Southern Australian Marine Brown Alga Sargassum Fallax. *Phytochemistry* 2009, 70 (2), 250–255. https://doi.org/10.1016/j. phytochem.2008.12.007.
[82] Cordell, G. Changing Strategies in Natural Products Chemistry. *Phytochemistry* 1995, 40 (6), 1585–1612. https://doi.org/10.1016/0031-9422(95)00444-C.
[83] Mandwal, A. Diverse Survival Functions of Secondary Metabolites in Nature. In *Secondary Metabolites - Trends and Reviews*, Vijayakumar, R., Selvapuram Sudalaimuthu Raja, S., Eds., IntechOpen, 2022. https://doi.org/10.5772/intechopen.101977.
[84] Bhatnagar, I., Kim, S.-K. Immense Essence of Excellence: Marine Microbial Bioactive Compounds. *Marine Drugs* 2010, 8 (10), 2673–2701. https://doi.org/10.3390/md8102673.
[85] Romanenko, L. A., Uchino, M., Kalinovskaya, N. I., Mikhailov, V. V. Isolation, Phylogenetic Analysis and Screening of Marine Mollusc-Associated Bacteria for Antimicrobial, Hemolytic and Surface Activities. *Microbiological Research* 2008, 163 (6), 633–644. https://doi.org/10.1016/j.micres.2006. 10.001.
[86] Bowman, J. P. Bioactive Compound Synthetic Capacity and Ecological Significance of Marine Bacterial Genus Pseudoalteromonas. *Marine Drugs* 2007, 5 (4), 220–241. https://doi.org/10.3390/ md504220.
[87] Khalifa, S. A. M., Elias, N., Farag, M. A., Chen, L., Saeed, A., Hegazy, M.-E. F., Moustafa, M. S., Abd El-Wahed, A., Al-Mousawi, S. M., Musharraf, S. G., Chang, F.-R., Iwasaki, A., Suenaga, K.,

Alajlani, M., Göransson, U., El-Seedi, H. R. Marine Natural Products: A Source of Novel Anticancer Drugs. *Marine Drugs* 2019, 17 (9), 491. https://doi.org/10.3390/md17090491.

[88] Kr, D. A Potential Source of Anti-Cancer Drugs: Marine Raw Materials. *IPCM* 2023, 7 (2), 1–7. https://doi.org/10.23880/ipcm-16000260.

[89] Saeed, A. F. U. H., Su, J., Ouyang, S. Marine-Derived Drugs: Recent Advances in Cancer Therapy and Immune Signaling. *Biomedicine & Pharmacotherapy* 2021, 134, 111091. https://doi.org/10.1016/j.biopha.2020.111091.

[90] Sugumaran, A., Pandiyan, R., Kandasamy, P., Antoniraj, M. G., Navabshan, I., Sakthivel, B., Dharmaraj, S., Chinnaiyan, S. K., Ashokkumar, V., Ngamcharussrivichai, C. Marine Biome-Derived Secondary Metabolites, a Class of Promising Antineoplastic Agents: A Systematic Review on Their Classification, Mechanism of Action and Future Perspectives. *Science of The Total Environment* 2022, 836, 155445. https://doi.org/10.1016/ j.scitotenv.2022.155445.

[91] Wang, L., Groβe, T., Stevens, H., Brinkhoff, T., Simon, M., Liang, L., Bitzer, J., Bach, G., Zeeck, A., Tokuda, H., Lang, S. Bioactive Hydroxyphenylpyrrole-Dicarboxylic Acids from a New Marine Halomonas Sp.: Production and Structure Elucidation. *Appl Microbiol Biotechnol* 2006, 72 (4), 816–822. https://doi.org/10.1007/s00253-006-0370-1.

[92] El-Garawani, I. M., El-Sabbagh, S. M., Abbas, N. H., Ahmed, H. S., Eissa, O. A., Abo-Atya, D. M., Khalifa, S. A. M., El-Seedi, H. R. A Newly Isolated Strain of Halomonas Sp. (HA1) Exerts Anticancer Potential via Induction of Apoptosis and G2/M Arrest in Hepatocellular Carcinoma (HepG2) Cell Line. *Sci Rep* 2020, 10 (1), 14076. https://doi.org/10.1038/s41598-020-70945-8.

[93] Ciavatta, M. L., Lefranc, F., Vieira, L. M., Kiss, R., Carbone, M., Van Otterlo, W. A. L., Lopanik, N. B., Waeschenbach, A. The Phylum Bryozoa: From Biology to Biomedical Potential. *Marine Drugs* 2020, 18 (4), 200. https://doi.org/10.3390/md18040200.

[94] Figuerola, B., Gordon, D. P., Cristobo, J. New Deep Cheilostomata (Bryozoa) Species from the Southwestern Atlantic: Shedding Light in the Dark. *Zootaxa* 2018, 4375 (2). https://doi.org/10.11646/zootaxa.4375.2.3.

[95] Figuerola, B., Avila, C. The Phylum Bryozoa as a Promising Source of Anticancer Drugs. *Marine Drugs* 2019, 17 (8), 477. https://doi.org/10.3390/md17080477.

[96] Bock, P. E., Gordon, D. P.Phylum Bryozoa Ehrenberg, 1831. In: Zhang, Z.-Q. (Ed.) Animal Biodiversity: An Outline of Higher-Level Classification and Survey of Taxonomic Richness (Addenda 2013) *Zootaxa* 2013, 3703 (1). https://doi.org/10.11646/zootaxa.3703.1.14.

[97] Jassim YA, Awadh EFA, Al-Amery SMH. A Review of General Properties of Blue-Green Algae (*Cyanobacteria*). BCS [Internet]. 2023 Apr 1 [cited 2024 Jun 29],2(2). Available from: https://journals.irapa.org/index.php/BCS/article/view/397.

[98] Gaysina LA, Saraf A, Singh P. Cyanobacteria in Diverse Habitats. In: *Cyanobacteria* [Internet]. Elsevier, 2019 [cited 2024 Jun 29]. p. 1–28. Available from: https://linkinghub.elsevier.com/retrieve/pii/B9780128146675000015.

[99] Aguilera, A., Haakonsson, S., Martin, M. V., Salerno, G. L., Echenique, R. O. Bloom-Forming Cyanobacteria and Cyanotoxins in Argentina: A Growing Health and Environmental Concern. *Limnologica* 2018, 69, 103–114. https://doi.org/10.1016/ j.limno.2017.10.006.

[100] Sánchez-Baracaldo, P. Origin of Marine Planktonic Cyanobacteria. *Sci Rep* 2015, 5 (1), 17418. https://doi.org/10.1038/srep17418.

[101] Robles-Bañuelos, B., Durán-Riveroll, L. M., Rangel-López, E., Pérez-López, H. I., González-Maya, L. Marine Cyanobacteria as Sources of Lead Anticancer Compounds: A Review of Families of Metabolites with Cytotoxic, Antiproliferative, and Antineoplastic Effects. *Molecules* 2022, 27 (15), 4814. https://doi.org/10.3390/ molecules27154814.

[102] Cai, W., Matthew, S., Chen, Q.-Y., Paul, V. J., Luesch, H. Discovery of New A- and B-Type Laxaphycins with Synergistic Anticancer Activity. *Bioorganic & Medicinal Chemistry* 2018, 26 (9), 2310–2319. https://doi.org/10.1016/j.bmc.2018.03.022.

[103] Prabakaran, G., Sampathkumar, P., Kavisri, M., Moovendhan, M. Extraction and Characterization of Phycocyanin from Spirulina Platensis and Evaluation of Its Anticancer, Antidiabetic and Antiinflammatory Effect. *International Journal of Biological Macromolecules* 2020, 153, 256–263. https://doi.org/10.1016/j.ijbiomac.2020.03.009.

[104] Shishido, T. K., Popin, R. V., Jokela, J., Wahlsten, M., Fiore, M. F., Fewer, D. P., Herfindal, L., Sivonen, K. Dereplication of Natural Products with Antimicrobial and Anticancer Activity from Brazilian Cyanobacteria. *Toxins* (Basel) 2019, 12 (1), 12. https://doi.org/10.3390/toxins12010012.

[105] Fathoni, I., Petitbois, J. G., Alarif, W. M., Abdel-Lateff, A., Al-Lihaibi, S. S., Yoshimura, E., Nogata, Y., Vairappan, C. S., Sholikhah, E. N., Okino, T. Bioactivities of Lyngbyabellins from Cyanobacteria of Moorea and Okeania Genera. *Molecules* 2020, 25 (17), 3986. https://doi.org/10.3390/molecules25173986.

[106] Wunder, A., Rothemund, M., Schobert, R. Synthesis and Anticancer Activity of the Proposed Structure of Caldoramide, an N-Peptidyltetramate from the Cyanobacterium Caldora Penicillata. *Tetrahedron* 2018, 74 (38), 5138–5142. https://doi.org/10.1016/j.tet. 2018.04.004.

[107] Fayyad, R., Mohammed Ali, A., Dwaish, A., Khayoon, A. Anticancer Activity of Spirulina Platensis Methanolic Extracts against L20b and Mcf7 Human Cancer Cell Lines. *Plant Archives* 2019, 19, 1419–1426.

[108] Elkomy, R. G., Ismail, M. M. Crude Sulfated Polysaccharides Extracted from Marine Cyanobacterium Oscillatoria Simplicissima with Evaluation Antioxidant and Cytotoxic Activities. *IJM* 2021. https://doi.org/10.18502/ijm.v13i4.6983.

[109] Safavi, M., Nowruzi, B., Estalaki, S., Shokri, M. Biological Activity of Methanol Extract from Nostoc Sp. N42 and Fischerella Sp. S29 Isolated from Aquatic and Terrestrial Ecosystems. *Inter J Algae* 2019, 21 (4), 373–391. https://doi.org/10.1615/InterJAlgae. v21.i4.80.

[110] Perera, R. M. T. D., Herath, K. H. I. N. M., Sanjeewa, K. K. A., Jayawardena, T. U. Recent Reports on Bioactive Compounds from Marine Cyanobacteria in Relation to Human Health Applications. *Life* 2023, 13 (6), 1411. https://doi.org/10.3390/life13061411.

[111] Zhang, H.-H., Zhang, X.-K., Si, R.-R., Shen, S.-C., Liang, T.-T., Fan, T.-T., Chen, W., Xu, L.-H., Han, B.-N. Chemical and Biological Study of Novel Aplysiatoxin Derivatives from the Marine Cyanobacterium Lyngbya Sp. *Toxins* 2020, 12 (11), 733. https://doi.org/10.3390/toxins12110733.

[112] Shiota, I., Iwasaki, A., Sumimoto, S., Tomoda, H., Suenaga, K. Caldorin, a New Polyketide from the Marine Cyanobacterium Caldora Penicillata. *Tetrahedron Letters* 2018, 59 (13), 1261–1263. https://doi.org/10.1016/j.tetlet.2018.02.046.

[113] Bouley, T. A., Machalaba, C., Keast, J., Gerwick, W. H., Fleming, L. E. Marine Biotechnology: A One Health Approach to Linking Life on Land to Life Underwater. In *Oceans and Human Health*, Elsevier, 2023, pp 149–180. https://doi.org/10.1016/B978-0-323-95227-9.00024-5.

[114] Cunliffe, M. Who Are the Marine Fungi? *Environmental Microbiology* 2023, 25 (1), 131–134. https://doi.org/10.1111/1462-2920.16240.

[115] Gonçalves, M. F. M., Esteves, A. C., Alves, A. Marine Fungi: Opportunities and Challenges. *Encyclopedia* 2022, 2 (1), 559–577. https://doi.org/10.3390/encyclopedia2010 037.

[116] Burgaud, G., Edgcomb, V., Hassett, B. T., Kumar, A., Li, W., Mara, P., Peng, X., Philippe, A., Phule, P., Prado, S., Quéméner, M., Roullier, C. Marine Fungi. In The Marine Microbiome, Stal, L. J., Cretoiu, M. S., Eds., *The Microbiomes of Humans, Animals, Plants, and the Environment*, Springer International Publishing: Cham, 2022, Vol. 3, pp 243–295. https://doi.org/10.1007/978-3-030-90383-1_5.

[117] Peng, X., Amend, A. S., Baltar, F., Blanco-Bercial, L., Breyer, E., Burgaud, G., Cunliffe, M., Edgcomb, V. P., Grossart, H., Mara, P., Masigol, H., Pang, K., Retter, A., Roberts, C., Van Bleijswijk, J., Walker, A. K., Whitner, S. Planktonic Marine Fungi: A Review. JGR *Biogeosciences* 2024, 129 (3), e2023JG007887. https://doi.org/10.1029/ 2023JG007887.

[118] Helber, S. B., Hoeijmakers, D. J. J., Muhando, C. A., Rohde, S., Schupp, P. J. Sponge Chemical Defenses Are a Possible Mechanism for Increasing Sponge Abundance on Reefs in Zanzibar. *PLoS ONE* 2018, 13 (6), e0197617. https://doi.org/10.1371/journal.pone. 0197617.

[119] Rohde, S., Nietzer, S., Schupp, P. J. Prevalence and Mechanisms of Dynamic Chemical Defenses in Tropical Sponges. *PLoS ONE* 2015, 10 (7), e0132236. https://doi.org/10.1371/ journal.pone.0132236.

[120] Lee, Y.-J., Cho, Y., Tran, H. N. K. Secondary Metabolites from the Marine Sponges of the Genus Petrosia: A Literature Review of 43 Years of Research. *Marine Drugs* 2021, 19 (3), 122. https://doi.org/10.3390/md19030122.

[121] Maslin, M., Gaertner-Mazouni, N., Debitus, C., Joy, N., Ho, R. Marine Sponge Aquaculture towards Drug Development: An Ongoing History of Technical, Ecological, Chemical Considerations and Challenges. *Aquaculture Reports* 2021, 21, 100813. https://doi.org/10.1016/j.aqrep.2021.100813.

[122] Varijakzhan, D., Loh, J.-Y., Yap, W.-S., Yusoff, K., Seboussi, R., Lim, S.-H. E., Lai, K.-S., Chong, C.-M. Bioactive Compounds from Marine Sponges: Fundamentals and Applications. *Marine Drugs* 2021, 19 (5), 246. https://doi.org/10.3390/md19050246.

[123] Hwang, C., Agulnik, M., Schulte, B. Prices and Trends in FDA-Approved Medications for Sarcomas. *Cancers* 2024, 16 (8), 1545. https://doi.org/10.3390/cancers16081545.

[124] Yun, C. W., Kim, H. J., Lee, S. H. Therapeutic Application of Diverse Marine-Derived Natural Products in Cancer Therapy. *Anticancer Res* 2019, 39 (10), 5261–5284. https://doi.org/10.21873/anticanres.13721.

[125] Calcabrini, C., Catanzaro, E., Bishayee, A., Turrini, E., Fimognari, C. Marine Sponge Natural Products with Anticancer Potential: An Updated Review. *Marine Drugs* 2017, 15 (10), 310. https://doi.org/10.3390/md15100310.

[126] Shubina LK, Makarieva TN, Dyshlovoy SA, Fedorov SN, Dmitrenok PS, Stonik VA. Three New Aaptamines from the Marine Sponge Aaptos sp. and Their Proapoptotic Properties. *Natural Product Communications*. 2010 Dec,5(12):1934578X1000501.

[127] Calcabrini, C., Catanzaro, E., Bishayee, A., Turrini, E., Fimognari, C. Marine Sponge Natural Products with Anticancer Potential: An Updated Review. *Marine Drugs* 2017, 15 (10), 310. https://doi.org/10.3390/md15100310.

[128] Youssef, D. T. A., Asfour, H. Z., Shaala, L. A. Psammaceratin A: A Cytotoxic Psammaplysin Dimer Featuring an Unprecedented (2Z,3Z)-2,3-Bis(Amino methylene)Succinamide Backbone from the Red Sea Sponge Pseudoceratina Arabica. *Marine Drugs* 2021, 19 (8), 433. https://doi.org/10.3390/md19080433.

[129] Chen, Y., Zhou, Q., Zhang, L., Zhong, Y., Fan, G., Zhang, Z., Wang, R., Jin, M., Qiu, Y., Kong, D. Stellettin B Induces Apoptosis in Human Chronic Myeloid Leukemia Cells via Targeting PI3K and Stat5. *Oncotarget* 2017, 8 (17), 28906–28921. https://doi.org/ 10.18632/oncotarget.15957.

[130] Cheng, S.-Y., Chen, N.-F., Lin, P.-Y., Su, J.-H., Chen, B.-H., Kuo, H.-M., Sung, C.-S., Sung, P.-J., Wen, Z.-H., Chen, W.-F. Anti-Invasion and Antiangiogenic Effects of Stellettin B through Inhibition of the Akt/Girdin Signaling Pathway and VEGF in Glioblastoma Cells. *Cancers* 2019, 11 (2), 220. https://doi.org/10.3390/cancers11020220.

[131] Santalova, E. A., Denisenko, V. A., Glazunov, V. P., Kalinovskii, A. I., Anastyuk, S. D., Stonik, V. A. Dibromotyrosine Derivatives from the Ethanol Extract of the Marine Sponge Aplysina Sp.: Structures, Transformations, and Origin. *Russ Chem Bull* 2011, 60 (3), 570–580. https://doi.org/10.1007/s11172-011-0088-9.

[132] Su, J.-H., Chen, Y.-C., El-Shazly, M., Du, Y.-C., Su, C.-W., Tsao, C.-W., Liu, L.-L., Chou, Y., Chang, W.-B., Su, Y.-D., Chiang, M., Yeh, Y.-T., Lu, M.-C. Towards the Small and the Beautiful: A Small Dibromotyrosine Derivative from Pseudoceratina Sp. Sponge Exhibits Potent Apoptotic Effect through Targeting IKK/NFκB Signaling Pathway. *Marine Drugs* 2013, 11 (9), 3168–3185. https://doi.org/10.3390/md11093168.

[133] Lee, M.-G., Liu, Y.-C., Lee, Y.-L., El-Shazly, M., Lai, K.-H., Shih, S.-P., Ke, S.-C., Hong, M.-C., Du, Y.-C., Yang, J.-C., Sung, P.-J., Wen, Z.-H., Lu, M.-C. Heteronemin, a Marine Sesterterpenoid-Type Metabolite, Induces Apoptosis in Prostate LNcap Cells via Oxidative and ER Stress Combined with the Inhibition of Topoisomerase II and Hsp90. *Marine Drugs* 2018, 16 (6), 204. https://doi.org/10.3390/md16060204.

[134] Park, C., Jung, J. H., Kim, N. D., Choi, Y. H. Inhibition of Cyclooxygenase-2 and Telomerase Activities in Human Leukemia Cells by Dideoxypetrosynol A, a Polyacetylene from the Marine Sponge Petrosia Sp. *Int J Oncol* 2007, 30 (1), 291–298.

[135] Cai, J., Shi, S.-Y., Cheng, F., Wei, M., Zou, K., Yu, X.-Q., Chen, J.-F. Polyacetylene Isomers Isolated from Bidens Pilosa L. Suppress the Metastasis of Gastric Cancer Cells by Inhibiting Wnt/β-Catenin and Hippo/YAP Signaling Pathways. *Molecules* 2023, 28 (4), 1837. https://doi.org/10.3390/molecules28041837.

[136] Möller, D., Geist, N., Gellert, M., Lillig, C. H., Langel, W. CRMP2: Extending the Structure Determination, 2016.Lenz KD, Klosterman KE, Mukundan H, Kubicek-Sutherland JZ. Macrolides: From Toxins to Therapeutics. *Toxins*. 2021 May 12,13(5):347.
[137] Miller, J. H., Singh, A. J., Northcote, P. T. Microtubule-Stabilizing Drugs from Marine Sponges: Focus on Peloruside A and Zampanolide. *Marine Drugs* 2010, 8 (4), 1059–1079. https://doi.org/10.3390/md8041059.
[138] Gnanambal. K, M. E., Lakshmipathy, S. V. Dictyoceratidan Poisons: Defined Mark on Microtubule-Tubulin Dynamics. *Life Sciences* 2016, 148, 229–240. https://doi.org/10. 1016/j.lfs.2016.02.034.
[139] Mohammed, H. A., Emwas, A.-H., Khan, R. A. Salt-Tolerant Plants, Halophytes, as Renewable Natural Resources for Cancer Prevention and Treatment: Roles of Phenolics and Flavonoids in Immunomodulation and Suppression of Oxidative Stress towards Cancer Management. *IJMS* 2023, 24 (6), 5171. https://doi.org/10.3390/ijms24065171.
[140] Flowers, T. J., Muscolo, A. Introduction to the Special Issue: Halophytes in a Changing World. *AoB PLANTS* 2015, 7. https://doi.org/10.1093/aobpla/plv020.
[141] Hashim, A. M., Alharbi, B. M., Abdulmajeed, A. M., Elkelish, A., Hozzein, W. N., Hassan, H. M. Oxidative Stress Responses of Some Endemic Plants to High Altitudes by Intensifying Antioxidants and Secondary Metabolites Content. *Plants* 2020, 9 (7), 869. https://doi.org/10.3390/plants9070869.
[142] Bozyel, M. E., Merdamert Bozyel, E., Canli, K., Altuner, E. M. Anticancer Uses of Medicinal Plants in Turkish Traditional Medicine. *Kahramanmaraş Sütçü İmam Üniversitesi Tarım ve Doğa Dergisi* 2019, 22, 465–484. https://doi.org/10.18016/ksutarimdoga.vi.550334.
[143] Arya, S. S., Devi, S., Ram, K., Kumar, S., Kumar, N., Mann, A., Kumar, A., Chand, G. Halophytes: The Plants of Therapeutic Medicine. In *Ecophysiology, Abiotic Stress Responses and Utilization of Halophytes,* Hasanuzzaman, M., Nahar, K., Öztürk, M., Eds., Springer Singapore: Singapore, 2019, pp 271–287. https://doi.org/10.1007/978-981-13-3762-8_13.
[144] Kong, C.-S., Kim, Y. A., Kim, H., Seo, Y. Evaluation of a Furochromone from the Halophyte Corydalis Heterocarpa for Cytotoxic Activity against Human Gastric Cancer (AGS) Cells. *Food Funct.* 2016, 7 (12), 4823–4829. https://doi.org/10.1039/C6FO01052K.
[145] Wagabi, H. *Marine Ecosystems Of Mangroves*, 2015.
[146] Martin, C., Almahasheer, H., Duarte, C. M. Mangrove Forests as Traps for Marine Litter. *Environmental Pollution* 2019, 247, 499–508. https://doi.org/10.1016/j.envpol. 2019.01.067.
[147] Nagelkerken, I., Blaber, S. J. M., Bouillon, S., Green, P., Haywood, M., Kirton, L. G., Meynecke, J.-O., Pawlik, J., Penrose, H. M., Sasekumar, A., Somerfield, P. J. The Habitat Function of Mangroves for Terrestrial and Marine Fauna: A Review. *Aquatic Botany* 2008, 89 (2), 155–185. https://doi.org/10.1016/j.aquabot.2007.12.007.
[148] Rog, S. M., Clarke, R. H., Cook, C. N. More than Marine: Revealing the Critical Importance of Mangrove Ecosystems for Terrestrial Vertebrates. *Diversity and Distributions* 2017, 23 (2), 221–230. https://doi.org/10.1111/ddi.12514.
[149] Das, Gouda, S., Mohanta, Y., Patra, J. K. Mangrove Plants: A Potential Source for Anticancer Drugs. *Indian J. Mar. Sci* 2015, 44.
[150] Ciavatta, M. L., Lefranc, F., Carbone, M., Mollo, E., Gavagnin, M., Betancourt, T., Dasari, R., Kornienko, A., Kiss, R. Marine Mollusk-Derived Agents with Antiproliferative Activity as Promising Anticancer Agents to Overcome Chemotherapy Resistance. *Medicinal Research Reviews* 2017, 37 (4), 702–801. https://doi.org/10.1002/med.21423.
[151] Santaniello, G., Nebbioso, A., Altucci, L., Conte, M. Recent Advancement in Anticancer Compounds from Marine Organisms: Approval, Use and Bioinformatic Approaches to Predict New Targets. *Marine Drugs* 2022, 21 (1), 24. https://doi.org/ 10.3390/md21010024.
[152] Padmanaban, D., Samuel, A., Sahayanathan, G. J., Raja, K., Chinnasamy, A. Anticancer Effect of Marine Bivalves Derived Polysaccharides against Human Cancer Cells. *Biocatalysis and Agricultural Biotechnology* 2022, 39, 102240. https://doi.org/ 10.1016/j.bcab.2021.102240.
[153] Gao, P., Khong, H. Y., Mao, W., Chen, X., Bao, L., Wen, X., Xu, Y. Tunicates as Sources of High-Quality Nutrients and Bioactive Compounds for Food/Feed and Pharmaceutical Applications: A Review. *Foods* 2023, 12 (19), 3684. https://doi.org/10. 3390/foods12193684.

[154] Lloret, J. Human Health Benefits Supplied by Mediterranean Marine Biodiversity. *Marine Pollution Bulletin* 2010, 60 (10), 1640–1646. https://doi.org/10.1016/j.marpolbul. 2010.07.034.
[155] Jumeri, Kim, S. M. Antioxidant and Anticancer Activities of Enzymatic Hydrolysates of Solitary Tunicate (Styela Clava). *Food Sci Biotechnol* 2011, 20 (4), 1075–1085. https://doi.org/10.1007/s10068-011-0146-y.
[156] Wang, L., Dong, C., Li, X., Han, W., Su, X. Anticancer Potential of Bioactive Peptides from Animal Sources. *Oncology Reports* 2017, 38 (2), 637–651. https://doi.org/10.3892/ or.2017.5778.
[157] Kaleemullah Khan, B., Praba, L. K., Abdul Jaffar Ali, H. Ascidians as Bioresources: An Anti-Inflammatory Activity of Colonial Ascidians Eudistoma Ovatum and Didemnum Perlucidum. *Journal of Biologically Active Products from Natu*re 2021, 11 (3), 254–268. https://doi.org/10.1080/22311866. 2021.1919208.
[158] Munn, C. B. Viruses as Pathogens of Marine Organisms—from Bacteria to Whales. *J. Mar. Biol. Ass.* 2006, 86 (3), 453–467. https://doi.org/10.1017/S002531540601335X.
[159] Gil, J. F., Mesa, V., Estrada-Ortiz, N., Lopez-Obando, M., Gómez, A., Plácido, J. Viruses in Extreme Environments, Current Overview, and Biotechnological Potential. *Viruses* 2021, 13 (1), 81. https://doi.org/10.3390/v13010081.
[160] Liao, W., Huang, L., Han, S., Hu, D., Xu, Y., Liu, M., Yu, Q., Huang, S., Wei, D., Li, P. Review of Medicinal Plants and Active Pharmaceutical Ingredients against Aquatic Pathogenic Viruses. *Viruses* 2022, 14 (6), 1281. https://doi.org/10.3390/v14061281.
[161] Middelboe, M., Brussaard, C. Marine Viruses: Key Players in Marine Ecosystems. *Viruses* 2017, 9 (10), 302. https://doi.org/10.3390/v9100302.
[162] Xie, J., Li, F., Cai, Y., Zhang, J., Zhang, Y., Zhai, Z., Su, Z., Chen, X., Lei, M., Liu, R., Li, W., Kang, D., Chen, X., Hong, A. SAIF Plays Anti-Angiogenesis via Blocking VEGF-VEGFR2-ERK Signal in Tumor Treatment. *Heliyon* 2023, 9 (7), e18240. https://doi.org/10.1016/j.heliyon.2023.e18240.
[163] Li, F., Lei, M., Xie, J., Guo, S., Li, W., Ren, X., Wang, T., Lin, S., Xie, Q., Chen, X. Discovery and Characteristics of a Novel Antitumor Cyclopeptide Derived from Shark. *Bioengineering* 2023, 10 (6), 674. https://doi.org/10.3390/bioengineering10060674.
[164] Rodrigues, T., Reker, D., Schneider, P., Schneider, G. Counting on Natural Products for Drug Design. *Nature Chem* 2016, 8 (6), 531–541. https://doi.org/10.1038/nchem.2479.
[165] Zhou, S., Huang, G. Retracted Article: The Synthesis and Biological Activity of Marine Alkaloid Derivatives and Analogues. *RSC Adv.* 2020, 10 (53), 31909–31935. https://doi.org/10.1039/D0 RA05856D.
[166] Choi, C., Cho, Y., Son, A., Shin, S.-W., Lee, Y.-J., Park, H. C. Therapeutic Potential of (−)-Agelamide D, a Diterpene Alkaloid from the Marine Sponge Agelas Sp., as a Natural Radiosensitizer in Hepatocellular Carcinoma Models. *Marine Drugs* 2020, 18 (10), 500. https://doi.org/10.3390/md18100500.
[167] Wang, L., Huang, Y., Huang, C., Yu, J., Zheng, Y., Chen, Y., She, Z., Yuan, J. A Marine Alkaloid, Ascomylactam A, Suppresses Lung Tumorigenesis via Inducing Cell Cycle G1/S Arrest through ROS/Akt/Rb Pathway. *Marine Drugs* 2020, 18 (10), 494. https://doi.org/10.3390/md18100494.
[168] Fedorov, S. N., Bode, A. M., Stonik, V. A., Gorshkova, I. A., Schmid, P. C., Radchenko, O. S., Berdyshev, E. V., Dong, Z. Marine Alkaloid Polycarpine and Its Synthetic Derivative Dimethylpolycarpine Induce Apoptosis in JB6 Cells Through P53- and Caspase 3-Dependent Pathways. *Pharm Res* 2004, 21 (12), 2307–2319. https://doi.org/10.1007/s11095-004-7683-5.
[169] Clement, J. A., Kitagaki, J., Yang, Y., Saucedo, C. J., O'Keefe, B. R., Weissman, A. M., McKee, T. C., McMahon, J. B. Discovery of New Pyridoacridine Alkaloids from Lissoclinum Cf. Badium That Inhibit the Ubiquitin Ligase Activity of Hdm2 and Stabilize P53. *Bioorganic & Medicinal Chemistry* 2008, 16 (23), 10022–10028. https://doi.org/10.1016/j.bmc.2008.10.024.
[170] Dassonneville, L., Wattez, N., Baldeyrou, B., Mahieu, C., Lansiaux, A., Banaigs, B., Bonnard, I., Bailly, C. Inhibition of Topoisomerase II by the Marine Alkaloid Ascididemin and Induction of Apoptosis in Leukemia Cells. *Biochemical Pharmacology* 2000, 60 (4), 527–537. https://doi.org/10.1016/S0006-2952(00)00351-8.
[171] Nicoletti, R., Trincone, A. Bioactive Compounds Produced by Strains of Penicillium and Talaromyces of Marine Origin. *Marine Drugs* 2016, 14 (2), 37. https://doi.org/10.3390/md14020037.

[172] Liu, R., Zhu, T., Li, D., Gu, J., Xia, W., Fang, Y., Liu, H., Zhu, W., Gu, Q. Two Indolocarbazole Alkaloids with Apoptosis Activity from a Marine-Derived Actinomycete Z2039-2. *Arch Pharm Res* 2007, 30 (3), 270–274. https://doi.org/10.1007/BF02977605.

[173] Zhou, S., Huang, G. Retracted Article: The Synthesis and Biological Activity of Marine Alkaloid Derivatives and Analogues. *RSC Adv.* 2020, 10 (53), 31909–31935. https://doi.org/10.1039/D0RA05856D.

[174] Pereira, L., Cotas, J. Therapeutic Potential of Polyphenols and Other Micronutrients of Marine Origin. *Marine Drugs* 2023, 21 (6), 323. https://doi.org/10.3390/md21060323.

[175] Mateos, R., Pérez-Correa, J. R., Domínguez, H. Bioactive Properties of Marine Phenolics. *Marine Drugs* 2020, 18 (10), 501. https://doi.org/10.3390/md18100501.

[176] Lomartire, S., Gonçalves, A. M. M. Marine Macroalgae Polyphenols as Potential Neuroprotective Antioxidants in Neurodegenerative Diseases. *Marine Drugs* 2023, 21 (5), 261. https://doi.org/10.3390/md21050261.

[177] Rudrapal, M., Khairnar, S. J., Khan, J., Dukhyil, A. B., Ansari, M. A., Alomary, M. N., Alshabrmi, F. M., Palai, S., Deb, P. K., Devi, R. Dietary Polyphenols and Their Role in Oxidative Stress-Induced Human Diseases: Insights Into Protective Effects, Antioxidant Potentials and Mechanism(s) of Action. *Front. Pharmacol.* 2022, 13, 806470. https://doi.org/10.3389/fphar.2022.806470.

[178] El-Beltagi, H. S., Mohamed, A. A., Mohamed, H. I., Ramadan, K. M. A., Barqawi, A. A., Mansour, A. T. Phytochemical and Potential Properties of Seaweeds and Their Recent Applications: A Review. *Marine Drugs* 2022, 20 (6), 342. https://doi.org/10.3390/ md20060342.

[179] Anand David, A., Arulmoli, R., Parasuraman, S. Overviews of Biological Importance of Quercetin: A Bioactive Flavonoid. *Phcog Rev* 2016, 10 (20), 84. https://doi.org/10.4103/ 0973-7847.194044.

[180] Hassan, S., Hamed, S., Almuhayawi, M., Hozzein, W., Selim, S., AbdElgawad, H. Bioactivity of Ellagic Acid and Velutin: Two Phenolic Compounds Isolated from Marine Algae. *Egypt. J. Bot.* 2020, 0 (0), 0–0. https://doi.org/10.21608/ejbo.2020.23778.1456.

[181] Khan, N., Syed, D. N., Ahmad, N., Mukhtar, H. Fisetin: A Dietary Antioxidant for Health Promotion. *Antioxidants & Redox Signaling* 2013, 19 (2), 151–162. https://doi.org/10.1089/ars.2012.4901.

[182] Afnan, Saleem, A., Akhtar, M. F., Sharif, A., Akhtar, B., Siddique, R., Ashraf, G. M., Alghamdi, B. S., Alharthy, S. A. Anticancer, Cardio-Protective and Anti-Inflammatory Potential of Natural-Sources-Derived Phenolic Acids. *Molecules* 2022, 27 (21), 7286. https://doi.org/10.3390/molecules27217286.

[183] Akbarian, M., Khani, A., Eghbalpour, S., Uversky, V. N. Bioactive Peptides: Synthesis, Sources, Applications, and Proposed Mechanisms of Action. *IJMS* 2022, 23 (3), 1445. https://doi.org/10.3390/ijms23031445.

[184] Kim, S.-K., Wijesekara, I. Development and Biological Activities of Marine-Derived Bioactive Peptides: A Review. *Journal of Functional Foods* 2010, 2 (1), 1–9. https://doi.org/10.1016/j.jff.2010.01.003.

[185] Rossino, G., Marchese, E., Galli, G., Verde, F., Finizio, M., Serra, M., Linciano, P., Collina, S. Peptides as Therapeutic Agents: Challenges and Opportunities in the Green Transition Era. *Molecules* 2023, 28 (20), 7165. https://doi.org/10.3390/molecules 28207165.

[186] Della Sala, G., Hochmuth, T., Teta, R., Costantino, V., Mangoni, A. Polyketide Synthases in the Microbiome of the Marine Sponge Plakortis Halichondrioides: A Metagenomic Update. *Marine Drugs* 2014, 12 (11), 5425–5440. https://doi.org/10.3390/md 12115425.

[187] Gingras, D., Batist, G., Béliveau, R. Æ-941 (Neovastat ®): A Novel Multifunctional Antiangiogenic Compound. *Expert Review of Anticancer Therapy* 2001, 1 (3), 341–347. https://doi.org/10.1586/14737140.1.3.341.

[188] Mita, A. C., Hammond, L. A., Bonate, P. L., Weiss, G., McCreery, H., Syed, S., Garrison, M., Chu, Q. S. C., DeBono, J. S., Jones, C. B., Weitman, S., Rowinsky, E. K. Phase I and Pharmacokinetic Study of Tasidotin Hydrochloride (ILX651), a Third-Generation Dolastatin-15 Analogue, Administered Weekly for 3 Weeks Every 28 Days in Patients with Advanced Solid Tumors. *Clinical Cancer Research* 2006, 12 (17), 5207–5215. https://doi.org/10.1158/1078-0432.CCR-06-0179.

[189] Hearn, B. R., Shaw, S. J., Myles, D. C. Microtubule Targeting Agents. In *Comprehensive Medicinal Chemistry II*, Elsevier, 2007, pp 81–110. https://doi.org/ 10.1016/B0-08-045044-X/00205-4.

[190] Cheung, R., Ng, T., Wong, J. Marine Peptides: Bioactivities and Applications. *Marine Drugs* 2015, 13 (7), 4006–4043. https://doi.org/10.3390/md13074006.

[191] Bradley, A. M., Devine, M., DeRemer, D. Brentuximab Vedotin: An Anti-CD30 Antibody–Drug Conjugate. *American Journal of Health-System Pharmacy* 2013, 70 (7), 589–597. https://doi.org/10.2146/ajhp110608.

[192] Advani, R. H., Lebovic, D., Chen, A., Brunvand, M., Goy, A., Chang, J. E., Hochberg, E., Yalamanchili, S., Kahn, R., Lu, D., Agarwal, P., Dere, R. C., Hsieh, H.-J., Jones, S., Chu, Y.-W., Cheson, B. D. Phase I Study of the Anti-CD22 Antibody–Drug Conjugate Pinatuzumab Vedotin with/without Rituximab in Patients with Relapsed/Refractory B-Cell Non-Hodgkin Lymphoma. *Clinical Cancer Research* 2017, 23 (5), 1167–1176. https://doi.org/10.1158/1078-0432.CCR-16-0772.

[193] Kuznetsov, G., TenDyke, K., Towle, M. J., Cheng, H., Liu, J., Marsh, J. P., Schiller, S. E. R., Spyvee, M. R., Yang, H., Seletsky, B. M., Shaffer, C. J., Marceau, V., Yao, Y., Suh, E. M., Campagna, S., Fang, F. G., Kowalczyk, J. J., Littlefield, B. A. Tubulin-Based Antimitotic Mechanism of E7974, a Novel Analogue of the Marine Sponge Natural Product Hemiasterlin. *Molecular Cancer Therapeutics* 2009, 8 (10), 2852–2860. https://doi.org/10.1158/1535-7163.MCT-09-0301.

[194] Petty, R., Anthoney, A., Metges, J.-P., Alsina, M., Gonçalves, A., Brown, J., Montagut, C., Gunzer, K., Laus, G., Iglesias Dios, J. L., Miguel-Lillo, B., Bohan, P., Salazar, R. Phase Ib/II Study of Elisidepsin in Metastatic or Advanced Gastroesophageal Cancer (IMAGE Trial). *Cancer Chemother Pharmacol* 2016, 77 (4), 819–827. https://doi.org/10.1007/s00280-016-2991-0.

[195] Ott, P. A., Pavlick, A. C., Johnson, D. B., Hart, L. L., Infante, J. R., Luke, J. J., Lutzky, J., Rothschild, N., Spitler, L., Cowey, C. L., Alizadeh, A., Salama, A., He, Y., Bagley, R. G., Zhang, J., Hamid, O. A Phase 2 Study of Glembatumumab Vedotin (GV), an Antibody-Drug Conjugate (ADC) Targeting gpNMB, in Advanced Melanoma. *Annals of Oncology* 2016, 27, vi393. https://doi.org/10.1093/annonc/mdw379.42.

[196] Alonso-Álvarez, S., Pardal, E., Sánchez-Nieto, D., Navarro, M., Caballero, M. D., Mateos, M. V., Martin, A. Plitidepsin: Design, Development, and Potential Place in Therapy. *DDDT 2017*, Volume11, 253–264. https://doi.org/10.2147/DDDT.S94165.

[197] Zheng, L.-H., Wang, Y.-J., Sheng, J., Wang, F., Zheng, Y., Lin, X.-K., Sun, M. Antitumor Peptides from Marine Organisms. *Marine Drugs* 2011, 9 (10), 1840–1859. https://doi.org/10.3390/md9101840.

[198] Bertin, M. J., Schwartz, S. L., Lee, J., Korobeynikov, A., Dorrestein, P. C., Gerwick, L., Gerwick, W. H. Spongosine Production by a Vibrio Harveyi Strain Associated with the Sponge Tectitethya Crypta. *J. Nat. Prod.* 2015, 78 (3), 493–499. https://doi.org/10.1021/ np5009762.

[199] Huang, R.-M., Chen, Y.-N., Zeng, Z., Gao, C.-H., Su, X., Peng, Y. Marine Nucleosides: Structure, Bioactivity, Synthesis and Biosynthesis. *Marine Drugs* 2014, 12 (12), 5817–5838. https://doi.org/10.3390/md12125817.

[200] Wu, L., Ye, K., Jiang, S., Zhou, G. Marine Power on Cancer: Drugs, Lead Compounds, and Mechanisms. *Marine Drugs* 2021, 19 (9), 488. https://doi.org/10.3390/md19090488.

[201] Kisor, D., Reilly. Profile of Nelarabine: Use in the Treatment of T-Cell Acute Lymphoblastic Leukemia. *OTT* 2009, 219. https://doi.org/10.2147/OTT.S4770.

[202] Ciccolini, J., Serdjebi, C., Peters, G. J., Giovannetti, E. Pharmacokinetics and Pharmacogenetics of Gemcitabine as a Mainstay in Adult and Pediatric Oncology: An EORTC-PAMM Perspective. *Cancer Chemother Pharmacol* 2016, 78 (1), 1–12. https://doi.org/10.1007/s00280-016-3003-0.

[203] Haque, N., Parveen, S., Tang, T., Wei, J., Huang, Z. Marine Natural Products in Clinical Use. *Marine Drugs* 2022, 20 (8), 528. https://doi.org/10.3390/md20080528.

[204] Napolitano, J. G., Daranas, A. H., Norte, M., Fernandez, J. J. Marine Macrolides, a Promising Source of Antitumor Compounds. *ACAMC* 2009, 9 (2), 122–137. https://doi.org/10.2174/187152009787313800.

[205] Mamada, S. S., Nainu, F., Masyita, A., Frediansyah, A., Utami, R. N., Salampe, M., Emran, T. B., Lima, C. M. G., Chopra, H., Simal-Gandara, J. Marine Macrolides to Tackle Antimicrobial Resistance of Mycobacterium Tuberculosis. *Marine Drugs* 2022, 20 (11), 691. https://doi.org/10.3390/md20110691.

[206] Konishi, H., Ochiai, T., Ikoma, H., Kubota, T., Ichikawa, D., Fujiwara, H., Okamoto, K., Sakakura, C., Sonoyama, T., Kokuba, Y., Sasaki, H., Matsui, T., Otsuji, E. Latrunculin A Has a Strong Anticancer

Effect in a Peritoneal Dissemination Model of Human Gastric Cancer in Mice. *Anticancer research* 2009, 29, 2091–2097.
[207] Mooberry, S. L., Hilinski, M. K., Clark, E. A., Wender, P. A. Function-Oriented Synthesis: Biological Evaluation of Laulimalide Analogues Derived from a Last Step Cross Metathesis Diversification Strategy. *Mol. Pharmaceutics* 2008, 5 (5), 829–838. https://doi.org/10.1021/mp800043n.
[208] Ludwiczuk, A., Skalicka-Woźniak, K., Georgiev, M. I. Terpenoids. In *Pharmacognosy*, Elsevier, 2017, pp 233–266. https://doi.org/10.1016/B978-0-12-802104-0.00011-1.
[209] Masyita, A., Mustika Sari, R., Dwi Astuti, A., Yasir, B., Rahma Rumata, N., Emran, T. B., Nainu, F., Simal-Gandara, J. Terpenes and Terpenoids as Main Bioactive Compounds of Essential Oils, Their Roles in Human Health and Potential Application as Natural Food Preservatives. *Food Chemistry*: X 2022, 13, 100217. https://doi.org/10.1016/ j.fochx.2022.100217.
[210] Gross, H., König, G. M. Terpenoids from Marine Organisms: Unique Structures and Their Pharmacological Potential. *Phytochem Rev* 2006, 5 (1), 115–141. https://doi.org/10. 1007/s11101-005-5464-3.
[211] Zhao, Q., Xue, Y., Liu, Z., Li, H., Wang, J., Li, Z., Wang, Y., Dong, P., Xue, C. Differential Effects of Sulfated Triterpene Glycosides, Holothurin A 1 , and 24-Dehydroechinoside A, on Antimetastasic Activity via Regulation of the MMP-9 Signal Pathway. *Journal of Food Science* 2010, 75 (9). https://doi.org/10.1111/j.1750-3841.2010.01837.x.
[212] Sajwani, F. Frondoside A Is a Potential Anticancer Agent from Sea Cucumbers. *J Can Res Ther* 2019, 15 (5), 953. https://doi.org/10.4103/jcrt.JCRT_1427_16.
[213] Wang, K., Chen, Y.-F., Yang, Y.-C. S. H., Huang, H.-M., Lee, S.-Y., Shih, Y.-J., Li, Z.-L., Whang-Peng, J., Lin, H.-Y., Davis, P. J. The Power of Heteronemin in Cancers. *J Biomed Sci* 2022, 29 (1), 41. https://doi.org/10.1186/s12929-022-00816-z.
[214] Hassan, H. M., Khanfar, M. A., Elnagar, A. Y., Mohammed, R., Shaala, L. A., Youssef, D. T. A., Hifnawy, M. S., El Sayed, K. A. Pachycladins A−E, Prostate Cancer Invasion and Migration Inhibitory Eunicellin-Based Diterpenoids from the Red Sea Soft Coral Cladiella Pachyclados. *J. Nat. Prod.* 2010, 73 (5), 848–853. https://doi.org/10.1021/np900787p.
[215] Shi, Z., Jain, S., Kim, I., Peng, X., Abraham, I., Youssef, D. T. A., Fu, L., El Sayed, K., Ambudkar, S. V., Chen, Z. Sipholenol A, a Marine-derived Sipholane Triterpene, Potently Reverses P-glycoprotein (ABCB1)-mediated Multidrug Resistance in Cancer Cells. *Cancer Science* 2007, 98 (9), 1373–1380. https://doi.org/10.1111/j.1349-7006.2007.00554.x.
[216] Grote, D., Shaker, K. H., Soliman, H. S. M., Hegazi, M. M., Seifert, K. Cembranoid Diterpenes from the Soft Corals Sarcophyton Sp. and Sarcophyton Glaucum. *Natural Product Communications* 2008, 3 (9), 1934578X0800300. https://doi.org/10.1177/ 1934578X0800300914.
[217] Aoki, S., Sanagawa, M., Watanabe, Y., Setiawan, A., Arai, M., Kobayashi, M. Novel Isomarabarican Triterpenes, Exhibiting Selective Anti-Proliferative Activity against Vascular Endothelial Cells, from Marine Sponge Rhabdastrella Globostellata. *Bioorganic & Medicinal Chemistry* 2007, 15 (14), 4818–4828. https://doi.org/10.1016/j.bmc.2007. 04.070.
[218] Kong, D., Yamori, T., Kobayashi, M., Duan, H. Antiproliferative and Antiangiogenic Activities of Smenospongine, a Marine Sponge Sesquiterpene Aminoquinone. *Marine Drugs* 2011, 9 (2), 154–161. https://doi.org/10.3390/md9020154.
[219] Rocha-Santos, T., Duarte, A. C. Introduction to the Analysis of Bioactive Compounds in Marine Samples. In *Comprehensive Analytical Chemistry*, Elsevier, 2014, Vol. 65, pp 1–13. https://doi.org/10.1016/B978-0-444-63359-0.00001-X.
[220] Hochmuth, T., Piel, J. Polyketide Synthases of Bacterial Symbionts in Sponges – Evolution-Based Applications in Natural Products Research. *Phytochemistry* 2009, 70 (15–16), 1841–1849. https://doi.org/10.1016/j.phytochem.2009.04.010.
[221] Ghareeb, M. A., Tammam, M. A., El-Demerdash, A., Atanasov, A. G. Insights about Clinically Approved and Preclinically Investigated Marine Natural Products. *Current Research in Biotechnology* 2020, 2, 88–102. https://doi.org/10.1016/j.crbiot.2020.09.001.
[222] Blunt, J. W., Copp, B. R., Keyzers, R. A., Munro, M. H. G., Prinsep, M. R. Marine Natural Products. *Nat. Prod. Rep.* 2015, 32 (2), 116–211. https://doi.org/10.1039/C4NP00144C.

[223] McBride, A., Butler, S. K. Eribulin Mesylate: A Novel Halichondrin B Analogue for the Treatment of Metastatic Breast Cancer. *American Journal of Health-System Pharmacy* 2012, 69 (9), 745–755. https://doi.org/10.2146/ajhp110237.

[224] Ida, H., Shimizu, T., Nishino, M., Nakamura, Y., Yazaki, S., Katsuya, Y., Sato, J., Koyama, T., Iwasa, S., Sudo, K., Kondo, S., Yonemori, K., Shitara, K., Shiono, S., Matsuoka, D., Yasuda, K., Otake, Y., Suzuki, T., Takase, T., Takashima, S., Yamaguchi, K., Semba, T., Yamamoto, N. Phase Ib/II Study of a Liposomal Formulation of Eribulin (E7389-LF) plus Nivolumab in Patients with Advanced Solid Tumors: Results from Phase Ib. *Cancer Research Communications* 2023, 3 (7), 1189–1199. https://doi.org/10.1158/2767-9764.CRC-22-0401.

[225] Isbrucker, R., Gunasekera, S., Longley, R. Structure-Activity Relationship Studies of Discodermolide and Its Semisynthetic Acetylated Analogs on Microtubule Function and Cytotoxicity. *Cancer Chemotherapy and Pharmacology* 2001, 48 (1), 29–36. https://doi.org/10.1007/s002800100287.

[226] Sudek, S., Lopanik, N. B., Waggoner, L. E., Hildebrand, M., Anderson, C., Liu, H., Patel, A., Sherman, D. H., Haygood, M. G. Identification of the Putative Bryostatin Polyketide Synthase Gene Cluster from " Candidatus Endobugula Sertula," the Uncultivated Microbial Symbiont of the Marine Bryozoan Bugula Neritina. *J. Nat. Prod.* 2007, 70 (1), 67–74. https://doi.org/10.1021/np060361d.

[227] Ruocco, N., Costantini, S., Guariniello, S., Costantini, M. Polysaccharides from the Marine Environment with Pharmacological, Cosmeceutical and Nutraceutical Potential. *Molecules* 2016, 21 (5), 551. https://doi.org/10.3390/molecules21050551.

[228] Khodse, V. B., Fernandes, L., Gopalkrishna, V. V., Bhosle, N. B., Fernandes, V., Matondkar, S. G. P., Bhushan, R. Distribution and Seasonal Variation of Concentrations of Particulate Carbohydrates and Uronic Acids in the Northern Indian Ocean. *Marine Chemistry* 2007, 103 (3–4), 327–346. https://doi.org/10.1016/j.marchem.2006.10.003.

[229] D'Ayala, G. G., Malinconico, M., Laurienzo, P. Marine Derived Polysaccharides for Biomedical Applications: Chemical Modification Approaches. *Molecules* 2008, 13 (9), 2069–2106. https://doi.org/10.3390/molecules13092069.

[230] Barbosa, J. D. S., Costa, M. S. S. P., Melo, L. F. M. D., Medeiros, M. J. C. D., Pontes, D. D. L., Scortecci, K. C., Rocha, H. A. O. In Vitro Immunostimulating Activity of Sulfated Polysaccharides from Caulerpa Cupressoides Var. Flabellata. *Marine Drugs* 2019, 17 (2), 105. https://doi.org/10.3390/md17020105.

[231] Li, B., Lu, F., Wei, X., Zhao, R. Fucoidan: Structure and Bioactivity. *Molecules* 2008, 13 (8), 1671–1695. https://doi.org/10.3390/molecules13081671.

[232] Sharma, S., Shekhar, S., Sharma, B., Jain, P. Decoding Glycans: Deciphering the Sugary Secrets to Be Coherent on the Implication. *RSC Adv.* 2020, 10 (56), 34099–34113. https://doi.org/10.1039/D0RA04471G.

[233] Bartnik, M., Facey, P. C. Glycosides. In *Pharmacognosy*, Elsevier, 2017, pp 101–161. https://doi.org/10.1016/B978-0-12-802104-0.00008-1.

[234] Ford, P. W., Gadepalli, M., Davidson, B. S. Halawanones A–D, New Polycyclic Quinones from a Marine-Derived Streptomycete. *J. Nat. Prod.* 1998, 61 (10), 1232–1236. https://doi.org/10.1021/np980126y.

[235] Olano, C., Méndez, C., Salas, J. A. Antitumor Compounds from Marine Actinomycetes. *Marine Drugs* 2009, 7 (2), 210–248. https://doi.org/10.3390/md7020210.

[236] Asolkar, R. N., Kirkland, T. N., Jensen, P. R., Fenical, W. Arenimycin, an Antibiotic Effective against Rifampin- and Methicillin-Resistant Staphylococcus Aureus from the Marine Actinomycete Salinispora Arenicola. *J Antibiot* 2010, 63 (1), 37–39. https://doi.org/10.1038/ja.2009.114.

[237] Maskey, R. P., Sevvana, M., Usón, I., Helmke, E., Laatsch, H. Gutingimycin: A Highly Complex Metabolite from a Marine Streptomycete. *Angew Chem Int Ed* 2004, 43 (10), 1281–1283. https://doi.org/10.1002/anie.200352312.

[238] (Wordeman, L., Vicente, J. J. Microtubule Targeting Agents in Disease: Classic Drugs, Novel Roles. *Cancers* 2021, 13 (22), 5650. https://doi.org/10.3390/cancers13225650.

[239] Wordeman, L., Vicente, J. J. Microtubule Targeting Agents in Disease: Classic Drugs, Novel Roles. *Cancers* 2021, 13 (22), 5650. https://doi.org/10.3390/cancers13225650.

[240] Lafanechère, L. The Microtubule Cytoskeleton: An Old Validated Target for Novel Therapeutic Drugs. *Front. Pharmacol.* 2022, 13, 969183. https://doi.org/10.3389/fphar. 2022.969183.
[241] Kaul, R., Risinger, A. L., Mooberry, S. L. Microtubule-Targeting Drugs: More than Antimitotics. *J. Nat. Prod.* 2019, 82 (3), 680–685. https://doi.org/10.1021/acs.jnatprod. 9b00105.
[242] Bertheloot, D., Latz, E., Franklin, B. S. Necroptosis, Pyroptosis and Apoptosis: An Intricate Game of Cell Death. *Cell Mol Immunol* 2021, 18 (5), 1106–1121. https://doi.org/10.1038/s41423-020-00630-3.
[243] Hussar, P. Apoptosis Regulators Bcl-2 and Caspase-3. *Encyclopedia* 2022, 2 (4), 1624–1636. https://doi.org/10.3390/encyclopedia2040111.
[244] Moffitt, K. L., Martin, S. L., Walker, B. From Sentencing to Execution – the Processes of Apoptosis. *Journal of Pharmacy and Pharmacology* 2010, 62 (5), 547–562. https://doi.org/10.1211/jpp.62.05.0001.
[245] Chaudhry, G.-S., Md Akim, A., Sung, Y. Y., Sifzizul, T. M. T. Cancer and Apoptosis: The Apoptotic Activity of Plant and Marine Natural Products and Their Potential as Targeted Cancer Therapeutics. *Front. Pharmacol.* 2022, 13, 842376. https://doi.org/10.3389/fphar.2022.842376.
[246] Danial, N. N. BCL-2 Family Proteins: Critical Checkpoints of Apoptotic Cell Death. *Clinical Cancer Research* 2007, 13 (24), 7254–7263. https://doi.org/10.1158/1078-0432.CCR-07-1598.
[247] Johnson, K. E., Wilgus, T. A. Vascular Endothelial Growth Factor and Angiogenesis in the Regulation of Cutaneous Wound Repair. *Advances in Wound Care* 2014, 3 (10), 647–661. https://doi.org/10.1089/wound.2013.0517.
[248] Everts, P. A., Lana, J. F., Onishi, K., Buford, D., Peng, J., Mahmood, A., Fonseca, L. F., Van Zundert, A., Podesta, L. Angiogenesis and Tissue Repair Depend on Platelet Dosing and Bioformulation Strategies Following Orthobiological Platelet-Rich Plasma Procedures: A Narrative Review. *Biomedicines* 2023, 11 (7), 1922. https://doi.org/10.3390/biomedicines11071922.
[249] Kumar, P., Kumar, S., Udupa, E. P., Kumar, U., Rao, P., Honnegowda, T. Role of Angiogenesis and Angiogenic Factors in Acute and Chronic Wound Healing. *Plast Aesthet Res* 2015, 2 (5), 243. https://doi.org/10.4103/2347-9264.165438.
[250] Shi, Z., Yao, C., Shui, Y., Li, S., Yan, H. Research Progress on the Mechanism of Angiogenesis in Wound Repair and Regeneration. *Front. Physiol.* 2023, 14, 1284981. https://doi.org/10.3389/fphys.2023.1284981.
[251] Simons, M., Alitalo, K., Annex, B. H., Augustin, H. G., Beam, C., Berk, B. C., Byzova, T., Carmeliet, P., Chilian, W., Cooke, J. P., Davis, G. E., Eichmann, A., Iruela-Arispe, M. L., Keshet, E., Sinusas, A. J., Ruhrberg, C., Woo, Y. J., Dimmeler, S. State-of-the-Art Methods for Evaluation of Angiogenesis and Tissue Vascularization: A Scientific Statement From the American Heart Association. *Circulation Research* 2015, 116 (11). https://doi.org/10.1161/RES.0000000000000054.
[252] Cargnello, M., Roux, P. P. Activation and Function of the MAPKs and Their Substrates, the MAPK-Activated Protein Kinases. *Microbiol Mol Biol Rev* 2011, 75 (1), 50–83. https://doi.org/10. 1128/MMBR.00031-10.
[253] Munshi, A., Ramesh, R. Mitogen-Activated Protein Kinases and Their Role in Radiation Response. *Genes & Cancer* 2013, 4 (9–10), 401–408. https://doi.org/10.1177/ 1947601913485414.
[254] Elkhadragy, L., Myers, A., Long, W. Role of the Atypical MAPK ERK3 in Cancer Growth and Progression. *Cancers* 2024, 16 (7), 1381. https://doi.org/10.3390/ cancers16071381.
[255] Kciuk, M., Gielecińska, A., Budzinska, A., Mojzych, M., Kontek, R. Metastasis and MAPK Pathways. *IJMS* 2022, 23 (7), 3847. https://doi.org/10.3390/ijms23073847.
[256] Soares-Silva, M., Diniz, F. F., Gomes, G. N., Bahia, D. The Mitogen-Activated Protein Kinase (MAPK) Pathway: Role in Immune Evasion by Trypanosomatids. *Front. Microbiol.* 2016, 7. https://doi.org/10.3389/fmicb.2016.00183.
[257] He, Y., Sun, M. M., Zhang, G. G., Yang, J., Chen, K. S., Xu, W. W., Li, B. Targeting PI3K/Akt Signal Transduction for Cancer Therapy. *Sig Transduct Target Ther* 2021, 6 (1), 425. https://doi.org/10.1038/ s41392-021-00828-5.
[258] Tatsuta, T., Hosono, M., Rotinsulu, H., Wewengkang, D. S., Sumilat, D. A., Namikoshi, M., Yamazaki, H. Lissoclibadin 1, a Polysulfur Aromatic Alkaloid from the Indonesian Ascidian Lissoclinum Cf. Badium , Induces Caspase-Dependent Apoptosis in Human Colon Cancer Cells and Suppresses Tumor

Growth in Nude Mice. *J. Nat. Prod.* 2017, 80 (2), 499–502. https://doi.org/10.1021/acs.jnatprod.6b01051.

[259] Dassonneville, L., Wattez, N., Baldeyrou, B., Mahieu, C., Lansiaux, A., Banaigs, B., Bonnard, I., Bailly, C. Inhibition of Topoisomerase II by the Marine Alkaloid Ascididemin and Induction of Apoptosis in Leukemia Cells. *Biochemical Pharmacology* 2000, 60 (4), 527–537. https://doi.org/10.1016/S0006-2952(00)00351-8.

[260] Gomes, N., Lefranc, F., Kijjoa, A., Kiss, R. Can Some Marine-Derived Fungal Metabolites Become Actual Anticancer Agents? *Marine Drugs* 2015, 13 (6), 3950–3991. https://doi.org/10.3390/md13063950.

[261] Fabre, S., Prudhomme, M., Rapp, M. Protein Kinase C Inhibitors, Structure—Activity Relationships in K252c-Related Compounds. *Bioorganic & Medicinal Chemistry* 1993, 1 (3), 193–196. https://doi.org/10.1016/S0968-0896(00)82121-5.

[262] Hoshino, S., Zhang, L., Awakawa, T., Wakimoto, T., Onaka, H., Abe, I. Arcyriaflavin E, a New Cytotoxic Indolocarbazole Alkaloid Isolated by Combined-Culture of Mycolic Acid-Containing Bacteria and Streptomyces Cinnamoneus NBRC 13823. *J Antibiot* 2015, 68 (5), 342–344. https://doi.org/10.1038/ja.2014.147.

[263] Pereira, A., Silva, L., Laranjeiro, C., Lino, C., Pena, A. Selected Pharmaceuticals in Different Aquatic Compartments: Part I—Source, Fate and Occurrence. *Molecules* 2020, 25 (5), 1026. https://doi.org/10.3390/molecules25051026.

[264] McGivern, J. G. Ziconotide: A Review of Its Pharmacology and Use in the Treatment of Pain. *Neuropsychiatric Disease and Treatment* 2007, 3 (1), 69–85. https://doi.org/10.2147/nedt.2007.3.1.69.

[265] Sawalha, Y., Maddocks, K. Profile of Polatuzumab Vedotin in the Treatment of Patients with Relapsed/Refractory Non-Hodgkin Lymphoma: A Brief Report on the Emerging Clinical Data. *OTT* 2020, Volume 13, 5123–5133. https://doi.org/10.2147/OTT.S219449.

[266] Singh, D., Dheer, D., Samykutty, A., Shankar, R. Antibody Drug Conjugates in Gastrointestinal Cancer: From Lab to Clinical Development. *Journal of Controlled Release* 2021, 340, 1–34. https://doi.org/10.1016/j.jconrel.2021.10.006.

[267] Rocha-Lima, C. M., Bayraktar, S., MacIntyre, J., Raez, L., Flores, A. M., Ferrell, A., Rubin, E. H., Poplin, E. A., Tan, A. R., Lucarelli, A., Zojwalla, N. A Phase 1 Trial of E7974 Administered on Day 1 of a 21-day Cycle in Patients with Advanced Solid Tumors. *Cancer* 2012, 118 (17), 4262–4270. https://doi.org/10.1002/cncr.27428.

[268] Coronado, C., Galmarini, C. M., Alfaro, V., Yovine, A. Elisidepsin. *Drug News Perspect* 2010, 35 (4), 287. https://doi.org/10.1358/dof.2010.35.4.1497635.

[269] Vijayakumar, S., Menakha, M. Pharmaceutical Applications of Cyanobacteria—A Review. *Journal of Acute Medicine* 2015, 5 (1), 15–23. https://doi.org/10.1016/j.jacme. 2015.02.004.

[270] Eghtedari, M., Jafari Porzani, S., Nowruzi, B. Anticancer Potential of Natural Peptides from Terrestrial and Marine Environments: A Review. *Phytochemistry Letters* 2021, 42, 87–103. https://doi.org/10.1016/j.phytol.2021.02.008.

[271] Mondal, A., Bose, S., Banerjee, S., Patra, J. K., Malik, J., Mandal, S. K., Kilpatrick, K. L., Das, G., Kerry, R. G., Fimognari, C., Bishayee, A. Marine Cyanobacteria and Microalgae Metabolites—A Rich Source of Potential Anticancer Drugs. *Marine Drugs* 2020, 18 (9), 476. https://doi.org/10.3390/md18090476.

[272] Michon, S., Cavelier, F., Salom-Roig, X. J. Synthesis and Biological Activities of Cyclodepsipeptides of Aurilide Family from Marine Origin. *Marine Drugs* 2021, 19 (2), 55. https://doi.org/10.3390/md19020055.

[273] Qamar, H., Hussain, K., Soni, A., Khan, A., Hussain, T., Chénais, B. Cyanobacteria as Natural Therapeutics and Pharmaceutical Potential: Role in Antitumor Activity and as Nanovectors. *Molecules* 2021, 26 (1), 247. https://doi.org/10.3390/molecules26010247.

[274] Suzuki, K., Mizuno, R., Suenaga, K., Teruya, T., Tanaka, N., Kosaka, T., Oya, M. B Isebromoamide, an Extract from L Yngbya Species, Induces Apoptosis through ERK and mTOR Inhibitions in Renal Cancer Cells. *Cancer Medicine* 2013, 2 (1), 32–39. https://doi.org/10.1002/cam4.53.

[275] Morita, M., Ogawa, H., Ohno, O., Yamori, T., Suenaga, K., Toyoshima, C. Biselyngbyasides, Cytotoxic Marine Macrolides, Are Novel and Potent Inhibitors of the Ca 2+ Pumps with a Unique

Mode of Binding. *FEBS Letters* 2015, 589 (13), 1406–1411. https://doi.org/10.1016/j.febslet.2015.04.056.
[276] Kim, J., Lee, S., Han, S., Lee, H.-Y. *Divergent Synthesis of Conidiogenones B–F and 12β-Hydroxyconidiogenone C. Chem 2023*, 9 (5), 1270–1280. https://doi.org/10.1016/ j.chempr.2023.01.018.
[277] .Kulkarni, Y. A., Garud, M. S., Gaud, R. S., Gaikwad, A. B. Recent Developments in Using Plant-Derived Natural Products as Tubulin Inhibitors for the Management of Cancer. In *Fruits, Vegetables, and Herbs*, Elsevier, 2016, pp 507–524. https://doi.org/10.1016/B978-0-12-802972-5.00024-X.
[278] Tiwari, K., Gupta, R. K. Bioactive Metabolites from Rare Actinomycetes. In *Studies in Natural Products Chemistry*, Elsevier, 2014, Vol. 41, pp 419–512. https://doi.org/10.1016/B978-0-444-63294-4.
[279] Yan, S., Zeng, M., Wang, H., Zhang, H. Micromonospora : A Prolific Source of Bioactive Secondary Metabolites with Therapeutic Potential. *J. Med. Chem.* 2022, 65 (13), 8735–8771. https://doi.org/10.1021/acs.jmedchem.2c00626.
[280] Cho, J. Y., Williams, P. G., Kwon, H. C., Jensen, P. R., Fenical, W. Lucentamycins A−D, Cytotoxic Peptides from the Marine-Derived Actinomycete Nocardiopsis Lucentensis. *J. Nat. Prod.* 2007, 70 (8), 1321–1328. https://doi.org/10.1021/np070101b.
[281] Mueller, A.-L., Brockmueller, A., Fahimi, N., Ghotbi, T., Hashemi, S., Sadri, S., Khorshidi, N., Kunnumakkara, A. B., Shakibaei, M. Bacteria-Mediated Modulatory Strategies for Colorectal Cancer Treatment. *Biomedicines* 2022, 10 (4), 832. https://doi.org/10.3390/biomedicines10040832.
[282] Cha, J. W., Park, J.-S., Sim, T., Nam, S.-J., Kwon, H. C., Del Valle, J. R., Fenical, W. Structure Assignment of Lucentamycin E and Revision of the Olefin Geometries of the Marine-Derived Lucentamycins. *J. Nat. Prod.* 2012, 75 (9), 1648–1651. https://doi.org/10.1021/np3003854.
[283] Pérez, M., Crespo, C., Schleissner, C., Rodríguez, P., Zúñiga, P., Reyes, F. Tartrolon D, a Cytotoxic Macrodiolide from the Marine-Derived Actinomycete Streptomyces Sp. MDG-04-17-069. *J. Nat. Prod.* 2009, 72 (12), 2192–2194. https://doi.org/10.1021/np9006603.
[284] Hillman, P. F., Lee, J. Y., Lee, J., Lee, J., Oh, D.-C., Fenical, W., Nam, S.-J. Lodopyridones D − G from a Marine-Derived Bacterium Saccharomonospora Sp. *Natural Product Research* 2023, 1–6. https://doi.org/10.1080/14786419.2023.2173192.
[285] Maloney, K. N., MacMillan, J. B., Kauffman, C. A., Jensen, P. R., DiPasquale, A. G., Rheingold, A. L., Fenical, W. Lodopyridone, a Structurally Unprecedented Alkaloid from a Marine Actinomycete. *Org. Lett.* 2009, 11 (23), 5422–5424. https://doi.org/10.1021/ol901997k.
[286] Serrano-Aparicio, N., Moliner, V., Świderek, K. On the Origin of the Different Reversible Characters of Salinosporamide A and Homosalinosporamide A in the Covalent Inhibition of the Human 20S Proteasome. *ACS Catal.* 2021, 11 (18), 11806–11819. https://doi.org/10.1021/acscatal.1c02614.
[287] Pokharkar, O., Zyryanov, G. V., Tsurkan, M. V. Natural Products from Marine Actinomycete Genus Salinispora Might Inhibit 3CLpro and PLpro Proteins of SARS-CoV-2: An In Silico Evidence. *Microbiology Research* 2023, 14 (4), 1907–1941. https://doi.org/10.3390/microbiolres14040130.
[288] Chhun, A., Sousoni, D., Del Mar Aguiló-Ferretjans, M., Song, L., Corre, C., Christie-Oleza, J. A. Phytoplankton Trigger the Production of Cryptic Metabolites in the Marine *Actinobacteria Salinispora Tropica.* May 20, 2020. https://doi.org/10.1101/2020.05.18.103358.
[289] Fenical, W., Jensen, P. R., Palladino, M. A., Lam, K. S., Lloyd, G. K., Potts, B. C. Discovery and Development of the Anticancer Agent Salinosporamide A (NPI-0052). *Bioorganic & Medicinal Chemistry* 2009, 17 (6), 2175–2180. https://doi.org/10.1016/j.bmc.2008.10.075.
[290] Gulder, T. A. M., Moore, B. S. Salinosporamide Natural Products: Potent 20 S Proteasome Inhibitors as Promising Cancer Chemotherapeutics. *Angew Chem Int Ed* 2010, 49 (49), 9346–9367. https://doi.org/10.1002/anie.201000728.
[291] Serrano-Aparicio, N., Moliner, V., Świderek, K. On the Origin of the Different Reversible Characters of Salinosporamide A and Homosalinosporamide A in the Covalent Inhibition of the Human 20S Proteasome. *ACS Catal.* 2021, 11 (18), 11806–11819. https://doi.org/10.1021/acscatal.1c02614.
[292] Yang, Q., Gao, W., Li, X., Li, X., Zhou, X., Li, W., Zhou, C., Luo, A., Liu, Z. Targeting ABCA1 via Extracellular Vesicle-Encapsulated Staurosporine as a Therapeutic Strategy to Enhance

Radiosensitivity. *Adv Healthcare Materials* 2024, 13 (16), 2400381. https://doi.org/10.1002/adhm.202400381.

[293] Zambrano, J. N., Williams, C. J., Williams, C. B., Hedgepeth, L., Burger, P., Dilday, T., Eblen, S. T., Armeson, K., Hill, E. G., Yeh, E. S. Staurosporine, an Inhibitor of Hormonally up-Regulated Neu-Associated Kinase. *Oncotarget* 2018, 9 (89), 35962–35973. https://doi.org/10.18632/oncotarget.26311.

[294] Hou, J., Liu, P., Qu, H., Fu, P., Wang, Y., Wang, Z., Li, Y., Teng, X., Zhu, W. Gilvocarcin HE: A New Polyketide Glycoside from Streptomyces Sp. *J Antibiot* 2012, 65 (10), 523–526. https://doi.org/10.1038/ja.2012.61.

[295] Karthikeyan, A., Joseph, A., Nair, B. G. Promising Bioactive Compounds from the Marine Environment and Their Potential Effects on Various Diseases. *Journal of Genetic Engineering and Biotechnology* 2022, 20 (1), 14. https://doi.org/10.1186/s43141-021-00290-4.

[296] Li, K., Cai, J., Su, Z., Yang, B., Liu, Y., Zhou, X., Huang, J., Tao, H. Glycosylated Natural Products From Marine Microbes. *Front. Chem.* 2020, 7, 879. https://doi.org/10.3389/fchem.2019.00879.

[297] Herzon, S. B. The Mechanism of Action of (−)-Lomaiviticin A. *Acc. Chem. Res.* 2017, 50 (10), 2577–2588. https://doi.org/10.1021/acs.accounts.7b00347.

[298] He, H., Ding, W.-D., Bernan, V. S., Richardson, A. D., Ireland, C. M., Greenstein, M., Ellestad, G. A., Carter, G. T. Lomaiviticins A and B, Potent Antitumor Antibiotics from Micromonospora l Omaivitiensis. *J. Am. Chem. Soc.* 2001, 123 (22), 5362–5363. https://doi.org/10.1021/ja010129o.

[299] Salem, S. Biosynthesis of Marineosin, a *Spiroaminal Undecylprodiginine Natural Product*, 2000. https://doi.org/10.15760/etd.936.

[300] Varghese, S., Jisha, M. S., Rajeshkumar, K. C., Gajbhiye, V., Alrefaei, A. F., Jeewon, R. Endophytic Fungi: A Future Prospect for Breast Cancer Therapeutics and Drug Development. *Heliyon* 2024, 10 (13), e33995. https://doi.org/10.1016/j.heliyon.2024.e33995.

[301] Wang, Y., Islam, Md. A., Davis, R. A., McAlpine, S. R. The Fungal Natural Product (1S,3S)-Austrocortirubin Induces DNA Damage in HCT116 Cells via a Mechanism Unique from Other DNA Damaging Agents. *Bioorganic & Medicinal Chemistry Letters* 2015, 25 (2), 249–253. https://doi.org/10.1016/j.bmcl.2014.11.055.

[302] Wang, Y., Zhang, F., Zhang, Y., Liu, J. O., Ma, D. Synthesis and Antitumor Activity of Cyclodepsipeptide Zygosporamide and Its Analogues. *Bioorganic & Medicinal Chemistry Letters* 2008, 18 (15), 4385–4387. https://doi.org/10.1016/j.bmcl.2008.06.066.

[303] Shin, H. J., Kim, T. S., Lee, H.-S., Park, J. Y., Choi, I.-K., Kwon, H. J. Streptopyrrolidine, an Angiogenesis Inhibitor from a Marine-Derived Streptomyces Sp. KORDI-3973. *Phytochemistry* 2008, 69 (12), 2363–2366. https://doi.org/10.1016/j.phytochem.2008.05.020.

[304] Li, W., Zhang, C., Zhang, H., Dong, R., Liu, J., Wang, C., Wang, M., Wang, Y., Wang, C., Zhang, Y., Shi, L., Xu, Y., Sun, L.-P. Design, Synthesis, and Anticancer Evaluation of Ammosamide B with Pyrroloquinoline Derivatives as Novel BRD4 Inhibitors. *Bioorganic Chemistry* 2022, 127, 105917. https://doi.org/10.1016/j.bioorg.2022.105917.

[305] Takemura, M., Sahara, T., Misawa, N. Violaxanthin: Natural Function and Occurrence, Biosynthesis, and Heterologous Production. *Appl Microbiol Biotechnol* 2021, 105 (16–17), 6133–6142. https://doi.org/10.1007/s00253-021-11452-2.

[306] Ahmed, S. A., Mendonca, P., Messeha, S. S., Soliman, K. F. A. Anticancer Effects of Fucoxanthin through Cell Cycle Arrest, Apoptosis Induction, and Angiogenesis Inhibition in Triple-Negative Breast Cancer Cells. *Molecules* 2023, 28 (18), 6536. https://doi.org/10.3390/molecules28186536.

[307] Long, Q., Xiao, X., Yi, P., Liu, Y., Varier, K. M., Rao, Q., Song, J., Qiu, J., Wang, C., Liu, W., Gajendran, B., He, Z., Liu, S., Li, Y. L20, a Calothrixin B Analog, Induces Intrinsic Apoptosis on HEL Cells through ROS/γ-H2AX/P38 MAPK Pathway. *Biomedicine & Pharmacotherapy* 2021, 137, 111336. https://doi.org/10.1016/j.biopha.2021.111336.

[308] Rechkoblit, O., Johnson, R. E., Buku, A., Prakash, L., Prakash, S., Aggarwal, A. K. Structural Insights into Mutagenicity of Anticancer Nucleoside Analog Cytarabine during Replication by DNA Polymerase η. *Sci Rep* 2019, 9 (1), 16400. https://doi.org/10.1038/s41598-019-52703-7.

[309] Pereira, R. B., Evdokimov, N. M., Lefranc, F., Valentão, P., Kornienko, A., Pereira, D. M., Andrade, P. B., Gomes, N. G. M. Marine-Derived Anticancer Agents: Clinical Benefits, Innovative Mechanisms, and New Targets. *Marine Drugs* 2019, 17 (6), 329. https://doi.org/10.3390/md17060329.

[310] Abramson, J. S., Straus, D. J., Bartlett, N. L., Burke, J. M., Lynch, R. C., Domingo Domenech, E., Hess, B., Schuster, S. R., Linhares, Y., Ramchandren, R., Gandhi, M., Mowat, R., Shah, H., Jurczak, W., Re, A., Hahn, U., Prince, H. M., Guo, W., Ho, L., Beck, R. C., Yasenchak, C. A., Lee, H. J. Brentuximab Vedotin, Nivolumab, Doxorubicin, and Dacarbazine (AN+AD) for Early-Stage Classical Hodgkin Lymphoma (SGN35-027 Part C). *Blood* 2023, 142 (Supplement 1), 611–611. https://doi.org/10.1182/blood-2023-172574.

[311] Gadducci, A., Cosio, S. Trabectedin and Lurbinectedin: Mechanisms of Action, Clinical Impact, and Future Perspectives in Uterine and Soft Tissue Sarcoma, Ovarian Carcinoma, and Endometrial Carcinoma. *Front. Oncol.* 2022, 12, 914342. https://doi.org/10.3389/fonc.2022.914342 .

[312] Leal, J., Martínez-Díez, M., García-Hernández, V., Moneo, V., Domingo, A., Bueren-Calabuig, J., Negri, A., Gago, F., Guillén-Navarro, M., Avilés, P., Cuevas, C., García-Fernández, L., Galmarini, C. PM01183, a New DNA Minor Groove Covalent Binder with Potent in Vitro and in Vivo Anti-tumour Activity. *British J Pharmacology* 2010, 161 (5), 1099–1110. https://doi.org/10.1111/j.1476-5381.2010.00945.x.

[313] Kauffmann-Guerrero, D., Huber, R. M. Orphan Drugs in Development for the Treatment of Small-Cell Lung Cancer: Emerging Data on Lurbinectedin. *LCTT 2020*, Volume 11, 27–31. https://doi.org/10.2147/LCTT.S239223.

[314] Monk, B. J., Lorusso, D., Italiano, A., Kaye, S. B., Aracil, M., Tanović, A., D'Incalci, M. Trabectedin as a Chemotherapy Option for Patients with BRCA Deficiency. *Cancer Treatment Reviews* 2016, 50, 175–182. https://doi.org/10.1016/j.ctrv.2016.09.009.

[315] Li, K., Tay, F. R., Yiu, C. K. Y. The Past, Present and Future Perspectives of Matrix Metalloproteinase Inhibitors. *Pharmacology & Therapeutics* 2020, 207, 107465. https://doi.org/10.1016/j.pharmthera.2019.107465.

[316] Miller, L. H., Maxa, K. L., Winter, S. S., Gossai, N. P. The Role of Nelarabine in the Treatment of T-Cell Acute Lymphoblastic Leukemia/Lymphoma: Challenges, Opportunities, and Future Directions. *Expert Review of Anticancer Therapy* 2023, 23 (12), 1229–1236. https://doi.org/10.1080/14737140.2023.2271662.

[317] Montillo, M., Ricci, Tedeschi. Fludarabine in the Treatment of Chronic Lymphocytic Leukemia: A Review. *TCRM* 2009, 187. https://doi.org/10.2147/TCRM.S3688.

[318] Fan, Y. Eribulin and Its Application as Anti-Cancer Treatment. *TNS* 2023, 6 (1), 86–93. https://doi.org/10.54254/2753-8818/6/20230176.

[319] Tilly, H., Morschhauser, F., Sehn, L. H., Friedberg, J. W., Trněný, M., Sharman, J. P., Herbaux, C., Burke, J. M., Matasar, M., Rai, S., Izutsu, K., Mehta-Shah, N., Oberic, L., Chauchet, A., Jurczak, W., Song, Y., Greil, R., Mykhalska, L., Bergua-Burgués, J. M., Cheung, M. C., Pinto, A., Shin, H.-J., Hapgood, G., Munhoz, E., Abrisqueta, P., Gau, J.-P., Hirata, J., Jiang, Y., Yan, M., Lee, C., Flowers, C. R., Salles, G. Polatuzumab Vedotin in Previously Untreated Diffuse Large B-Cell Lymphoma. *N Engl J Med* 2022, 386 (4), 351–363. https://doi.org/10.1056/NEJMoa2115304 .

[320] S. Hanna, PharmD, Bcps, Bcop, K. Advancements in Therapy for Bladder Cancer: Enfortumab Vedotin. *JADPRO* 2020, 11 (4). https://doi.org/10.6004/jadpro.2020.11.4.8.

[321] Gallogly, M. M., Lazarus, H. M., Cooper, B. W. Midostaurin: A Novel Therapeutic Agent for Patients with FLT3-Mutated Acute Myeloid Leukemia and Systemic Mastocytosis. *Therapeutic Advances in Hematology* 2017, 8 (9), 245–261. https://doi.org/10.1177/2040620717721459.

[322] Robles-Bañuelos, B., Durán-Riveroll, L. M., Rangel-López, E., Pérez-López, H. I., González-Maya, L. Marine Cyanobacteria as Sources of Lead Anticancer Compounds: A Review of Families of Metabolites with Cytotoxic, Antiproliferative, and Antineoplastic Effects. *Molecules* 2022, 27 (15), 4814. https://doi.org/10.3390/molecules27154814.

[323] Lassiter, G., Bergeron, C., Guedry, R., Cucarola, J., Kaye, A. M., Cornett, E. M., Kaye, A. D., Varrassi, G., Viswanath, O., Urits, I. Belantamab Mafodotin to Treat Multiple Myeloma: A Comprehensive Review of Disease, Drug Efficacy and Side Effects. *Current Oncology* 2021, 28 (1), 640–660. https://doi.org/10.3390/curroncol28010063.

[324] Ferlay, J., Colombet, M., Soerjomataram, I., Parkin, D. M., Piñeros, M., Znaor, A., Bray, F. Cancer Statistics for the Year 2020: An Overview. *Intl Journal of Cancer* 2021, 149 (4), 778–789. https://doi.org/10.1002/ijc.33588.

[325] Bray, F., Ferlay, J., Soerjomataram, I., Siegel, R. L., Torre, L. A., Jemal, A. Global Cancer Statistics 2018: GLOBOCAN Estimates of Incidence and Mortality Worldwide for 36 Cancers in 185 Countries. *CA A Cancer J Clinicians* 2018, 68 (6), 394–424. https://doi.org/10.3322/caac.21492.

[326] Tan, S. C., Poh, W. T., Yong, A. C. H., Chua, E. W., Ooi, D. J., Mahmud, R., Thiagarajan, M., Stanslas, J. Challenges and Strategies for Improving Access to Cancer Drugs in Malaysia: Summary of Opinions Expressed at *the 2nd MACR International Scientific Conference* 2022. CMAR 2023, Volume 15, 851–862. https://doi.org/10.2147/CMAR.S420890 .

[327] Avhad, A. B., Bhangale, C. J. Marine Natural Products and Derivatives. *RPS Pharmacy and Pharmacology Reports* 2023, 2 (2), rqad008. https://doi.org/10.1093/rpsppr/rqad008.

[328] Rasmi, Y., Kirboğa, K. K., Rao, T., Ali, M., Ahmed, M. Z. Marine Natural Products as a Bioresource for Cosmeceuticals. In Bioprospecting of Natural Sources for Cosmeceuticals, Kathuria, D., Sharma, A., Verma, M., Nayik, G. A., Eds., *Royal Society of Chemistry*, 2024, pp 223–275. https://doi.org/10.1039/9781837672288-00223.

[329] Lindequist, U. Marine-Derived Pharmaceuticals - Challenges and Opportunities. *Biomolecules & Therapeutics* 2016, 24 (6), 561–571. https://doi.org/10.4062/biomolther.2016.181.

[330] Ozyerli-Goknar, E., Bagci-Onder, T. Epigenetic Deregulation of Apoptosis in Cancers. *Cancers* 2021, 13 (13), 3210. https://doi.org/10.3390/cancers13133210.

[331] Chinnadurai, R. K., Khan, N., Meghwanshi, G. K., Ponne, S., Althobiti, M., Kumar, R. Current Research Status of Anti-Cancer Peptides: Mechanism of Action, Production, and Clinical Applications. *Biomedicine & Pharmacotherapy* 2023, 164, 114996. https://doi.org/10.1016/j.biopha.2023.114996.

Chapter 9

Neurological Disorders and Beyond: Marine Metabolites for Brain Health

N. K. Mujawar[1,*]
and S. B. Patil[2]
[1]Department of Pharmaceutics, Women College of Pharmacy, Peth Vadgaon, Kolhapur, Maharashtra, India
[2]Department of Pharmacology, Dr. Shivajirao Kadam College of Pharmacy,
Kasbe Digraj, Sangli, Maharashtra, India

Abstract

Worldwide, public health is seriously threatened by neurological illnesses for which there are few efficient therapies. Marine metabolites have been the focus of recent studies as possible therapeutic agents for enhancing brain health and treating neurological diseases. An overview of the developing area of marine-derived chemicals in neuroscience is given in this abstract, with particular attention to their therapeutic potential, mechanisms of action, and future possibilities. Among the many bioactive metabolites produced by marine creatures are algae, sponges, and microbes, all showing promise for neurological impacts. Many pharmacological actions of these substances include neuroprotection, anti-inflammatory, antioxidant, and neurotrophic qualities, which are crucial for preserving brain function and preventing neurodegeneration. In preclinical research and early clinical trials, several marine-derived chemicals have demonstrated benefits for neurological diseases, including stroke, Parkinson's disease, and Alzheimer's disease. Fish oil's omega-3 fatty acids, for instance, have been shown to have neuroprotective and cognitive-enhancing properties. Furthermore, there is promise for marine polyphenols and peptides to regulate neurotransmitter systems and lower neuroinflammation. Technological developments in metabolomics, bioinformatics, and synthetic biology have quickened the search for and creation of neuroactive substances obtained from the sea. Moreover, the sustainable use of marine resources and developing marine metabolites into clinically feasible medicines need the cooperation of marine biologists, pharmacologists, and neurologists. In conclusion, novel substances for improving brain health and treating neurological diseases are promisingly found in marine metabolites. Identifying safe and efficient treatments for the increasing global burden of neurological disorders is highly possible with continued study in this area.

---

* Corresponding Author's Email: nigarwcop2017@gmail.com

In: Healing from the Depths Marine Metabolites in Disease Management
Editors: Pranav Kumar Prabhakar and Tarun Kumar Upadhyay
ISBN: 979-8-89530-832-5

Keywords: neurological disorders, marine metabolites, brain health, neuroactive compounds, therapeutic agents

1. Introduction of Neurological Disorders

Neurological disorders encompass a broad range of medical conditions that hinder the effective functioning of the nervous system. This complex network comprises the brain, spinal cord, and an extensive array of nerves that spread throughout the body, controlling functions such as muscle movement, sensory perception, cognitive tasks, and emotional regulation. Disruption in any part of this system can lead to significant impairment of vital bodily functions, often resulting in serious ramifications for a person's physical and mental health [1].

A variety of causative factors contribute to neurological disorders. These include genetic mutations (such as those responsible for Huntington's disease), prenatal or birth-related injuries, infections like meningitis or encephalitis, autoimmune responses (as seen in multiple sclerosis), trauma to the brain or spinal cord, exposure to toxic substances, and even lifestyle and environmental stressors. These factors may act alone or in combination, making diagnosis and treatment particularly challenging [2].

The symptoms of neurological disorders can vary significantly depending on the region of the nervous system involved and the nature of the disorder. Sensory disruptions such as numbness, tingling, or pain may indicate peripheral nerve damage. Motor symptoms, including muscle weakness, paralysis, or involuntary movements like tremors, point to involvement of motor pathways. Cognitive and behavioural manifestations, such as memory loss, confusion, depression, or hallucinations, are common in disorders affecting the cortex or deeper structures of the brain [3].

Neurological disorders are broadly classified based on their underlying pathology and the regions they affect. Neurodegenerative diseases, such as Alzheimer's and Parkinson's, involve progressive loss of neuron function. Seizure disorders, including epilepsy, result from abnormal electrical activity in the brain. Vascular disorders, such as stroke, occur due to interruption of the blood supply. Other classifications include demyelinating diseases (e.g., multiple sclerosis) and headache syndromes, like migraines [4-5].

Diagnosis is often a multistep process involving a detailed medical history, neurological examinations, neuroimaging (such as MRI and CT scans), and, in some cases, electrophysiological studies like EEG or nerve conduction tests. Laboratory tests and genetic screening may be employed to confirm certain conditions [6].

Management strategies vary depending on the type and severity of the condition. Common treatments include medications (e.g., anticonvulsants, dopamine agonists), rehabilitation therapies (physical, occupational, and speech), psychotherapy, and surgical interventions (such as deep brain stimulation for Parkinson's disease). For many disorders, long-term care with a multidisciplinary team—neurologists, therapists, psychiatrists, and social workers—is essential. This integrated approach enhances symptom control and improves quality of life [7].

2. Types of Neurological Disorders

Neurological disorders refer to a wide variety of ailments that impact the neurological system, which includes the brain, spinal cord, and peripheral nerves. Each neurological condition may necessitate distinct diagnosis, treatment, and care approaches. Understanding these categories enables healthcare practitioners to properly define and handle the specific requirements of neurological disease patients, resulting in more successful care and treatment outcomes. These disorders can be classified as different based on numerous criteria shown in Figure 1 below.

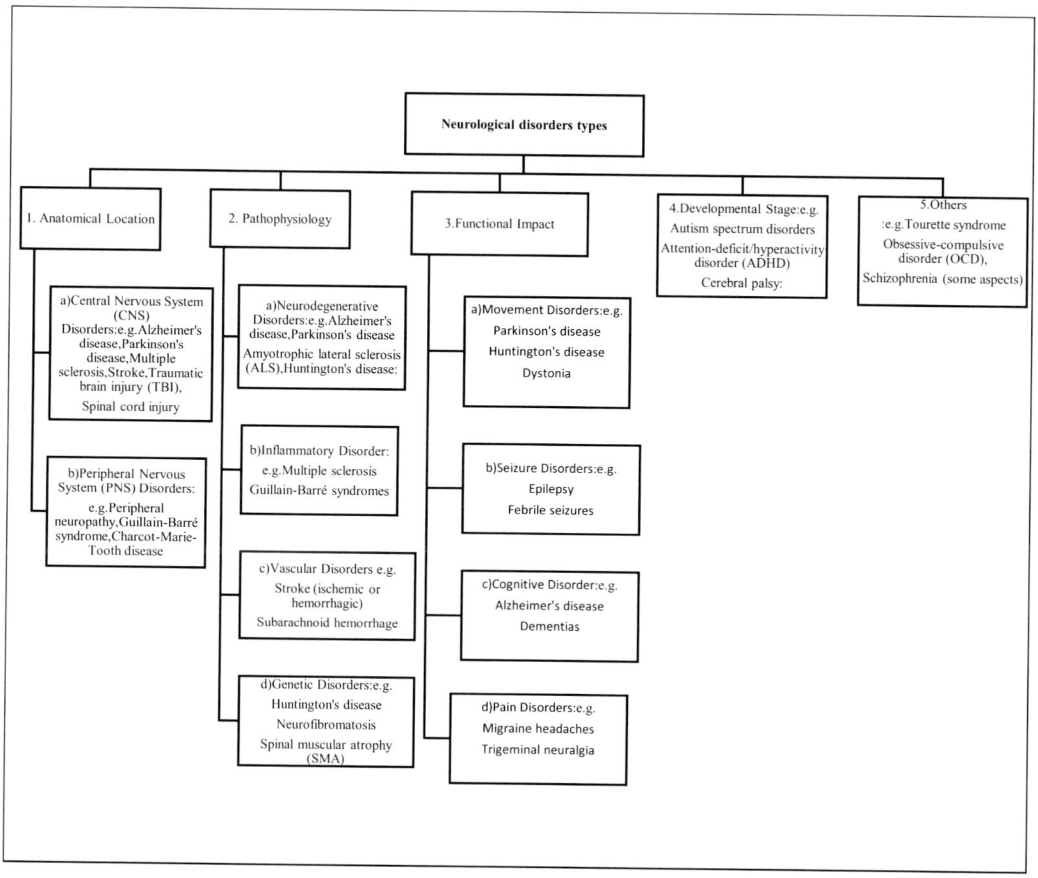

Figure 1. Types of Neurological disorders. Neurological disorders are a broad category of ailments affecting the neurological system, which includes the brain, spinal cord, and peripheral nerves. These disorders can be classified into groups based on criteria, such as disease introduction, causes, symptoms, and therapies, detailed in Table 1 below.

3. Neurological Disorders Challenges in Treatment

Because of their complexity and the variety of symptoms and courses among people, neurological illnesses provide unique treatment problems. Some of the significant challenges are:

Table 1. Neurological disorders

Sr. No.	Neurological Disorder	Introduction	Causes	Symptoms	Treatments
1	Alzheimer's Disease	A progressive neurodegenerative disorder affecting memory as well cognition.	Genetic factors (e.g., APOE gene), brain changes (amyloid plaques, neurofibrillary tangles), ageing.	Memory loss, confusion, difficulty in problem-solving, disorientation, and mood changes.	Medications (cholinesterase inhibitors, memantine), cognitive therapy, behavioral interventions, and support services [8].
2	Parkinson's Disease	A neurodegenerative disorder affecting movement due to dopamine deficiency.	Genetic predisposition, environmental factors, ageing.	Tremors, rigidity, bradykinesia (slow movement), postural instability, freezing episodes.	Medications (levodopa, dopamine agonists), deep brain stimulation, physical therapy, speech therapy, occupational therapy [9].
3	Multiple Sclerosis (MS)	The autoimmune disease causes demyelination of nerve fibres in the central nervous system.	Genetic susceptibility, environmental factors (e.g., viral infections), immune system dysfunction.	Fatigue, numbness or tingling, muscle weakness, vision problems, cognitive impairment, bladder dysfunction.	Disease-modifying therapies (interferons, monoclonal antibodies), corticosteroids, physical therapy, occupational therapy, speech therapy, and symptomatic treatments (for specific symptoms like spasticity) [10].
4	Stroke	Sudden interruption of blood flow to the brain causes brain cell damage and neurological deficits.	Ischemic (blood clot) or hemorrhagic (bleeding) causes risk factors like hypertension and diabetes.	Sudden numbness or weakness (usually on one side of the body), confusion, trouble speaking or understanding, severe headache.	Thrombolytic therapy (alteplase), mechanical thrombectomy (clot retrieval), rehabilitation (physical therapy, speech therapy, occupational therapy), prevention and management of risk factors [11].
5	Traumatic Brain Injury (TBI)	Brain damage caused by an external force, resulting in cognitive, physical, and emotional impairments	Falls, motor vehicle accidents, sports injuries, assaults, blast injuries (military settings).	Headache, dizziness, memory problems, difficulty concentrating, mood changes, sensory deficits.	Emergency medical care, monitoring and stabilization, imaging (CT scan, MRI), rehabilitation (physical therapy, speech therapy, occupational therapy), and medications (for symptoms like headaches or seizures) [12].
6	Spinal Cord Injury	Damage to the spinal cord results in loss of function below the level of injury, potentially including paralysis.	Trauma (e.g., accidents, falls, sports injuries), violence, diseases (e.g., spinal tumours, infections).	Loss of sensation or movement, paralysis, changes in reflexes, pain, bowel and bladder dysfunction, and respiratory problems (depending on the level and severity of injury).	Emergency medical treatment, stabilization, surgery (if applicable), rehabilitation (physical therapy, occupational therapy, speech therapy), assistive devices (wheelchairs, braces), and medications (for pain and spasticity) [13].
7	Peripheral Neuropathy	A condition involving peripheral nerve damage leads to pain, numbness, tingling, and muscle weakness.	Diabetes, infections (e.g., HIV, Lyme disease), autoimmune diseases (e.g., Guillain-Barré syndrome), toxins.	Burning or shooting pain, tingling or numbness, muscle weakness, sensitivity to touch, coordination difficulties.	Treatment of underlying conditions (e.g., glycaemic control for diabetes), medications (e.g., pain relievers, antidepressants, anticonvulsants), and physical therapy [14].
8	Guillain-Barré Syndrome (GBS)	An autoimmune disorder where the immune system attacks peripheral nerves, leading to muscle weakness and paralysis.	It often follows viral or bacterial infections (e.g., Campylobacter jejuni, Zika virus) and autoimmune response.	Ascending muscle weakness, tingling or numbness, difficulty with facial movements, respiratory weakness (in severe cases).	Supportive care (monitoring for respiratory function, physical therapy), immunoglobulin therapy, plasmapheresis (to remove antibodies) [15].

Sr. No.	Neurological Disorder	Introduction	Causes	Symptoms	Treatments
9	Charcot-Marie-Tooth Disease (CMT)	An inherited genetic disorder affecting peripheral nerves, resulting in muscle weakness and sensory loss.	Genetic mutations affect proteins in peripheral nerves (e.g., myelin or axon proteins).	Muscle weakness (typically in lower legs and feet), foot deformities, loss of muscle tissue, and sensory loss.	Physical therapy (to maintain mobility and strength), orthopaedic devices (e.g., braces), surgical interventions (for severe deformities), and genetic counselling [16].
10	Amyotrophic Lateral Sclerosis (ALS)	Progressive neurodegenerative disease affecting motor neurons in the brain and spinal cord.	Genetic mutations (e.g., in the SOD1 gene), environmental factors, oxidative stress.	Muscle weakness, twitching, difficulty speaking, swallowing, and breathing, and eventual paralysis.	Supportive care (respiratory support, feeding tubes), medications (riluzole, edaravone), physical therapy, and assistive devices [17].
11	Huntington's Disease	Genetic disorders cause progressive breakdown of nerve cells in the brain.	Inherited genetic mutation (HTT gene), affects neurotransmitter levels and cellular energy production.	Involuntary movements (chorea), cognitive decline, psychiatric symptoms (e.g., depression, anxiety).	Medications (antipsychotics, antidepressants), supportive care, genetic counselling, and clinical trials [18].
12	Subarachnoid Haemorrhage	Bleeding into the space between the brain and the surrounding membrane (subarachnoid space).	Ruptured aneurysm, trauma, arteriovenous malformation (AVM), high blood pressure.	Sudden severe headache (often described as "thunderclap"), nausea, vomiting, neck stiffness, and sensitivity to light.	Emergency medical intervention, surgical repair (clipping or coiling of aneurysm), supportive care (pain management, monitoring) [19].
13	Neurofibro-matosis	Genetic disorders cause tumours to grow on nerve tissue, leading to various neurological and non-neurological complications	Inherited genetic mutation (NF1 or NF2 gene) affects tumour suppressor genes.	Neurofibromas (tumours on nerves), skin changes (café-au-lait spots), bone deformities, and learning disabilities.	Monitoring for complications, surgery (for tumours causing complications), medications (for pain and seizures), and genetic counselling [20].
14	Spinal Muscular Atrophy (SMA)	Genetic disorders cause progressive muscle weakness and atrophy due to the loss of motor neurons in the spinal cord.	Genetic mutations affecting survival motor neuron 1 (SMN1) gene.	Muscle weakness, poor muscle tone, respiratory difficulties, difficulty swallowing.	Supportive care (respiratory support, feeding assistance), medications (nusinersen, gene therapy), and physical therapy [21].
15	Dystonia	Neurological movement disorder causes involuntary muscle contractions and repetitive twisting movements.	Primary (genetic mutations), secondary to brain injury, stroke, infections, or drugs, idiopathic (cause unknown).	Involuntary muscle contractions (sustained or intermittent), twisting or repetitive movements, abnormal postures (e.g., torsion dystonia, cervical dystonia), tremors, and pain.	Medications (muscle relaxants, botulinum toxin injections), physical therapy, occupational therapy, speech therapy, deep brain stimulation (for severe cases), surgery (selective denervation or deep brain stimulation), and lifestyle adjustments [22].
16	Epilepsy	Chronic neurological disorder characterized by recurrent seizures due to abnormal brain activity.	Genetic predisposition, brain injury (e.g., trauma, stroke), infections (e.g., meningitis), developmental disorders, tumours, vascular abnormalities, idiopathic (unknown cause).	Seizures (various types: generalized tonic-clonic, absence, focal), loss of consciousness, convulsions, unusual sensations or emotions, temporary confusion, and staring spells.	Antiepileptic drugs (AEDs), ketogenic diet, surgery (for epilepsy focus removal), vagus nerve stimulation, responsive neurostimulation (RNS), deep brain stimulation (DBS) in select cases, lifestyle modifications (e.g., sleep, stress management) [23].

Table 1. (Continued)

Sr. No.	Neurological Disorder	Introduction	Causes	Symptoms	Treatments
17	Febrile Seizures	Seizures triggered by fever are most common in young children.	A rapid rise in body temperature, typically due to viral infections.	Seizures (brief, usually generalized tonic-clonic), fever (often over 100.4°F or 38°C), loss of consciousness, shaking or jerking limbs.	Management of fever (antipyretics), reassurance, observation, treatment of underlying illness (if present), education, and guidance for caregivers [24].
18	Dementias	Group of conditions characterized by cognitive decline and memory loss severe enough to interfere with daily life.	Alzheimer's disease (most common), vascular dementia, Lewy body dementia, frontotemporal dementia, mixed dementia (combination of Alzheimer's and vascular), and traumatic brain injury.	Memory loss, difficulty in problem-solving, confusion, disorientation, personality changes, mood swings, withdrawal from social activities.	Medications (cholinesterase inhibitors, memantine), cognitive therapy, behavioral interventions (e.g., structured routines, memory aids), support services (caregiver support, adult day care), and lifestyle modifications (e.g., diet, exercise) [25].
19	Migraine Headaches	Other symptoms often accompany recurrent severe headaches.	Genetic predisposition, neurovascular dysfunction, triggers (e.g., stress, certain foods, hormonal changes).	Throbbing headache (often unilateral), nausea, vomiting, sensitivity to light and sound, aura (visual disturbances).	Acute medications (triptans, NSAIDs), preventive medications (beta-blockers, antiepileptic drugs), lifestyle modifications (e.g., avoiding triggers, regular sleep), and complementary therapies (acupuncture, biofeedback) [26].
20	Trigeminal Neuralgia	Chronic pain affecting the trigeminal nerve, causing intense facial pain.	Compression of the trigeminal nerve by blood vessels, tumour, or MS lesions, demyelination of nerve fiber tumour, or MS lesions, demyelination of nerve fibres.	Intense, stabbing or electric shock-like pain in the face (typically on one side), triggered by light touch, eating, talking, or even a breeze.	Medications (anticonvulsants, muscle relaxants), nerve blocks, surgery (microvascular decompression, rhizotomy), stereotactic radiosurgery (Gamma Knife), complementary therapies (acupuncture), lifestyle modifications (e.g., diet) [27].
21	Attention-Deficit/ Hyperactivity Disorder (ADHD)	A neurodevelopmental disorder characterized by inattention, hyperactivity, and impulsivity.	Genetic factors, brain structure and function differences, and environmental factors (e.g., prenatal exposure to toxins and maternal smoking).	Inattention, hyperactivity, impulsivity, difficulty organizing tasks, forgetfulness, excessive talking, restlessness.	Stimulant medications (e.g., methylphenidate, amphetamines), non-stimulant medications (e.g., atomoxetine), behavioral therapies (e.g., parent training, cognitive behavioral therapy), educational interventions (e.g., classroom accommodations) [28].
22	Cerebral Palsy	Non-progressive neurological disorder affecting movement and posture is usually caused by brain damage before or during birth.	Brain injury or abnormal development in utero or early infancy (e.g., oxygen deprivation, infections, genetic mutations).	Spasticity (muscle stiffness), poor coordination, involuntary movements, speech and swallowing difficulties, and joint contractures.	Physical therapy, occupational therapy, speech therapy, medications (muscle relaxants, botulinum toxin injections), orthopaedic devices (e.g., braces, walkers), and surgery (for severe cases or to correct deformities) [29].

Sr. No.	Neurological Disorder	Introduction	Causes	Symptoms	Treatments
23	Tourette Syndrome	A neurological disorder characterized by repetitive, involuntary movements and vocalizations (tics).	Genetic predisposition, abnormalities in neurotransmitter systems (e.g., dopamine).	Motor tics (e.g., eye blinking, head jerking), vocal tics (e.g., throat clearing, swearing), behavioural symptoms (e.g., compulsions, impulsivity).	Behavioural therapies (e.g., habit reversal training, cognitive behavioral therapy), medications (antipsychotics, alpha-2 adrenergic agonists), deep brain stimulation (for severe cases), and supportive therapies (education, support groups) [30].
24	Obsessive-Compulsive Disorder (OCD)	Obsessive thoughts and compulsive behaviours characterize chronic mental health disorders.	Genetic predisposition, brain structure and function differences, environmental factors (e.g., stress, childhood trauma).	Obsessions (persistent, unwanted thoughts), compulsions (repetitive behaviors or mental acts), anxiety.	Cognitive behavioral therapy (CBT), exposure and response prevention (ERP), medications (selective serotonin reuptake inhibitors SSRIs, clomipramine), and lifestyle adjustments (stress management, regular exercise) [31].
25	Schizophrenia (some aspects)	A chronic psychiatric disorder affecting thoughts, emotions, and behaviour.	Genetic predisposition, abnormalities in brain structure and function (e.g., neurotransmitter imbalances), environmental factors (e.g., prenatal exposure to infections, stress).	Delusions (false beliefs), hallucinations (seeing or hearing things that aren't there), disorganized thinking, reduced emotional expression, and social withdrawal.	Antipsychotic medications (typical and atypical), psychotherapy (e.g., cognitive behavioral therapy, family therapy), supportive services (housing, vocational rehabilitation), and lifestyle modifications (exercise, healthy diet) [32].

1. Neurological illnesses: Neurological illnesses vary in their causes, symptoms, and clinical manifestations. This variety makes it difficult to establish universal therapeutic techniques that work for all patients.
2. Difficulty in Diagnosis: Multiple neurological illnesses might present with overlapping symptoms or atypical features, causing delays or misdiagnoses. This can impede early intervention and proper therapy initiation.
3. Limited Understanding of Pathophysiology: Despite improvements in neuroscience, the fundamental processes of many neurological illnesses remain unclear. This hinders the development of focused medicines that address the underlying cause of the disease [33].
4. Neuroplasticity and Recovery: The nervous system's ability to adapt and reorganize differs by disease and individual. Promoting recovery or preventing further decline frequently necessitates therapies that successfully harness or increase neural plasticity.
5. Drug Delivery Challenges: Drugs targeting the central nervous system (CNS) may struggle to pass the blood-brain barrier and reach therapeutic concentrations in affected areas. This complicates pharmaceutical effectiveness and may result in systemic side effects.
6. Long-term Management and Care: Chronic and progressive neurological illnesses require regular monitoring, therapy changes, and carer and patient support [34].
7. Multidisciplinary Care: Neurological problems may interact with other medical conditions, requiring coordinated care from neurologists, psychiatrists, and physiatrists to fully address complicated clinical demands.
8. Psychosocial Impact: Neurological illnesses can affect cognitive function, behaviour, and emotional well-being, creating significant issues for patients and families. Addressing these issues is critical for achieving comprehensive treatment outcomes [35].
9. Cost and Access to Care: Treating neurological illnesses can be costly, including drugs, therapies, assistive equipment, and surgical procedures. Access to specialized treatment, especially in rural or disadvantaged locations, can be limited.
10. Research and Innovation: Neurological illness therapy relies on continuous research and innovation. We need funding and cross-disciplinary collaboration to further our understanding and treatment of these complicated illnesses.

Addressing these issues necessitates a diverse approach that includes scientific research, professional expertise, patient advocacy, and public health programs. Collaboration among stakeholders, including healthcare professionals, researchers, policymakers, and advocacy groups, is critical to improving outcomes and quality of life for those with neurological illnesses [36].

4. Marine Metabolites: An Emerging Frontier

Marine metabolites are chemical substances many organisms produce, including algae, bacteria, sponges, corals, and even some fish. These metabolites have received much attention recently for their potential medicinal, agricultural, and industrial applications. Here is a summary of why marine metabolites are considered an emerging frontier [37]:

1. Biodiversity and Novelty: Marine settings are rich in biodiversity, with creatures adapting to severe conditions such as high pressure, cold temperatures, and low light. This diversity helps produce distinct and novel metabolites not found in terrestrial species.
2. Biological Activities: Marine metabolites have powerful antibacterial, antiviral, anticancer, anti-inflammatory, and antioxidant capabilities. These bioactive chemicals can potentially be used as novel medicines and therapeutic agents [38].
3. Drug Discovery Potential: Marine metabolites have the potential for drug discovery and development. Chemicals such as cytarabine (from a Caribbean sponge) and ziconotide (from a cone snail) have already been turned into therapeutically viable medications, highlighting the medicinal potential of marine-derived chemicals.
4. Ecological Significance: These metabolites play essential roles in marine species' defence against predators, rivals, and pathogens. Understanding their ecological functions can provide information on ecosystem dynamics and biodiversity conservation [39].
5. Biotechnological Applications: Marine metabolites have biotechnological applications beyond pharmaceuticals, such as biopesticides, growth stimulants in agriculture, and antioxidants and anti-ageing compounds in cosmetics.
6. Sustainable Resource: Marine creatures can be a sustainable supply of metabolites when fished appropriately or created through aquaculture and biotechnology. This alleviates the demand for land resources and ecosystems [40].

5. Challenges and Future Directions

1. Supply and Extraction: Challenges include sustainable procurement of marine species, efficient extraction technologies, and scaling up production to satisfy commercial demands without affecting natural populations.
2. Bioactivity and Mechanisms: Understanding marine metabolites' bioactivity and potential adverse effects is vital for their successful use in therapeutics [38].
3. Regulatory Challenges: Developing marine-derived pharmaceuticals requires navigating complex and time-consuming approval channels.
4. Technological Advances: Analytical techniques like genomics, metabolomics, and bioinformatics are crucial for rapidly finding and characterizing novel marine compounds.
5. Collaboration and financing: Scientists, industry, and policymakers work together to advance research, secure funding, and apply discoveries.

Marine metabolites are a rich source of bioactive chemicals with numerous applications in various industries. Continued study and innovation in this field promise to discover novel medications, biotechnological products, and ecological insights, all of which will benefit human health and the environment [41].

6. Exploration of Marine Resources

Marine resource exploration includes sustainably discovering, assessing, and utilizing varied biological and mineral resources in oceans and seas. Here's an overview of significant features and considerations while researching marine resources:

A. Biological resources
   1) Biodiversity Hotspots: Ocean biodiversity hotspots include microbes, algae, marine plants, sponges, corals, and fish/marine animals. These organisms generate many bioactive chemicals with potential applications in medicines, biotechnology, and food products.
   2) Drug Discovery: Marine species include numerous bioactive compounds with potential pharmacological use. Examples include antiviral chemicals from marine algae, anticancer drugs from marine sponges, and pain-relieving peptides from cone snails. Exploration entails bioprospecting missions to locate and characterize these chemicals [42].
   3) Aquaculture and Fisheries: Sustainable aquaculture involves raising marine organisms for food, pharmaceuticals, and bioproducts. It attempts to alleviate the burden on wild fish populations while satisfying rising worldwide demand for seafood and other marine products.
   4) Biotechnology Applications: Marine biomaterials and bioproducts have applications in biotechnology, including enzymes, biofuels, biomaterials, and bioremediation. For example, microorganisms from deep-sea hydrothermal vents create enzymes with specialized characteristics for harsh circumstances [43].

B. Mineral and Energy Resources
   1) Mineral Exploration: Oceans include abundant mineral resources, including manganese nodules, polymetallic sulfides, and cobalt-rich crusts. These minerals have potential applications in electronics, renewable energy technology (such as battery metals), and industry [44].
   2) Energy Exploration: Marine energy sources include offshore wind, wave, tidal, and ocean thermal energy conversion (OTEC). These renewable energy sources can potentially improve global energy security and reduce climate change.
   3) Deep-Sea Mining: This emerging business has environmental and regulatory challenges. Companies are developing ways to extract minerals while reducing environmental effects and adhering to international standards [45].

C. Challenges and Considerations
   1) Environmental Impact: Exploiting marine resources can harm habitats, biodiversity, and ecosystems. Sustainable practices and environmental impact assessments are critical for mitigating these effects [46].
   2) Regulatory Frameworks: International rules and regulations control marine resource exploration and exploitation, such as the United Nations Convention on the Law of the Sea (UNCLOS) and regional accords. Compliance with these frameworks is critical to responsible ocean stewardship.
   3) Technological Innovation: Remote sensing, robotics, autonomous vehicles, and deep-sea exploration technologies are advancing marine resource exploration.

These technologies advance efficiency, safety, and environmental monitoring [47].

4) Economic Viability: Assessing the economic viability of marine resource projects requires examining resource abundance, extraction costs, market demand, and regulatory compliance. Public-private partnerships and investment are critical to finance exploration projects.
5) Sustainability and Governance: Collaboration among stakeholders, such as governments, industry, research institutions, and local populations, is essential for sustainable marine resource management. Integrated ocean management approaches encourage conservation, equitable benefits, and long-term sustainability. Exploration of marine resources includes a variety of biological, mineral, and energy prospects, each with its own set of challenges and considerations. Sustainable practices, technological innovation, and international collaboration are critical for maximizing potential benefits while minimizing environmental and social consequences on marine ecosystems [48].

7. Importance of Marine Metabolites in Neuroscience

Marine metabolites are essential in neuroscience because of their distinct chemical structures and bioactive capabilities, which could be used to treat neurological illnesses and better understand brain function. Marine metabolites, generated from marine animals such as sponges, algae, and mollusks, contain bioactive chemicals with various pharmacological actions. These metabolites are especially useful in neuroscience for various reasons:

1. Neuroprotective Properties: Marine metabolites can protect neurons from oxidative stress, inflammation, and neurodegeneration. Examples include marine algae chemicals that protect against neurotoxicity.
2. Potential for medication Development: Marine metabolites offer new medication options for neurological disorders like Alzheimer's, Parkinson's disease, and stroke. Compounds such as bromotyrosine derivatives from marine sponges have been demonstrated to prevent neurodegenerative processes [49].
3. Neuroregenerative Potential: Marine chemicals can promote neuroregeneration, development, and adaptability. These qualities are essential for mending injured neural tissues in diseases such as spinal cord injury and traumatic brain injury.
4. Modulation of Neurotransmission: Marine metabolites can modulate neurotransmitter systems, affecting synaptic transmission and neural communication. This skill is important for illnesses that include neurotransmitter abnormalities, such as epilepsy and mood disorders [50].
5. Anti-inflammatory and Antioxidant Effects: Marine metabolites can reduce neuroinflammation and oxidative stress linked to neurodegenerative disorders and brain ageing.
6. Targeting Neurological Pathways: Marine bioactive chemicals can target specific neurological pathways or disease-related cellular targets. This focused strategy

improves efficacy while potentially lowering adverse effects compared to traditional treatments [51].

7. Biotechnological Applications: Advances in biotechnology allow for the manufacture and modification of marine metabolites to improve their pharmacological characteristics and bioavailability, making them potentially effective therapeutic agents for neurological diseases.
8. Environmental Sustainability: Using marine metabolites in brain research and medication development improves sustainable bioprospecting and marine conservation. Responsible sourcing and extraction procedures help to protect marine biodiversity and ecosystems.

Marine metabolites are an invaluable and diverse resource for neuroscience research and drug discovery. Their distinct chemical structures and biological activity present intriguing opportunities for creating novel therapeutics to treat neurological diseases and promote brain health. Continued investigation and use of marine biodiversity are promising for enhancing neuroscience and meeting unmet medical needs in neurology [52].

8. Pharmacological Activities of Marine Metabolites

Marine metabolites have a diverse spectrum of pharmacological actions, making them important sources of bioactive chemicals for various medical and commercial purposes. Here are some essential pharmacological actions of marine metabolites:

1. Anticancer and Antitumor Properties

The pharmacological activity of marine metabolites, with a focus on their anticancer and antitumor properties:

Mechanisms of Action:

a) Cytotoxicity: Marine metabolites can specifically target and kill cancer cells while sparing healthy cells.
b) Examples include halichondrins (derived from sea sponges), which affect microtubule activity in cancer cells, resulting in mitotic arrest and apoptosis [53].
c) Apoptosis Induction:

Marine metabolites can induce apoptosis (programmed cell death) in cancer cells via many mechanisms, including mitochondrial malfunction and caspase activation. Bryostatins, for example, promote apoptosis and prevent tumour growth by activating protein kinase C (PKC) pathways [54].

2. Anti-angiogenic Effects:

Marine metabolites can block angiogenesis, critical for tumour growth and metastasis. Compounds such as fucoidans (produced from brown algae) inhibit vascular endothelial growth factor (VEGF), lowering tumour vascularization and proliferation [55].

a) DNA Damage and Repair Inhibition:

Specific metabolites cause DNA damage and inhibit repair pathways in cancer cells, resulting in genomic instability and cell death. Trabectedin (ET-743), derived from the marine tunicate Ecteinascidiaturbinata, affects DNA repair pathways, increasing its cytotoxic effects on tumors [56].

3. Anti-inflammatory and Immunomodulatory Activities

Marine metabolites have potent anti-inflammatory and immunomodulatory properties, critical to their therapeutic potential in various disorders. Here is a summary of their principal pharmacological activity in various domains:

a) Anti-inflammatory Activity:
   1. Inhibition of Pro-inflammatory Mediators:

Marine metabolites can suppress pro-inflammatory cytokines and mediators, including TNF-α, IL-1, IL-6, and prostaglandins. Fucoidans, derived from brown algae, have been demonstrated to suppress the expression of pro-inflammatory cytokines and enzymes implicated in inflammation [57].

b) Modulation of Inflammatory Signaling Pathways:

They may disrupt the NF-κB and MAPK pathways, essential for inflammatory responses. For example, compounds found in marine species such as sponges and corals affect these pathways, reducing inflammatory reactions [58].

c) Antioxidant Effects:

Marine metabolites have significant antioxidant qualities that reduce free radicals and oxidative stress linked to chronic inflammation. This characteristic is essential for preventing oxidative damage caused by ageing and chronic diseases. For example, carotenoids from marine algae and polyphenols from marine plants work as antioxidants, shielding cells from oxidative damage and inflammation [59].

4. Antimicrobial and Antiviral Activities

Marine metabolites have garnered attention for their various pharmacological properties, particularly in treating microbial infections and viral illnesses. A summary of their primary antibacterial and antiviral characteristics is discussed below:

Antimicrobial Activities:

a) Antibacterial Effects:

Marine metabolites have potent antibacterial activity against various bacteria, including resistant forms. Peptides from marine creatures such as sponges and algae have antibacterial properties by disrupting bacterial cell membranes or interfering with critical bacterial functions.

b) Antifungal Effects:

Marine metabolites show antifungal properties against pathogenic fungi that cause diseases in humans and plants. Examples include compounds from marine fungi and algae, which limit fungal growth by targeting fungal cell walls or intracellular processes [60].

c) Antiparasitic Effects:

Marine metabolites have potential as antiparasitic agents for protozoan parasites and helminths. Examples include compounds found in marine sponges and mollusks that have activity against parasites that cause diseases like malaria and schistosomiasis [61].

d) Antiviral Activities:

Antiviral Effects: Marine metabolites have antiviral activity against several human viruses, including HSV, influenza, and HIV. Examples include sulfated polysaccharides from marine algae, which restrict viral entrance into host cells or interfere with viral reproduction mechanisms.

e) Anti-HIV Activity:

Marine metabolites have demonstrated anti-HIV action by blocking viral enzymes and preventing viral attachment to host cells. Compounds identified from marine sponges and tunicates have potential as HIV entry or reverse transcriptase inhibitors [62].

9. Neuroprotective and Neurological Activities

Marine metabolites have demonstrated promising pharmacological activity, notably in neuroprotection and neurological health. Here are some critical areas where these metabolites have shown significant effects discussed below:

1) Neuroprotection

Marine metabolites have been shown to protect neurons against various types of injury and degeneration. This includes [63]:

a) Antioxidant Activity:

Marine substances with significant antioxidant capabilities can reduce oxidative stress in neurons. Polyphenols, carotenoids, and peptides derived from marine sources are among examples.

b) Anti-inflammatory Effects:

Marine metabolites with anti-inflammatory characteristics can help reduce neuroinflammation, a prevalent component in neurodegenerative disorders [64].

c) Anti-apoptotic Effects:

Marine chemicals have anti-apoptotic effects, preventing neuronal cell death and maintaining brain function and structure [65].

2) Neurological activity:

Aside from neuroprotection, marine metabolites influence a variety of brain processes.

a) Neurotransmitter Modulation:

Certain chemicals can modulate neurotransmitter systems, potentially impacting mood, cognition, and behaviour [49].

b) Neurogenesis Promotion:

Specific metabolites induce neurogenesis, the development of new neurons in the brain. This is especially significant in disorders involving neuronal loss, such as Alzheimer's and Parkinson's disease.

c) Enhancement of Cognitive Function:

Research indicates marine metabolites may improve cognitive function, including memory and learning. This may have implications for addressing cognitive impairment caused by ageing or neurological illnesses [66].

3) Examples of Marine Metabolites:

The following chemicals and sources have been studied:

1. Omega-3 Fatty Acids:

Found in abundance in marine fish oils, omega-3 fatty acids such as EPA (eicosapentaenoic acid) and DHA (docosahexaenoic acid) are critical for brain health and have been related to neuroprotection and cognitive performance.

2. Phlorotannins:

Phlorotannins are polyphenolic chemicals found in brown algae (Phaeophyceae) that have antioxidant and anti-inflammatory activities, which could be advantageous for neuroprotection [67].

3. Astaxanthin:

Astaxanthin is a carotenoid found in marine creatures such as microalgae and krill. It has strong antioxidant properties and has shown promise in protecting neurons from oxidative stress.

Finally, marine metabolites are a rich bioactive chemical source with considerable neuroprotective and neurological health benefits. Further investigation into these chemicals may yield novel therapeutic options for treating neurodegenerative illnesses and increasing general brain function [68].

4) Other Pharmacological Activities

1. Anti-diabetic properties:

Marine metabolites provide interesting pathways for treating diabetes and its consequences via various processes. They include improving glucose metabolism and insulin sensitivity, as shown in substances such as fucoidans and fucoxanthin found in marine algae. These metabolites, such as omega-3 fatty acids, also have anti-inflammatory properties that can help reduce insulin resistance caused by chronic inflammation. Furthermore, their high antioxidant qualities, as observed in carotenoids and polyphenols from marine sources such as algae, aid in reducing oxidative stress on pancreatic cells and peripheral tissues. Marine metabolites help treat diabetes by suppressing α-glucosidase and α-amylase enzymes. This regulates postprandial glucose levels. They also show promise in controlling lipid metabolism and providing renoprotective actions, which may reduce kidney damage caused by diabetic nephropathy. While much of the research is still preliminary, marine metabolites represent a rich supply of natural chemicals with substantial promise for developing supplementary therapies for diabetes treatment and preventing complications.

2. Cardiovascular Benefits:

Marine metabolites have been examined for their cardiovascular benefits, including as vasodilation, antiplatelet activity, and cholesterol-lowering properties. Omega-3 fatty acids, such as those found in marine fish oils, are widely known for their cardiovascular advantages, which include a lower risk of heart disease and stroke [69].

3. Antioxidant and Free Radical Scavenging Activities:

Marine metabolites have high antioxidant and free radical scavenging activities, protecting cells from oxidative stress and damage. These substances improve cellular health and may help avoid chronic diseases linked to oxidative stress, such as cardiovascular disease and neurological disorders.

4. Immunomodulatory Effects:

Specific marine metabolites can affect immunological responses. This can help with autoimmune illnesses, allergies, and other immune system dysregulation conditions [70].

5. Skin Health and Cosmeceutical Applications:

Marine metabolites have moisturizing, anti-ageing, and wound-healing capabilities, promising them for skincare and cosmeceutical products. Compounds such as marine collagen, peptides, and polysaccharides have improved skin suppleness, reduced wrinkles, and promoted wound healing [71].

Overall, marine metabolites are a valuable and diversified source of bioactive chemicals with pharmacological properties that extend beyond neuroprotection and neurological health. Further research and exploration of these chemicals may result in new therapeutic agents and therapies for various human health issues.

6. Industrial and biotechnological applications:

Aside from their therapeutic effects, marine metabolites are rapidly recognized for their broad industrial and biotechnological applications [72]. A complete analysis of these applications is discussed below:

1. Enzyme Inhibition: Marine metabolites can inhibit enzymes involved in illness and industrial activities. Examples include proteases, lipases, and cellulases. These enzyme inhibitors can be used in medications to target specific disease pathways or in manufacturing processes to improve efficiency and product quality [73].
2. Biodegradability: Marine-derived polymers and biopolymers have biodegradable characteristics, making them ideal for eco-friendly applications. These materials are used in biodegradable plastics, biomaterials (such as wound dressings), and other sustainable goods where minimizing environmental impact is crucial.
3. Cosmetics and Skincare: Marine metabolites like collagen, peptides, and polysaccharides are commonly used in cosmetics and skincare products for their moisturizing, anti-ageing, and wound-healing benefits. Astaxanthin, a carotenoid derived from marine sources, is another prominent chemical known for its antioxidant properties, which protect the skin from UV damage and improve general skin health [74].
4. Nutraceuticals and Functional Foods: Dietary supplements containing omega-3 fatty acids (EPA and DHA) from marine fish oils have been shown to improve cardiovascular health and reduce inflammation. Marine-derived antioxidants, polyphenols, and bioactive peptides are also used in functional meals and supplements to promote general health and wellness [67].
5. Aquaculture and Animal Feed: Marine metabolites can improve feed quality and increase growth rates in fish and shellfish. Marine-derived enzymes and probiotics are important in aquatic species' digestive health and nutrient absorption, promoting production efficiency and sustainability [75].

6. Environmental Bioremediation: Marine metabolites can help clean up contaminants in marine environments. Marine-derived enzymes and microbes are being examined for their capacity to break down oil spills, poisons, and other contaminants, promoting environmental sustainability and ecosystem health [76].
7. Biopharmaceuticals and Drug Development:Marine metabolites are significant sources of new chemicals for drug research. Marine-derived peptides, alkaloids, and natural products are being studied for their potential therapeutic applications in cancer, infectious illnesses, and metabolic disorders [77].
8. Industrial Processes:

Marine enzymes are increasingly used in industrial operations like biofuel production, textile manufacture, and food processing. These enzymes have unique qualities such as heat stability and specificity, making them attractive tools in biocatalysis and green chemistry projects that aim to reduce environmental impact and improve process efficiency.

Marine metabolites are diverse and valuable chemicals with substantial commercial, biotechnological, and pharmaceutical applications. They are critical in improving skincare products, optimizing aquaculture methods, and promoting environmental sustainability through bioremediation. Their numerous applications include biodegradable materials, nutraceuticals, and functional foods loaded with omega-3 fatty acids and antioxidants. Furthermore, these metabolites are essential in drug research, providing potential leads for treating various disorders. Continued research and development are critical for fully realizing their benefits, emphasizing the need to study marine biodiversity to successfully utilize these bioactive substances for therapeutic and economic uses [78].

10. Preclinical Studies: Exploring Therapeutic Potential with Marine-Derived Compounds

Marine-derived chemicals hold significant therapeutic promise in various medical fields because of their unique biochemical properties and broad biological impacts. Preclinical studies continually reveal their potential in various therapeutic applications, demonstrating their ability to treat various diseases and disorderseffectively [79].

1) Anticancer Properties:

Marine-derived chemicals show promise as anticancer agents. Compounds derived from sea sponges, algae, and microbes have been shown to have cytotoxic effects on cancer cells. These chemicals may operate through various methods, including apoptosis induction, angiogenesis inhibition, and signalling pathway regulation.

Examples include compounds obtained from marine species such as sponges, algae, and bacteria, which have shown vigorous anticancer activity in preclinical models. Compounds derived from marine bryozoans, such as Bryostatin-1, have shown promise in reducing tumour growth by modifying protein kinase C (PKC) pathways. Similarly, marine-derived peptides and alkaloids are cytotoxic to numerous cancer cell lines, indicating they could be used as novel anticancer medicines [80].

2) Anti-inflammatory and Immunomodulatory Effects:

Marine metabolites have anti-inflammatory and immunomodulatory qualities that can help treat illnesses like arthritis, inflammatory bowel disease, and autoimmune disorders. Some chemicals also have immunomodulatory properties, which regulate immune responses and may reduce autoimmune reactions.

Examples: Marine metabolites, especially omega-3 fatty acids (EPA and DHA), have been widely researched for their anti-inflammatory potential. Preclinical studies indicate that EPA and DHA can reduce inflammatory responses by suppressing NF-κB activation and pro-inflammatory cytokine production. Other marine-derived chemicals, such as marine polysaccharides and bioactive peptides, have been shown to modulate immune cell activity and cytokine signalling pathways [81].

3) Neuroprotective and Neurological Benefits:

Marine metabolites have been studied for their neuroprotective properties, especially in illnesses such as Alzheimer's, Parkinson's, and stroke. Compounds such as omega-3 fatty acids, marine peptides, and antioxidants can protect neurons from oxidative stress, inflammation, and apoptosis, improving brain health and cognitive performance.

Examples: Marine-derived omega-3 fatty acids, particularly DHA, have been shown to have neuroprotective properties in preclinical models of neurodegenerative disorders. DHA improves neuronal membrane integrity, lowers neuroinflammation, and increases synaptic plasticity and neurogenesis. Furthermore, substances such as astaxanthin, a carotenoid derived from marine sources, have been shown in animal experiments to exhibit antioxidant characteristics that protect neurons from oxidative stress while improving cognitive performance [63].

4) Cardiovascular Health:

Omega-3 fatty acids from marine sources, like EPA and DHA, have been linked to improved cardiovascular health. They aid in lowering triglycerides, blood pressure, and general cardiac function, lowering the risk of cardiovascular disease.

Examples: Omega-3 fatty acids produced from marine fish oils, EPA, and DHA have been widely researched for their cardiovascular effects in preclinical models. These chemicals lower triglycerides, lower blood pressure, prevent platelet aggregation and improve endothelial function. Preclinical studies have also shown that they can lessen the risk of cardiovascular disorders such as coronary artery disease and stroke [82].

5) Antimicrobial and Antiviral Activities:

Marine metabolites have strong antibacterial and antiviral effects against a variety of diseases. These chemicals may be used as replacements or adjuncts to traditional antibiotics and antiviral medications, helping to address drug-resistant diseases and developing viral outbreaks. In preclinical research, marine-derived peptides and natural compounds showed vigorous antibacterial and antiviral activity. Peptides from marine creatures have extensive antibacterial activity against bacteria, fungi, and viruses. These chemicals may provide novel approaches to fighting drug-resistant bacteria and developing infectious illnesses [83].

6) Wound Healing and Tissue Regeneration:

Marine-derived substances such as peptides and polysaccharides have been shown to promote wound healing and tissue regeneration. They boost tissue regeneration, increase collagen synthesis, and speed up wound closure, making them useful in dermatology and wound care.

Examples: Preclinical models were used to explore the wound-healing characteristics of marine-derived polysaccharides, peptides, and growth factors. These chemicals stimulate angiogenesis, collagen synthesis, and epithelialization, resulting in faster wound closure and tissue restoration. Marine-derived collagen and elastin peptides have also been demonstrated to promote skin regeneration and reduce scar formation [84].

7) Metabolic Disorders and Diabetes:

Marine metabolites can help manage metabolic illnesses, including diabetes and obesity. Fucoxanthin and fucoidans from marine algae have been investigated for their capacity to regulate glucose metabolism, increase insulin sensitivity, and minimize fat buildup.

Examples: Marine metabolites such as fucoxanthin from algae and fucoidans from seaweed have shown promise in preclinical investigations for treating metabolic disorders like obesity and diabetes. In animal models, these drugs modulate lipid metabolism, increase insulin sensitivity, and diminish adipocyte differentiation and inflammation [85].

8) Bone Health:

Marine-derived calcium and bioactive peptides have been studied for their potential to promote bone health and prevent osteoporosis. These chemicals may increase bone mineral density while inhibiting bone resorption, providing prospective treatment alternatives for skeletal health [86]. Examples: Preclinical studies have demonstrated that marine-derived calcium and bioactive peptides can help promote bone health and prevent osteoporosis. These chemicals improve bone mineral density, decrease osteoclast activity, and stimulate osteoblast function, indicating potential applications in bone regeneration and fracture healing.

In conclusion, marine-derived chemicals provide a valuable reservoir of bioactive molecules for various therapeutic uses. Continued research and development provide optimism for improving therapies in various medical specializations, solving unmet clinical requirements, and even revolutionizing healthcare through sustainable and effective therapeutic innovations. Preclinical studies provide persuasive data supporting the therapeutic potential of marine-derived chemicals for various health disorders. These findings highlight the importance of continued research and clinical trials to confirm their efficacy and safety, allowing for the development of novel medicines based on marine biodiversity [87].

11. Exploring Marine-Derived Compounds in Early Clinical Trials for Neurological Disorders

Early clinical studies using marine-derived chemicals for neurological illnesses represent a new area in biomedical research, utilizing the varied biochemical features of marine species to treat neurological conditions. These trials are aimed at determining the safety, efficacy, and potential mechanisms of action of various marine-derived chemicals in treating illnesses such as Alzheimer's disease, Parkinson's disease, stroke, and other neurodegenerative conditions.

1. Compounds and Sources:

Omega-3 Fatty Acids Omega-3 fatty acids derived from marine fish oils, including EPA (eicosapentaenoic acid) and DHA (docosahexaenoic acid), have been proven in preclinical tests to have neuroprotective properties. Early clinical trials seek to validate these discoveries and determine their potential to reduce disease progression or ameliorate symptoms in neurodegenerative illnesses [88].

a) Marine Peptides and Proteins:

Some peptides and proteins obtained from marine species have distinct biological properties that may help brain health. These chemicals are being studied for their capacity to regulate neuroinflammatory responses, increase neuroplasticity, and protect neurons from oxidative damage [52].

b) Marine Natural Products:

Compounds such as marine alkaloids and polyphenols have antioxidant and anti-inflammatory characteristics that are important in preventing neurological diseases. Early clinical trials seek to assess their efficacy in minimizing neuronal damage and enhancing cognitive function in people with neurological diseases [67].

2. Mechanisms of Action:

a) Neuroprotection:

Many marine-derived chemicals have neuroprotective properties by lowering oxidative stress, suppressing inflammation, and encouraging neuronal survival and regeneration. These systems are crucial in countering neurodegenerative processes that cause neuron loss and cognitive impairment [89].

b) Neurotransmission Modulation:

Some substances alter neurotransmitter systems, such as increasing dopamine or acetylcholine levels, which are disturbed in Parkinson's and Alzheimer's diseases. These impacts could enhance neural transmission and cognitive function [50].

c) Anti-amyloid and Tau Effects:

The formation of amyloid-beta plaques and tau tangles in the brain characterizes Alzheimer's disease. Marine-derived chemicals are being investigated for their ability to prevent or inhibit the synthesis of these pathogenic proteins, reducing disease progression [90].

3. Design and progress of clinical trials:

Early clinical trials typically include Phase I and II studies on safety, dosage determination, and preliminary efficacy assessments. These trials use small patient populations to assess tolerability, pharmacokinetics (how the body processes the chemical), and preliminary therapeutic outcomes.

a) Biomarkers and Imaging: Early studies rely heavily on biomarkers and imaging tools to assess illness progression and therapy response objectively. Biomarkers such as amyloid and tau levels in CSF fluid or imaging scans can provide information on the compound's effects on disease pathology and brain function [91].

4. Challenges and Future Directions:

Early clinical studies have challenges such as patient recruitment, determining suitable dosing regimens, and managing side effects. Neurological illnesses may have complicated underlying mechanisms, necessitating careful choice of endpoints and outcome metrics to assess therapeutic efficacy appropriately.

Future directions include moving promising medicines from early stages to Phase III trials to demonstrate efficacy and safety in broader patient populations. Furthermore, investigating combination medicines and personalized medicine approaches based on genetic and biomarker profiles could improve treatment outcomes.

To summarize, early clinical studies of marine-derived chemicals for neurological illnesses reflect a blossoming sector with the potential to bring new therapeutic alternatives. These trials seek to translate promising preclinical results into viable medicines that can slow disease development, alleviate symptoms, and improve patients' quality of life.

12. Exploring Marine-Derived Compounds: Case Studies in Alzheimer's, Parkinson's, and Stroke Therapeutics

The therapeutic potential of marine-derived chemicals in the setting of Alzheimer's disease, Parkinson's disease, and stroke entails investigating natural molecules produced from marine creatures for their ability to treat these neurological disorders. Here's a detailed look at their uses in each case:

1. Alzheimer's Disease:

Alzheimer's disease is characterized by the deposition of amyloid-beta plaques and tau tangles in the brain, which causes progressive cognitive deterioration. Marine-derived chemicals are being studied for their ability to target these pathogenic processes and produce neuroprotective benefits.

a) Omega-3 Fatty Acids (EPA and DHA):

Found abundantly in marine fish oils, EPA and DHA have demonstrated neuroprotective qualities in preclinical tests. They may assist in reducing inflammation, improving synaptic plasticity, and boosting neuronal survival, thereby reducing the progression of Alzheimer's disease [92].

b) Marine Polyphenols and Antioxidants:

Astaxanthin, derived from marine algae and crustaceans, has strong antioxidant properties. Antioxidants aid in reducing oxidative stress, which contributes to brain damage in Alzheimer's disease [68].

c) Neuroinflammatory Modulators:

Some marine peptides and proteins have shown promise in modifying the neuroinflammatory responses associated with Alzheimer's disease. By lowering inflammation, these chemicals may protect neurons and improve cognitive function [52].

2. Parkinson's Disease:

The loss of dopamine-producing neurons in the brain causes motor symptoms such as tremors, stiffness, and bradykinesia. Marine-derived chemicals are being investigated for their ability to preserve dopaminergic neurons and reduce symptoms.

a) Neuroprotective Peptides:

Specific peptides obtained from marine species have been shown to exhibit neuroprotective properties in preclinical Parkinson's disease models. These peptides may help neurons survive and reduce oxidative stress, contributing to dopaminergic neuron degeneration [49].

b) Dopaminergic Modulation:

Compounds that modify dopamine receptors or neurotransmitter levels may help treat motor complaints. Marine-derived chemicals are being explored for their ability to increase dopamine synthesis or receptor sensitivity [93].

c) Anti-inflammatory properties:

Neuroinflammation, like Alzheimer's disease, contributes to Parkinson's pathogenesis. Marine-derived antioxidants and anti-inflammatory chemicals may aid in reducing inflammation in the brain, thereby reducing disease progression [57].

3. Stroke:

A stroke disrupts blood flow to the brain, causing neuronal damage and cognitive and motor deficits. Marine-derived chemicals are being investigated for their neuroprotective and regenerative benefits after stroke [49].

a) Neuroregenerative Factors:

In preclinical stroke models, several marine chemicals stimulate neurogenesis and neural repair pathways. These chemicals may improve brain plasticity and functional recovery following a stroke [50].

b) Anti-thrombotic Effects:

Marine-derived chemicals with anti-thrombotic qualities may help prevent or dissolve blood clots, lowering the severity of clot-induced ischemic strokes [94].

c) Anti-inflammatory and antioxidant actions:

Inflammation and oxidative stress worsen brain damage after a stroke. Marine chemicals with anti-inflammatory and antioxidant characteristics may help protect neurons from subsequent injury and promote recovery [95].

In conclusion, marine-derived chemicals offer intriguing treatment options for Alzheimer's disease, Parkinson's disease, and stroke due to their neuroprotective, anti-inflammatory, antioxidant, and perhaps neurodegenerative capabilities. Early preclinical and clinical investigations are critical for determining efficacy, safety, and mechanisms of action, paving the way for future therapeutic breakthroughs in neurology.

13. Pharmacological Potential of Marine-Derived Compounds: Mechanisms of Action

The mechanisms of action of marine-derived chemicals, specifically omega-3 fatty acids, marine peptides, and polyphenols, in terms of their therapeutic benefits.

1. Omega-3 Fatty Acids' Neuroprotective Effects

Omega-3 fatty acids, including EPA (eicosapentaenoic acid) and DHA (docosahexaenoic acid), are vital nutrients found in marine fish oils. They are well-known for their neuroprotective effects, essential for brain health and function. The mechanisms of action are:

a) Anti-inflammatory Effects:

EPA and DHA prevent neuroinflammation by suppressing the generation of pro-inflammatory cytokines and signalling molecules like NF-κB. Chronic neuroinflammation is linked to neurodegenerative illnesses such as Alzheimer's and Parkinson's, causing neuronal damage and cognitive impairment.

b) Enhanced Synaptic Plasticity:

DHA is a key structural component of neuronal membranes. It promotes synaptic plasticity, which is necessary for learning, memory formation, and neural communication. DHA increases synaptic development and maintains neuronal membrane integrity, which improves cognitive performance.

c) Antioxidant Properties:

Omega-3 fatty acids have antioxidant properties, scavenging free radicals and decreasing oxidative stress in the brain. Oxidative stress causes neuronal damage and is involved in the etiology of neurodegenerative diseases. Omega-3 fatty acids prevent neurons from degenerating prematurely by decreasing oxidative damage.

d) Neuroprotection and Cell Survival:

EPA and DHA support neuronal survival and prevent apoptotic cell death. They facilitate the generation of neurotrophic factors, such as brain-derived neurotrophic factor (BDNF), which promotes neuronal growth, differentiation, and survival. This neuroprotective effect is critical in preventing neurodegeneration and maintaining cognitive function [89].

2. Marine Peptides:
   a) Marine peptides modulate neurotransmitter systems:

Marine peptides produced by several marine creatures have demonstrated the potential for altering neurotransmitter systems in the brain. These peptides exert their effects via several mechanisms:

b) Neurotransmitter Release and Receptor Activity:

Marine peptides can improve neurotransmitter release and receptor function, improving neurotransmission and signalling. For example, peptides from marine algae or invertebrates may function as agonists or antagonists at specific neurotransmitter receptors, affecting synaptic transmission [50].

c) Neurotransmitter Metabolism:

Marine peptides influence neurotransmitter metabolism, including dopamine, serotonin, and acetylcholine. By altering enzyme activity involved in neurotransmitter production and breakdown, these peptides can influence neurotransmitter levels and equilibrium in the brain [96].

d) Neuroprotective Effects:

Marine peptides can reduce neuronal excitotoxicity, oxidative stress, and inflammation. They may increase neural resilience to stressors and boost neuronal survival during neurodegeneration.

e) Modulation of Neuroplasticity:

Neuroplasticity is the brain's ability to reorganize itself through new synaptic connections and pathways based on learning and experience. Marine peptides may promote neuroplasticity by increasing synaptic remodelling and promoting adaptive alterations in neural networks.

3. Polyphenols: Reduced Neuroinflammation

Polyphenols produced from marine sources, such as algae and marine plants, have potent anti-inflammatory characteristics that can improve neurological health. Their modes of action include:

a) Inhibition of Pro-inflammatory Pathways:

Marine polyphenols block pro-inflammatory pathways in microglia and astrocytes, immune cells in the central nervous system. They inhibit the generation of inflammatory cytokines (such as TNF-alpha and IL-6) and chemokines that contribute to neuroinflammation.

b) Antioxidant Activity:

Polyphenols protect neurons from oxidative injury by scavenging free radicals and reactive oxygen species (ROS). Polyphenols reduce oxidative stress, which protects neuronal membranes, proteins, and DNA from oxidative injury, preserving neuronal function and viability [59].

c) Modulating Glial Activation:

Glial cells, such as microglia and astrocytes, play a critical role in neuroinflammation and protection. Marine polyphenols regulate glial cell activation, encouraging the transition from a pro-inflammatory M1 to an anti-inflammatory M2 phenotype. This change decreases neuroinflammatory responses while promoting tissue repair and regeneration in the brain [97].

d) Enhancement of Blood-Brain Barrier Integrity:

Polyphenols improve blood-brain barrier integrity by strengthening the selectively permeable barrier that governs substance transit between the bloodstream and the brain. Polyphenols limit immune cell and inflammatory mediator penetration into the brain by strengthening the blood-brain barrier, thereby decreasing neuroinflammation [98].

Therefore, omega-3 fatty acids, marine peptides, and polyphenols produced from marine sources have various modes of action that contribute to their neuroprotective and therapeutic effects. These chemicals show promise for treating and managing neurological illnesses because they target inflammation, support neurotransmitter systems, and promote neuronal health and function. Continued study is required to understand their mechanisms further and maximize their therapeutic potential in clinical settings.

14. Advances in Technology Driving Discovery Marine-Derived Compounds

These technical breakthroughs and their impact on the identification and understanding of marine-derived compounds:

1. Metabolomics:

Metabolomics is a subfield of omics sciences that studies tiny compounds (metabolites) found in biological systems. Metabolomics is critical for identifying and characterizing the vast array of metabolites produced by marine species. These compounds frequently contain novel chemical structures and bioactivities, making them attractive in medicinal, agricultural, and industrial applications.

a) Application of Marine Compounds:

Metabolomics enables researchers to examine the metabolic profiles of marine creatures thoroughly. Scientists can detect and measure metabolites found in marine samples using techniques including mass spectrometry (MS) and nuclear magnetic resonance (NMR) spectroscopy. This method aids in the discovery of novel bioactive chemicals with potential therapeutic applications, such as antibiotics, anticancer medicines, and anti-inflammatory drugs.

b) Challenges and Advancements:

The complexity and variety of marine habitats and the requirement for advanced analytical tools and bioinformatics methodologies to analyze massive datasets provide challenges in marine metabolomics research. Recent advances in metabolomics technology, such as better analytical instrument resolution and sensitivity and advanced data processing methods, have greatly improved the ability to detect and characterize marine-derived chemicals [99].

2. Bioinformatics:

Bioinformatics refers to using computational tools and techniques to analyze biological data. Bioinformatics is essential in numerous ways when dealing with marine-derived compounds [100]:

a) Genomic and metagenomic analysis:

Bioinformatics techniques analyze genomic and metagenomic data collected from marine creatures and microbial ecosystems. This aids in discovering biosynthetic gene clusters that produce bioactive substances, predicting their chemical structures, and understanding their evolutionary origins.

b) Database Mining:

Bioinformatics databases like MarinLit, AntiMarin, and NORMAN store information on marine-derived compounds, their structures, biological functions, and sources. These databases help researchers identify new chemicals by giving reference data for comparison and analysis [101].

c) Molecular Docking and Simulation:

Molecular docking and simulations predict interactions between marine-derived chemicals and biological targets (e.g., enzymes and receptors). This aids in understanding the mechanism of action of bioactive chemicals and optimizing their medicinal efficacy.

d) Integrate with Other Technologies

Bioinformatics works with metabolomics and other omics disciplines to provide a comprehensive picture of marine ecosystems and their potential for bioactive chemical discovery. Researchers can discover new chemicals and understand their ecological roles and biotechnological uses by combining genomic data and metabolite profiles [100].

3. Synthetic Biology:

Synthetic biology is designing and engineering biological systems or components for novel uses. Synthetic biology offers creative ideas to:

a) Production Optimization:

Synthetic biology approaches are utilized to improve the production of bioactive chemicals by marine species or their microbial symbionts. This involves metabolic engineering to improve biosynthetic pathways, genetic alteration of host species to maximize yield, and route rebuilding in heterologous hosts [102].

b) Pathway Discovery:

Synthetic biology uses bioinformatics and genomic data to uncover and build biosynthetic pathways that produce bioactive chemicals. This method makes it easier to synthesize complicated natural products that would otherwise be difficult to obtain [100].

c) Bioengineering novel compounds:

Synthetic biology facilitates the development of new bioactive chemicals by developing and synthesizing non-natural routes or changing existing ones. This includes combinatorial biosynthesis, which combines enzymes from various pathways to create new chemical compounds with desired biological activity.

d) Future directions:

Advances in synthetic biology tools, such as CRISPR-Cas9 genome editing and high-throughput screening approaches, are broadening the scope of marine-derived drug discovery and use. These methods allow for the development of genetically altered organisms optimized for bioactive compound production and the quick screening of compound libraries for medicinal and commercial applications.

In conclusion, metabolomics, bioinformatics, and synthetic biology are critical for further discovering and understanding marine-derived chemicals. These technologies complement one another by allowing for complete metabolite profile analysis, the use of genetic data for pathway discovery, and the engineering of biological systems for increased compound production and novel bioactivity. They work together to pave the way for pharmacological, agricultural, and industrial breakthroughs based on the tremendous biochemical variety of marine ecosystems [103].

15. Marine-Derived Compounds: Collaborative Efforts for Translation

Collaboration among marine biologists, pharmacologists, and neuroscientists is critical for turning discoveries about marine-derived chemicals into practical uses. Each profession provides distinct knowledge and views that are required at various phases of the translation process:

1. The Role of Marine Biologists:

Marine biologists play an essential role in early discovering and characterising marine-derived chemicals. Their specialties include maritime ecosystems, biodiversity, and the ecological roles of marine creatures. This is how they contribute [104].

a) Bio-Prospecting and Sampling:

Maritime biologists investigate maritime ecosystems to gather samples from sponges, corals, algae, and microbes. They discover possible sources of bioactive chemicals using ecological and taxonomic data.

b) Isolation and ID:

Marine scientists use taxonomic and ecological expertise to isolate and identify organisms that create interesting chemicals. They use field surveys, specimen collection, and laboratory research to extract bioactive compounds.

c) Ecological context:

Understanding the ecological context of marine creatures is critical for successful bioprospecting. Marine biologists examine the impact of harvesting on marine ecosystems and strive for ethical and conservation-minded techniques.

d) In collaboration with other disciplines:

Marine biologists work with pharmacologists and neuroscientists to identify chemicals with therapeutic promise and understand their ecological activities, guiding future research and development [105].

2. Pharmacologists' involvement:

Pharmacologists investigate the pharmacological properties and potential therapeutic applications of marine-derived substances. Their contributions are critical in bridging the gap between basic research and therapeutic applications.

a) Bioactivity Screening:

Pharmacologists use screening assays to determine the biological activity of marine-derived substances. This includes evaluating antibacterial, anticancer, anti-inflammatory, and other related bioactivities utilizing in vitro and in vivo models. Understanding how marine-derived chemicals interact with biological targets, such as receptors and enzymes, is essential for understanding their modes of action. Pharmacologists study these pathways using biochemical, molecular, and pharmacological tools [106].

b) Drug development:

Pharmacologists work with medicinal chemists and bioengineers to improve the chemical structures of bioactive chemicals for efficacy, selectivity, and pharmacokinetic features. This includes developing analogues and derivatives using structure-activity relationship (SAR) research.

16. Preclinical and Clinical Investigations

Pharmacologists conduct preclinical investigations (animal models) and assist with clinical trials to assess promising marine-derived drugs' safety, effectiveness, and pharmacokinetics. These investigations are critical for turning findings into possible therapeutics [107].

1. Contribution by Neuroscientists:

Neuroscientists study the impact of marine-derived chemicals on the nervous system and their possible applications in treating neurological illnesses. Their expertise is crucial for investigating neuropharmacological elements.

a) Neuroprotective and Regenerative Effects:

Neuroscientists are studying whether marine-derived chemicals have neuroprotective effects, such as shielding neurons from injury or boosting neurodegeneration in circumstances like stroke or neurodegenerative illnesses.

b) Behavioral and cognitive effects:

The study of how marine-derived substances affect behaviour and cognition aids in the identification of new treatments for neurological and psychiatric illnesses. This includes testing the impact on memory, learning, mood control, and other brain functions [108].

c) Neurochemical mechanisms:

Neuroscientists study how marine-derived chemicals interact with neurotransmitter systems, ion channels, and other molecular targets of the nervous system. This information is critical for determining their therapeutic potential.

d) In collaboration with other disciplines:

Neuroscientists work with pharmacologists and clinicians to create research that evaluates the neurological effects of marine-derived chemicals in both laboratory and clinical settings. Their findings help to assess chemical efficacy and safety thoroughly.

To summarise, collaboration among marine biologists, pharmacologists, and neuroscientists is critical for turning discoveries of marine-derived chemicals into therapeutic applications. By combining skills in biodiversity, pharmacology, and neurology, researchers can uncover, characterize, and produce bioactive chemicals with potential therapeutic advantages for human health and well-being. These interdisciplinary interactions allow for the efficient and ethical research of marine biodiversity while also boosting medical and pharmacological innovation [52].

17. Marine-Derived Compounds: Future Directions and Implications

The future paths and consequences of marine-derived chemicals are varied, including the sustainable exploitation of marine resources, therapeutic use and translation, and tackling the worldwide burden of neurological illnesses.

1. Sustainable exploitation of maritime resources:

As research into marine-derived substances continues, it is critical to maintain sustainable methods in their exploitation:

a) Conservation and ethical practices:

Understanding and protecting marine biodiversity are essential components of sustainable bioprospecting strategies. This includes implementing conservation actions to conserve fragile marine habitats and animals that produce bioactive chemicals [109].

b) Biotechnological Approaches:

Biotechnology advances, such as synthetic biology and genomic-guided bioprospecting, can help to ensure long-term utilization. Cultured marine microorganisms and bioreactor systems help to lessen the need for wild harvesting [110].

c) Regulatory frameworks:

Creating and implementing regulatory frameworks at the national and international levels is critical. These frameworks should balance scientific exploration and environmental conservation and ensure that local residents benefit pretty.

d) Education and awareness:

Promoting public and stakeholder understanding of the value of marine biodiversity and the importance of sustainable practices is critical. Education can instil appropriate attitudes toward marine resources and aid conservation efforts [111].

2. Clinical Application and Translation:

Marine-derived chemicals show great potential for therapeutic uses in numerous medical sectors.

a) Drug Discovery & Development:

Continued research aims to uncover novel bioactive chemicals with potential therapeutic applications, such as antimicrobials, anticancer medicines, and anti-inflammatory drugs [112].

b) Translation to Therapeutics:

Moving from research to clinical application necessitates extensive testing in both preclinical models and clinical trials. Collaboration among researchers, pharmaceutical companies, and regulatory authorities is critical for navigating the arduous journey from bench to bedside.

c) Precise Medicine

Marine-derived chemicals may help with personalized treatment by targeting specific biological pathways or genetic profiles linked to disease. This can lead to more effective and personalized therapy for each patient.

d) Emerging diseases and resistance:

With the growth of antimicrobial resistance and increasing infectious illnesses, marine-derived chemicals provide new opportunities for creating alternative treatments against resistant pathogens [113].

3. Addressing the Global Burden of Neurological Diseases:

Neurological illnesses are a massive worldwide health concern, and marine-derived chemicals may provide potential solutions:

a) Neuroprotective and Regenerative Effects:

Compounds identified in marine sources may have neuroprotective characteristics, slow disease development or reducing neurological damage in illnesses such as Alzheimer's disease, Parkinson's disease, and stroke.

b) Cognitive enhancement:

Some marine-derived chemicals show potential for improving cognitive functions like memory and learning. These effects can be beneficial to people suffering from cognitive impairment or neurodegenerative diseases [114].

c) Therapeutic innovation:

Marine-derived chemicals may serve as novel medicines for neurological illnesses that lack viable treatments. Research efforts are centred on understanding their mechanisms of action and optimizing their neuropharmacological characteristics.

d) Translated Research:

Collaborations among neuroscientists, pharmacologists, and physicians are essential for converting preclinical findings into clinical trials and, eventually, authorized therapies. This multidisciplinary approach speeds up the development of new medicines for neurological diseases.

Therefore, future possibilities for marine-derived chemicals include balancing scientific inquiry with environmental sustainability, increasing clinical uses through rigorous research and development, and addressing critical global health concerns such as neurological illnesses. Researchers can contribute to medical innovation and conservation initiatives by appropriately utilizing marine biodiversity's bioactive potential, assuring long-term benefits for human health and marine ecosystems [64].

Conclusion

Marine-derived chemicals offer potential opportunities for drug development and therapeutic innovation due to their varied and bioactive metabolites. The incredible richness of marine species provides a rich source of molecules with potential applications in various medical sectors, including infectious diseases, neurological disorders, and more. The potential of marine metabolites stems from their distinct chemical structures and biological activity, which frequently display novel modes of action not found in terrestrial species. These chemicals have shown promise in preclinical tests for combating drug-resistant infections, alleviating inflammatory conditions, and even protecting and regenerating neural structures. Furthermore, advances in biotechnology, pharmacology, and bioinformatics are helping to build safe and effective marine-based medicines. These disciplines allow researchers to methodically investigate, identify, characterize, and optimize bioactive chemicals found in marine species, opening the door for clinical applications. Notably, the sustainable use of marine resources is critical to the ongoing discovery and development of marine-derived medicines. Conservation efforts, ethical bio-prospecting methods, and regulatory frameworks are essential in maintaining marine ecosystems while maximizing their potential medical value. In essence, the future of marine-derived chemicals seems bright for tackling global health issues and boosting human well-being. By appropriately utilizing the ocean's biochemical diversity, we can uncover new therapeutics that provide hope to patients while contributing to medical research advancement.

References

[1] Hellier JL, editor. *The Brain, the Nervous System, and Their Diseases*: [3 volumes]. Bloomsbury Publishing USA; 2014 Dec 16.

[2] Han VX, Patel S, Jones HF, Dale RC. Maternal immune activation and neuroinflammation in human neurodevelopmental disorders. *Nature Reviews Neurology*. 2021 Sep;17(9):564-79.

[3] Nagaratnam N, Nagaratnam K, Cheuk G, Nagaratnam N, Nagaratnam K, Cheuk G. Neurological disorders and related problems in the elderly. *Diseases in the Elderly: Age-Related Changes and Pathophysiology*. 2016:151-213.

[4] Hirtz D, Thurman DJ, Gwinn-Hardy K, Mohamed M, Chaudhuri AR, Zalutsky R. How common are the "common" neurologic disorders? *Neurology*. 2007 Jan 30;68(5):326-37.

[5] Lima AA, Mridha MF, Das SC, Kabir MM, Islam MR, Watanobe Y. A comprehensive survey on the detection, classification, and challenges of neurological disorders. *Biology*. 2022 Mar 18;11(3):469.

[6] Frishberg BM. The utility of neuroimaging in the evaluation of headache in patients with normal neurologic examinations. *Neurology*. 1994 Jul;44(7):1191-.

[7] Cup EH, Pieterse AJ, Knuijt S, Hendricks HT, van Engelen BG, Oostendorp RA, van der Wilt GJ. Referral of patients with neuromuscular disease to occupational therapy, physical therapy and speech therapy: usual practice versus multidisciplinary advice. *Disability and rehabilitation*. 2007 Jan 1;29(9):717-26.

[8] Breijyeh Z, Karaman R. Comprehensive review on Alzheimer's disease: causes and treatment. *Molecules*. 2020 Dec 8;25(24):5789.

[9] Bloem BR, Okun MS, Klein C. Parkinson's disease. *The Lancet*. 2021 Jun 12;397(10291):2284-303.

[10] Huang WJ, Chen WW, Zhang X. Multiple sclerosis: Pathology, diagnosis and treatments. *Experimental and therapeutic medicine*. 2017 Jun 1;13(6):3163-6.

[11] Yaghi S, Eisenberger A, Willey JZ. Symptomatic intracerebral hemorrhage in acute ischemic stroke after thrombolysis with intravenous recombinant tissue plasminogen activator: a review of natural history and treatment. *JAMA neurology*. 2014 Sep 1;71(9):1181-5.

[12] Capizzi A, Woo J, Verduzco-Gutierrez M. Traumatic brain injury: an overview of epidemiology, pathophysiology, and medical management. *Medical Clinics*. 2020 Mar 1;104(2):213-38.

[13] Ahuja CS, Fehlings MG. *Spinal cord injury. Principles of Neurological Surgery*, ed. 2018 Jan 1;4.

[14] Iqbal Z, Azmi S, Yadav R, Ferdousi M, Kumar M, Cuthbertson DJ, Lim J, Malik RA, Alam U. Diabetic peripheral neuropathy: epidemiology, diagnosis, and pharmacotherapy. *Clinical therapeutics*. 2018 Jun 1;40(6):828-49.

[15] Jasti AK, Selmi C, Sarmiento-Monroy JC, Vega DA, Anaya JM, Gershwin ME. Guillain-Barré syndrome: causes, immunopathogenic mechanisms and treatment. *Expert review of clinical immunology*. 2016 Nov 1;12(11):1175-89.

[16] d'Ydewalle C, Benoy V, Van Den Bosch L. Charcot-Marie-Tooth disease: emerging mechanisms and therapies. *The international journal of biochemistry & cell biology*. 2012 Aug 1;44(8):1299-304.

[17] Orsini M, Oliveira AB, Nascimento OJ, Reis CH, Leite MA, de Souza JA, Pupe C, de Souza OG, Bastos VH, de Freitas MR, Teixeira S. Amyotrophic lateral sclerosis: new perspectives and update. *Neurology international*. 2015 Sep 24;7(2):5885.

[18] Stoker TB, Mason SL, Greenland JC, Holden ST, Santini H, Barker RA. Huntington's disease: Diagnosis and management. *Practical neurology*. 2022 Feb 1;22(1):32-41.

[19] Lawton MT, Vates GE. Subarachnoid hemorrhage. *New England Journal of Medicine*. 2017 Jul 20;377(3):257-66.

[20] Sabatini C, Milani D, Menni F, Tadini G, Esposito S. Treatment of neurofibromatosis type 1. *Current treatment options in neurology*. 2015 Jun; 17:1-1.

[21] Ramdas S, Servais L. New treatments in spinal muscular atrophy: an overview of currently available data. *Expert opinion on pharmacotherapy*. 2020 Feb 11;21(3):307-15.

[22] Jinnah HA, Factor SA. Diagnosis and treatment of dystonia. *Neurologic clinics*. 2015 Feb 1;33(1):77-100.

[23] Falco-Walter J. *Epilepsy—definition, classification, pathophysiology, and epidemiology*. InSeminars in neurology 2020 Dec (Vol. 40, No. 06, pp. 617-623). Thieme Medical Publishers, Inc.

[24] Smith DK, Sadler KP, Benedum M. Febrile seizures: risks, evaluation, and prognosis. *American family physician*. 2019 Apr 1;99(7):445-50.

[25] Bhushan I, Kour M, Kour G, Gupta S, Sharma S, Yadav A. Alzheimer's disease: Causes & treatment– A review. *Ann Biotechnol*. 2018;1(1):1002.

[26] Peters GL. Migraine overview and summary of current and emerging treatment options. *Am J Manag Care*. 2019 Jan 1;25(2 Suppl): S23-34.

[27] Silva M, Ouanounou A. Trigeminal neuralgia: etiology, diagnosis, and treatment. *SN Comprehensive Clinical Medicine*. 2020 Sep; 2:1585-92.

[28] Fawns T. Attention deficit and hyperactivity disorder. *Primary Care: Clinics in Office Practice*. 2021 Sep 1;48(3):475-91.

[29] Sadowska M, Sarecka-Hujar B, Kopyta I. Cerebral palsy: current opinions on definition, epidemiology, risk factors, classification and treatment options. *Neuropsychiatric disease and treatment*. 2020 Jun 12:1505-18.

[30] Dina CZ, Porta M. *Understanding Tourette Syndrome: A guide to symptoms, management and treatment*. Routledge; 2019 Jun 14.

[31] Stein DJ, Costa DL, Lochner C, Miguel EC, Reddy YJ, Shavitt RG, van den Heuvel OA, Simpson HB. Obsessive-compulsive disorder. *Nature reviews Disease primers*. 2019 Aug 1;5(1):52.

[32] McCutcheon RA, Marques TR, Howes OD. Schizophrenia—an overview. *JAMA psychiatry*. 2020 Feb 1;77(2):201-10.

[33] Mavroudis I, Kazis D, Kamal FZ, Gurzu IL, Ciobica A, Pădurariu M, Novac B, Iordache A. Understanding Functional Neurological Disorder: Recent Insights and Diagnostic Challenges. *International Journal of Molecular Sciences*. 2024 Apr 18;25(8):4470.

[34] Cramer SC, Sur M, Dobkin BH, O'Brien C, Sanger TD, Trojanowski JQ, Rumsey JM, Hicks R, Cameron J, Chen D, Chen WG. *Harnessing neuroplasticity for clinical applications*. Brain. 2011 Jun 1;134(6):1591-609.

[35] Chambers SK, Grassi L, Hyde MK, Holland J, Dunn J. Integrating psychosocial care into neuro-oncology: challenges and strategies. *Frontiers in oncology*. 2015 Feb 23; 5:41.

[36] Birbeck GL, Hanna MG, Griggs RC. Global opportunities and challenges for clinical neuroscience. *JAMA*. 2014 Apr 23;311(16):1609-10.

[37] Singh H, Parida A, Debbarma K, Ray DP, Banerjee P. Common marine organisms: A novel source of medicinal compounds. *International Journal of Bioresource Science*. 2020;7(2):39-49.

[38] Kiuru P, D'Auria MV, Muller CD, Tammela P, Vuorela H, Yli-Kauhaluoma J. Exploring marine resources for bioactive compounds. *Planta medica*. 2014 Sep;80(14):1234-46.

[39] Mans DR. Exploring the global animal biodiversity in the search for new drugs–marine invertebrates. *J Transl Sci*. 2016;2(3):170-9.

[40] NEGREANU-PIRJOL BS, NEGREANU-PIRJOL T, POPOVICIU AP. *MARINE ALGAE AN IMPORTANT RESOURCE FOR BIOECONOMY AND BIOMEDICAL APPLICATIONS*. A REVIEW.

[41] Ma H, Hou G. Study on the evolution of collaborative innovation in marine economy considering the participation of financial institutions and two types of cooperation. *Heliyon*. 2024 Feb 29;10(4).

[42] Fernando IP, Jayawardena TU, Wu J. Marine proteins and peptides: Production, biological activities, and potential applications. *Food Innovation and Advances*. 2023;2(2):69-84.

[43] Bouley TA, Machalaba C, Keast J, Gerwick WH, Fleming LE. Marine biotechnology: A One Health approach to linking life on land to life underwater. *In Oceans and Human Health* 2023 Jan 1 (pp. 149-180). Academic Press.

[44] Wang X, Müller WE. Marine biominerals: perspectives and challenges for polymetallic nodules and crusts. *Trends in Biotechnology*. 2009 Jun 1;27(6):375-83.

[45] Bravo ME, Brandt MI, van der Grient JM, Dahlgren TG, Esquete P, Gollner S, Jones DO, Levin LA, McClain CR, Narayanaswamy BE, Sutton T. Insights from the management of offshore energy resources: Toward an ecosystem-services based management approach for deep-ocean industries. *Frontiers in Marine Science*. 2023 Jan 12; 9:994632.

[46] Arora NK, Fatima T, Mishra I, Verma M, Mishra J, Mishra V. Environmental sustainability: challenges and viable solutions. *Environmental Sustainability*. 2018 Dec 1; 1:309-40.

[47] Levin LA, Bett BJ, Gates AR, Heimbach P, Howe BM, Janssen F, McCurdy A, Ruhl HA, Snelgrove P, Stocks KI, Bailey D. Global observing needs in the deep ocean. *Frontiers in Marine Science*. 2019 May 29; 6:241.

[48] Field JG, Hempel G, Summerhayes CP, editors. Oceans 2020: *science, trends, and the challenge of sustainability*. Island Press; 2013 Apr 22.

[49] Silva J, Alves C, Soledade F, Martins A, Pinteus S, Gaspar H, Alfonso A, Pedrosa R. Marine-derived components: can they be a potential therapeutic approach to Parkinson's disease? *Marine drugs*. 2023 Aug 16;21(8):451.

[50] Küpeli Akkol E, Tatlı Çankaya I, Şeker Karatoprak G, Carpar E, Sobarzo-Sánchez E, Capasso R. Natural compounds as medical strategies in the prevention and treatment of psychiatric disorders seen in neurological diseases. *Frontiers in Pharmacology*. 2021 May 13; 12:669638.

[51] Barbosa M, Valentão P, Andrade PB. Bioactive compounds from macroalgae in the new millennium: implications for neurodegenerative diseases. *Marine drugs*. 2014 Sep 25;12(9):4934-72.

[52] Silva M, Seijas P, Otero P. Exploitation of marine molecules to manage alzheimer's disease. *Marine Drugs*. 2021 Jun 28;19(7):373.

[53] CS Souza A, de Fatima A, B da Silveira R, Z Justo G. Seek and destroy: the use of natural compounds for targeting the molecular roots of cancer. *Current drug targets*. 2012 Jul 1;13(8):1072-82.

[54] Hood KA, Bäckström BT, Northcote PT, Berridge MV. The novel cytotoxic sponge metabolite peloruside A, structurally similar to bryostatin-1, has unique bioactivity independent of protein kinase C. *Anti-cancer drug design*. 2001 Apr 1;16(2):155-66.

[55] Ustyuzhanina NE, Bilan MI, Ushakova NA, Usov AI, Kiselevskiy MV, Nifantiev NE. Fucoidans: Pro- or antiangiogenic agents? Glycobiology. 2014 Dec 1;24(12):1265-74.

[56] Jimenez PC, Wilke DV, Branco PC, Bauermeister A, Rezende-Teixeira P, Gaudêncio SP, Costa-Lotufo LV. Enriching cancer pharmacology with drugs of marine origin. *British Journal of Pharmacology*. 2020 Jan;177(1):3-27.

[57] Augustyniak A, Przyborska J, McMahon H. Differential effects of Fucus vesiculosus fucoidan on fibroblast and macrophage cell lines inflammatory activation. *Results in Chemistry*. 2024 Jan 1; 7:101443.

[58] Lee TH, Chen JL, Chang CH, Tsai MM, Tseng HC, Chang YC, Shanmugam V, Hsieh HL. A Brain-Protective Sterol from Soft Coral Inhibits Lipopolysaccharide-Induced Matrix Metalloproteinase-9-Mediated Astrocytic Migration. *Biomedicines*. 2024 Jan 19;12(1):226.

[59] Chandrasekhar T, Riazunnisa K, Vijaya Lakshmi D, Anu Prasanna V, Veera Bramhachari P. Exploration of Bioactive Functional Molecules from Marine Algae: Challenges and Applications in Nutraceuticals. *Marine Bioactive Molecules for Biomedical and Pharmacotherapeutic Applications*. 2024 Jan 24:187-96.

[60] Supanekar SP, Naik MS, Sarwade VV. Marine Resources for Marine-Derived Drugs: A Review. *UTTAR PRADESH JOURNAL OF ZOOLOGY*. 2024 Mar 9;45(5):175-86.

[61] Cheng W, Huang Y, Gao H, Bold B, Zhang T, Yang D. Marine Natural Products as Novel Treatments for Parasitic Diseases.

[62] Saini R, Ali MI, Pant M, Warghane A. Current Status of Potential Antiviral Drugs Derived from Plant, Marine, and Microbial Sources. *Anti-Infective Agents*. 2024 Apr 1;22(2):61-73.

[63] Bălașa AF, Chircov C, Grumezescu AM. Marine biocompounds for neuroprotection—A review. *Marine Drugs*. 2020 May 31;18(6):290.

[64] Naik B, Richa S, Bharadwaj S, Mishra S, Kumar V, Kumar V, Saris PE, Gupta AK, Mishra R, Gupta U, Rustagi S. Utilizing marine algal metabolites to fight neurodegenerative diseases. *Frontiers in Marine Science*. 2024 May 24; 11:1370839.

[65] Ercolano G, De Cicco P, Ianaro A. New drugs from the sea: Pro-apoptotic activity of sponges and algae-derived compounds. *Marine drugs*. 2019 Jan 7;17(1):31.

[66] Othman NF, Singh GK, Hazalin NA. *Roles of Nutraceuticals in Regulating Neurotransmitters in Dementia*.

[67] Tanna B, Mishra A. Metabolites unravel nutraceutical potential of edible seaweeds: An emerging source of functional food. *Comprehensive reviews in food science and food safety*. 2018 Nov;17(6):1613-24.

[68] Miyashita K. Marine antioxidants: polyphenols and carotenoids from algae. *Antioxidants and functional components in aquatic foods*. 2014 Jun 2:219-35.

[69] Pereira L, Valado A. Algae-Derived Natural Products in Diabetes and Its Complications—Current Advances and Future Prospects. *Life*. 2023 Aug 29;13(9):1831.

[70] Pradhan B, Nayak R, Patra S, Jit BP, Ragusa A, Jena M. Bioactive metabolites from marine algae as potent pharmacophores against oxidative stress-associated human diseases: A comprehensive review. *Molecules*. 2020 Dec 23;26(1):37.

[71] Fonseca S, Amaral MN, Reis CP, Custódio L. Marine natural products as innovative cosmetic ingredients. *Marine Drugs*. 2023 Mar 8;21(3):170.

[72] Carvalho LG, Pereira L. Review of marine algae as source of bioactive metabolites: A marine biotechnology approach. *Marine Algae*. 2014 Sep 26:195-227.

[73] Parte S, Sirisha VL, D'Souza JS. Biotechnological applications of marine enzymes from algae, bacteria, fungi, and sponges. *In Advances in food and nutrition research* 2017 Jan 1 (Vol. 80, pp. 75-106). Academic Press.

[74] Alves A, Sousa E, Kijjoa A, Pinto M. Marine-derived compounds with potential use as cosmeceuticals and nutricosmetics. *Molecules.* 2020 May 29;25(11):2536.

[75] Leal MC, Rocha RJ, Rosa R, Calado R. Aquaculture of marine non-food organisms: what, why and how? *Reviews in Aquaculture.* 2018 Jun;10(2):400-23.

[76] Hazaimeh MD, Ahmed ES. Bioremediation perspectives and progress in petroleum pollution in the marine environment: a review. *Environmental Science and Pollution Research.* 2021 Oct;28(39):54238-59.

[77] Anjum K, Abbas SQ, Akhter N, Shagufta BI, Shah SA, Hassan SS. Emerging biopharmaceuticals from bioactive peptides derived from marine organisms. *Chemical biology & drug design.* 2017 Jul;90(1):12-30.

[78] Senadheera TR, Hossain A, Shahidi F. Marine bioactives and their application in the food industry: A review. *Applied Sciences.* 2023 Nov 6;13(21):12088.

[79] Liang X, Luo D, Luesch H. Advances in exploring the therapeutic potential of marine natural products. *Pharmacological research.* 2019 Sep 1; 147:104373.

[80] Rai T, Kaushik N, Malviya R, Sharma PK. A review on marine source as anticancer agents. *Journal of Asian Natural Products Research.* 2024 Apr 2;26(4):415-51.

[81] Cutolo EA, Campitiello R, Caferri R, Pagliuca VF, Li J, Agathos SN, Cutolo M. Immunomodulatory Compounds from the Sea: From the Origins to a Modern Marine Pharmacopoeia. *Marine Drugs.* 2024 Jul;22(7):304.

[82] Burri L, Hoem N, Banni S, Berge K. Marine omega-3 phospholipids: metabolism and biological activities. *International journal of molecular sciences.* 2012 Nov 21;13(11):15401-19.

[83] Sukmarini L. Antiviral peptides (AVPs) of marine origin as propitious therapeutic drug candidates for the treatment of human viruses. *Molecules.* 2022 Apr 19;27(9):2619.

[84] Prasathkumar M, Sadhasivam S. Chitosan/Hyaluronic acid/Alginate and an assorted polymers loaded with honey, plant, and marine compounds for progressive wound healing—Know-how. *International Journal of Biological Macromolecules.* 2021 Sep 1; 186:656-85.

[85] Sharma PP, Baskaran V. Polysaccharide (laminaran and fucoidan), fucoxanthin and lipids as functional components from brown algae (Padina tetrastromatica) modulates adipogenesis and thermogenesis in diet-induced obesity in C57BL6 mice. *Algal Research.* 2021 Apr 1; 54:102187.

[86] Carletti A, Gavaia PJ, Cancela ML, Laizé V. Metabolic bone disorders and the promise of marine osteoactive compounds. *Cellular and Molecular Life Sciences.* 2024 Dec;81(1):11.

[87] Senthilkumar K, Venkatesan J, Kim SK. Marine derived natural products for osteoporosis. *Biomedicine & Preventive Nutrition.* 2014 Jan 1;4(1):1-7.

[88] Calder PC. Marine omega-3 fatty acids and inflammatory processes: Effects, mechanisms and clinical relevance. *Biochimica et Biophysica Acta (BBA)-Molecular and Cell Biology of Lipids.* 2015 Apr 1;1851(4):469-84.

[89] Lopes PA, Bandarra NM, Martins SV, Martinho J, Alfaia CM, Madeira MS, Cardoso C, Afonso C, Paulo MC, Pinto RM, Guil-Guerrero JL. Markers of neuroprotection of combined EPA and DHA provided by fish oil are higher than those of EPA (Nannochloropsis) and DHA (Schizochytrium) from microalgae oils in Wistar rats. *Nutrition & metabolism.* 2017 Dec; 14:1-7.

[90] Martins M, Silva R, MM Pinto M, Sousa E. Marine natural products, multitarget therapy and repurposed agents in Alzheimer's disease. *Pharmaceuticals.* 2020 Sep 11;13(9):242.

[91] Hu D, Jin Y, Hou X, Zhu Y, Chen D, Tai J, Chen Q, Shi C, Ye J, Wu M, Zhang H. Application of marine natural products against Alzheimer's disease: past, present and future. *Marine Drugs.* 2023 Jan 5;21(1):43.

[92] Ajith TA. A recent update on the effects of omega-3 fatty acids in Alzheimer's disease. *Current clinical pharmacology.* 2018 Nov 1;13(4):252-60.

[93] Seong SH, Paudel P, Choi JW, Ahn DH, Nam TJ, Jung HA, Choi JS. Probing multi-target action of phlorotannins as new monoamine oxidase inhibitors and dopaminergic receptor modulators with the potential for treatment of neuronal disorders. *Marine drugs.* 2019 Jun 24;17(6):377.

[94] Pan N, Li ZC, Li ZH, Chen SH, Jiang MH, Yang HY, Liu YS, Hu R, Zeng YW, Dai LH, Liu L. Antiplatelet and antithrombotic effects of isaridin E isolated from the marine-derived fungus via downregulating the PI3K/Akt signaling pathway. *Marine Drugs*. 2021 Dec 24;20(1):23.

[95] Lee JC, Hou MF, Huang HW, Chang FR, Yeh CC, Tang JY, Chang HW. Marine algal natural products with anti-oxidative, anti-inflammatory, and anti-cancer properties. *Cancer cell international.* 2013 Dec; 13:1-7.

[96] Grosso C, Valentão P, Ferreres F, Andrade PB. Bioactive marine drugs and marine biomaterials for brain diseases. *Marine drugs*. 2014 May 2;12(5):2539-89.

[97] Liu B, Zhang Y, Yang Z, Liu M, Zhang C, Zhao Y, Song C. ω-3 DPA protected neurons from neuroinflammation by balancing microglia M1/M2 polarizations through inhibiting NF-κB/MAPK p38 signaling and activating neuron-BDNF-PI3K/AKT pathways. *Marine drugs*. 2021 Oct 20;19(11):587.

[98] Fakhri S, Yarmohammadi A, Yarmohammadi M, Farzaei MH, Echeverria J. Marine natural products: Promising candidates in the modulation of gut-brain axis towards neuroprotection. *Marine Drugs*. 2021 Mar 19;19(3):165.

[99] Letertre MP, Dervilly G, Giraudeau P. Combined nuclear magnetic resonance spectroscopy and mass spectrometry approaches for metabolomics. *Analytical chemistry*. 2020 Nov 6;93(1):500-18.

[100] Ambrosino L, Tangherlini M, Colantuono C, Esposito A, Sangiovanni M, Miralto M, Sansone C, Chiusano ML. Bioinformatics for marine products: An overview of resources, bottlenecks, and perspectives. *Marine drugs*. 2019 Oct 11;17(10):576.

[101] Kennedy J, Flemer B, Jackson SA, Lejon DP, Morrissey JP, O'gara F, Dobson AD. Marine metagenomics: new tools for the study and exploitation of marine microbial metabolism. *Marine drugs*. 2010 Mar;8(3):608-28.

[102] Mao X, Liu Z, Sun J, Lee SY. Metabolic engineering for the microbial production of marine bioactive compounds. Biotechnology Advances. 2017 Dec 1;35(8):1004-21.

[103] Shuikan AM, Hozzein WN, Alshuwaykan RM, Arif IA. *Metabolomics and genetic engineering for secondary metabolites discovery. Secondary Metabolites: Trends and Reviews*; IntechOpen: London, UK. 2022 Mar 31:71.

[104] Guilloux B. *Marine Genetic Resources, R&D and the Law 1: Complex Objects of Use*. John Wiley & Sons; 2018 Jun 21.

[105] Rocha-Martin J, Harrington C, Dobson AD, O'Gara F. Emerging strategies and integrated systems microbiology technologies for biodiscovery of marine bioactive compounds. *Marine drugs*. 2014 Jun 10;12(6):3516-59.

[106] Glaser KB, Mayer AM. A renaissance in marine pharmacology: from preclinical curiosity to clinical reality. *Biochemical pharmacology*. 2009 Sep 1;78(5):440-8.

[107] Newman DJ, Cragg GM. Advanced preclinical and clinical trials of natural products and related compounds from marine sources. *Current medicinal chemistry*. 2004 Jul 1;11(13):1693-713.

[108] Giacobbe J, Benoiton B, Zunszain P, Pariante CM, Borsini A. The anti-inflammatory role of omega-3 polyunsaturated fatty acids metabolites in pre-clinical models of psychiatric, neurodegenerative, and neurological disorders. *Frontiers in Psychiatry*. 2020 Feb 28; 11:450833.

[109] Hunt B, Vincent AC. Scale and sustainability of marine bioprospecting for pharmaceuticals. AMBIO: *A Journal of the Human Environment*. 2006 Mar;35(2):57-64.

[110] Jain A, Tailor V. Emerging Trends of Biotechnology in Marine Bioprospecting: A New Vision. *Marine Niche: Applications in Pharmaceutical Sciences: Translational Research*. 2020:1-36.

[111] Martins A, Vieira H, Gaspar H, Santos S. Marketed marine natural products in the pharmaceutical and cosmeceutical industries: Tips for success. *Marine drugs*. 2014 Feb 17;12(2):1066-101.

[112] Altmann KH. Drugs from the oceans: Marine natural products as leads for drug discovery. *Chimia*. 2017 Oct 25;71(10):646-.

[113] Ibrahim HA. Future Perspectives for Developing Marine Drug Discovery. *Marine Ecosystems: A Unique Source of Valuable Bioactive Compounds*. 2023 Nov 6; 3:394-423.

[114] Hannan MA, Dash R, Haque MN, Mohibbullah M, Sohag AA, Rahman MA, Uddin MJ, Alam M, Moon IS. Neuroprotective potentials of marine algae and their bioactive metabolites: Pharmacological insights and therapeutic advances. *Marine drugs*. 2020 Jul 1;18(7):347.

Chapter 10

Unveiling Marine Biosurfactants: Genomics, Metabolomics, and Biotechnological Insights

Suthar Harish[1,*]
Kalaria Rishee[2]
and Radhey Shyam Kaushal[3]

[1]Department of Post Harvest Technology, ASPEE College of Horticulture, Navsari Agricultural University, Navsari, Gujarat, India
[2]ASPEE Shakilam Biotechnology Institute, Navsari Agricultural University, Surat, Gujarat, India
[3]Department of Life Sciences - Biotechnology, Parul Institute of Applied Sciences and Biophysics and Structural Biology Laboratory, Research and Development Cell, Parul University, Gujarat, India

Abstract

The marine environment is a key factor in determining the survival of several living organisms found in marine habitats. It displays a wide range of environments in which marine life has developed with a great deal of structural and functional diversity. The marine habitat contains a wealth of bioactive compounds that could be applied to a wide range of biotechnological applications. One of the most important things is to preserve the marine ecosystem. It is negatively impacted by the dangerous occurrence of a petroleum spill. Many marine microorganisms make biosurfactants to protect themselves against such occurrences.

Marine biosurfactants help in bioremediation of hydrocarbon waste by reducing the interfacial tension between oil and water, forming a stable emulsion, and using hydrocarbons as substrates. In the past two decades, development of advanced techniques that draw molecular investigations at the genomic, metagenomic, transcriptomic, metatranscriptomic, proteomic, and metabolomic levels is imperative. These techniques made it possible to study diverse biological systems, to identify the different types of biosurfactants produced by microorganisms, and to understand the critical molecular mechanisms involved in the synthesis and functioning of biosurfactants. This chapter includes the study of different types of marine biosurfactants, applications of bioinformatics and computational tools for the molecular structure of biosurfactants, and interpretation of the structure-function relationship. In the chapter, proper exploitation of

* Corresponding Author's Email: harish.s@nau.in

In: Healing from the Depths Marine Metabolites in Disease Management
Editors: Pranav Kumar Prabhakar and Tarun Kumar Upadhyay
ISBN: 979-8-89530-832-5

the information has been done which can be useful in developing marine biosurfactant based innovative applications in different fields.

Keywords: biosurfactants, bioinformatics, metagenomics, bioactive compounds

1. Marine Biosurfactants

Life first emerged in the oceans ~1.5 billion years ago. Since then, the structure and functions of marine organisms have diversified, making the marine environment the largest and most diverse ecosystem on Earth. It occupies ~70% of the planet's surface. It includes a wide range of environmental conditions such as the intense cold of the polar seas and the intense heat and pressure of deep-sea hydrothermal vents (Danovaro et al., 2011).

Marine habitats have different environmental conditions, such as salinity, pressure, temperature, light and nutritional composition, and provide an ideal place for the evolution of every living creatures in the sea. Thus, the marine ecosystem is one of the most potential sources of diverse organisms and novel bioactive compounds (Barzkar et al., 2023). It is a treasure of diverse forms of naturally occurring polysaccharides, peptides, proteins, and fatty acids with unique properties. Proteins from marine habitats are reported to be used as antibacterial, foaming, and gel-forming agents (Ghosh et al., 2022). Scientists are working to identify chemically diverse and novel compounds that have great commercial and social values. Marine microorganisms, such as bacteria, fungi and microalgae, are reported to produce a variety of metabolites having anti-viral, anti-bacterial, antioxidant, anti-tumor activity. Further, research on the application of biotechnological methods to synthesize and modify marine compounds to improve their functional properties is ongoing. The surfaceactive agents produced by microorganisms are called biosurfactants. They are used to reduce the surface tension between two immiscible solvents such as oil and water. They are also used as emulsifiers to form a stable emulsion. As their origin is biological, they are biodegradable and eco-friendly in nature. Hence, they can be the most suitable alternative to hazardous chemical surfactants (Poomalai et al., 2024). Intra-cellular or extracellular biosurfactants are produced by microorganisms in the early stationary phase or late log of the growth cycle. Biosurfactants are a diverse group of amphipathic molecules having hydrophilic (polar) and hydrophobic (non-polar) moieties. They are of two types, low molecular weight biosurfactants and high molecular weight biosurfactants. Low-molecular-weight biosurfactants lower the surface and interfacial tensions between two immiscible surfaces (Air-Water, Oil-Water, etc.) while high-molecular-weight biosurfactants bind tightly to surfaces and act as emulsifiers.

Different types of biosurfactants with different activities and structures are produced by various microorganisms. Microbes produce these biosurfactants to support growth. Production of biosurfactant helps microorganisms to survive and grow in the competitive environment in specific habitats. For this reason, some biosurfactants possess antibacterial and antifungal properties. Some biosurfactants help microbes to use hydrophobic material as a substrate by increasing its solubility in water, while other biosurfactants help in the attachment-detachment of microbes to surfaces. Further, it is also reported that biosurfactants play an important role in pathogenesis, quorum sensing, and biofilm formation. Bioemulsifiers can be transferred horizontally from one bacterium to another, thereby changing the recipient's surface properties and interactions with the environment (Ron and Rosenberg, 2001). Many researchers have identified unique biosurfactant molecules from marine habitats for innovation in health, nutrition, and agriculture. These are the reasons for the immense interest in investigating the marine environment (Ambrosino et al., 2019).

Currently, over 2000 unique biosurfactant structures are known. They vary in their structural, biological, and physicochemical characteristics. The potential of biosurfactants is determined by properties such as minimal Critical Micelle Concentrations (CMC), efficient surface tension reduction, metal ion complexation, eco-friendly nature, and great bioactivity potential. A low CMC indicates that the biosurfactant can function at very low concentration as compared to the chemical surfactants (Bhadoriya et al., 2013; Hausmann et al., 2014). Further, biosurfactants are produced from renewable resources, have a low eco-toxicity and excellent biological degradability that adds credit to the application of biosurfactant for environmental sustainability (Johann et al., 2016; Poremba et al., 1991). Biosurfactants are widely used as emulsifiers, detergents, and wetting agents for hydrophobic surfaces. They are also utilized as medicines in food as well as in cosmetics (Banat et al., 2000; Liang et al., 2019).

The current chapter contains updated information about a noticeable rise in research activity for the application of biosurfactants in different fields (Nikolova and Gutierrez, 2021). Numerous biosurfactants produced by microorganisms are of interest to the biotechnological and pharmaceutical industries. Among the most well-known examples of biosurfactants are surfactin produced by *Bacillus subtilis*, lichenisyn produced by *B. licheniformis*, pumilacidin by *B. pumilus*, rhamnolipids produced by *Pseudomonas aeruginosa*, sophorolipid produced by *Paracoccus marcusii* and galactolipid produced by *Cyberlindnera saturnus* (Kubicki et al., 2019).

For the development of innovative products, investigations are mainly focused on the identification of species that have the potential to produce novel bioactive metabolites. Entrepreneurs are making significant investments in advanced research that have the potential to efficiently reduce the time generally taken to identify and develop new drugs (Iskar et al., 2012; Whittaker, 2003). This thrust developed surprising interest in the field of bioinformatics (Ortega et al., 2012). Bioinformatics provides methods for effectively extracting information that adds value from omics experimental data and ~~for~~ modeling. Tools of bioinformatics are successfully used to speed up the identification of biologically active candidates, their structures and mode of actions on other living systems (Ambrosino et al., 2017; Esposito et al., 2016).

The ability of the species to produce biosurfactant compounds is encoded by its genome. Hence, the identification of novel biosurfactants and the molecular mechanism involved in their synthesis begin with genome or transcriptome sequencing. Introduction of sophisticated methodologies, such as next-generation sequencing (NGS) technologies, made it possible to use the "test and then isolate" in place of the "isolate and then test" (Trindade et al., 2015). In recent years, the development of cost-effective NGS has greatly benefited biology in general and marine biotechnology. Expansion of the sequencing projects and use of omics approaches such as proteomics and metabolomics provided great support in understanding the structure and functionality of molecules (Hartmann et al., 2014; Lacerda and Reardon, 2009; Lauritano and Ianora, 2018). Technological developments have now made it possible to isolate organisms after novel compounds have been identified and the metabolic pathways leading to these compounds have been elucidated. Alternatively, only the genes encoding the compounds in question can be isolated and subsequently expressed in heterologous hosts for controlled production. Furthermore, biotechnological instruments to improve and control the synthesis of natural molecules have significantly improved in the wake of recent developments. For instance, cutting-edge experimental methods like genome editing have just been made available and are extensively used to alter the genome sequence in specific regions of interest (Lin et al., 2019). One notable example of a unique technique is the use of CRISPR/Cas9 machinery to create mutant genomes that specifically target genes (Doudna and Charpentier, 2014). Furthermore, bioinformatics offers tools for these methods that can anticipate CRISPR/Cas9 targets, even in new or incomplete genome sequences (Heigwer et al., 2014; Montague et al., 2014; Stemmer et al., 2015).

2. Applications and Resources of Bioinformatics in Marine Omics

Exploring new and valuable chemicals from marine resources, such as metabolites, enzymes, and other molecules, provides further understaning of the molecular processes underlying their synthesis, and functional characteristics can be accomplished through various methods and techniques. These techniques range from metabolomics, which is used to comprehend the phenotypic effects of genome expression, to whole-genome sequencing. This stand-alone research line also serves as a reference baseline for additional omics techniques, like transcriptomics and proteomics, which are used to investigate the functional activity of species or tissues. Their community-level counterparts in meta-omics, such as metagenomics and metatranscriptomics, are capable of addressing comparable problems (Dai and Shen, 2022).

2.1. Genomics and Transcriptomics

High-throughput technologies are fast and less expensive. The emergence of novel technologies, such as NGS approaches, has facilitated the transition from the less effective Sanger methodology to the sequencing of large quantities of DNA fragmentsbecause genomics and transcriptomic approaches combine to locate and quickly annotate genes producing interesting substances. Nucleic acid sequencing techniques are also widely employed in marine biotechnology, particularly in the quest for new marine pharmaceuticals (Xie et al., 2017; Rivera-de-Torre et al., 2019). The regulatory mechanisms govern biosynthetic gene clusters, potentially identifying the kind of post-translational modification involved in the synthesis of proteins. In order to complement these investigations, various other NGS-based sequencing techniques are becoming more and more popular, such as small RNA, epigenome, and single cell sequencing (Amos et al., 2017).

2.2. Metagenomics and Metatranscriptomics

Our understanding of marine ecology is advancing significantly as a result of the intensively studied field of metagenomics. The study of the genetic and genomic composition of microbes isolated from mixed populations in a certain habitat is known as metagenomics (Handelsman, 2004). The taxonomic and functional description of the specific environment can be done using this method. One example is the detection of proteorhodopsin, which led to the discovery of new feeding methods on ocean surfaces (Béjà et al., 2000; Béjà et al., 2001). Building libraries and expanding knowledge for specific marine life segments became feasible with the advent of bulk DNA and RNA sequencing technology (Roumpeka et al., 2017; Teeling and Glockner, 2012).

Furthermore, to fulfill the demands of the ever-increasing complexity of the study and the necessity to stay up to date with the fast-expanding body of reference material, bioinformatic approaches are constantly updated and improved. According to Bremges and McHardy (2018), the implementation of programs like Critical Assessment of Metagenome Interpretation (CAMI) encouraged the validation of best practices for achieving reliable and consistent results. Recent investigations of the metabolic capabilities of genomes rebuilt from metagenomes have been conducted on thousands of genomes or their fragments from different marine ecosystems (Dombrowski et al., 2018; Tully et al., 2018; Seitz et al., 2019). Indeed, genome mining can help find new compounds (Machado et al., 2015), while comparative genomics can help determine ecological or evolutionary trends (Coutinho et al., 2018).

2.3. Proteomics and Structural Biology

The primary players in the functional operations that a biological system performs are proteins. They respond to changes in the environment as well as the emergence of internal or external stimuli (Feder and Walser, 2005). Proteomics is a system-based view of how organisms mount their molecular reactions, with the goal of identifying and quantifying proteins. Peptides and poisons from sea anemones, sea sponges, cone snails, and cyanobacteria that are beneficial for biomedical research were found via proteomic investigations. Like all omics methodologies, proteomics research depends on bioinformatic tools to facilitate both the processing of unprocessed data, and the utilization of the results generated. One of the most extensive sources of information about protein sequences is InterPro, a database that includes various classifications of features related to proteins (Finn et al., 2017). For instance, protein family information from Pfam, the protein families database, along with more in-depth explanations like protein domains or sequence conserved signatures, can be found using InterPro (El-Gebali et al., 2019). Protein structure prediction is a fundamental method to highlight conformational aspects of molecules of biotechnological interest. Examples of this type of work include defining enzymatic action useful for biotechnological applications or clarifying structural features related to environmental adaptation, such as warm- or cold-adapted mechanisms that confer thermostability in extremophilic enzymes. By offering an *in-silico* calculation of the stability of enzymes and protein complexes even before conducting in vitro tests, molecular dynamics simulations facilitate the assessment of the biotechnological potential of molecules of interest. The primary hindrance to proteomics research is the structural complexity of the biomolecules and physiological processes that are associated with it (De Santi et al., 2014; Russell, 2000). This complexity specifically affects the resolution of the protein structure data produced by crystallographic or NMR experiments in structural biology applications. Further, it has an impact on all the downstream bioinformatics processes such as homology modeling, molecular dynamics, and drug design techniques (Cruickshank, 1999).

2.4. Metabolomics

The metabolome of an organism actually represents the phenotypic responses of the organism to a wide range of physiological and environmental stimuli, and it directly correlates with gene expression and the corresponding creation of metabolites (Bundy et al., 2005). Reactions of organisms to changing conditions frequently involve metabolic remodeling and the regulation of certain metabolite levels, which may serve as indicators of a particular response (e.g., biotic or abiotic stressors). In this regard, metabolomics aids in the assessment of how climate change affects marine life, revealing potential roles for marine systems in reducing the effects of global warming (Brierley and Kingsford, 2009; Baltar et al., 2019). The antiSMASH server enables genome-wide identification, annotation, and analysis of gene clusters related to secondary metabolite biosynthesis. Global Natural Products Social (GNPS) molecular networking represents an open-access resource for sharing of raw, processed, or identified tandem mass (MS/MS) spectrometry data. Both these bioinformatics resources utilized in such efforts are very noteworthy (Blin et al., 2019; Wang et al., 2016). The accuracy with which the biological functions of the gathered bioactive chemicals and metabolites are defined is the foundation for all the resources that are offered. Although metabolite identification remains a key bottleneck in untargeted metabolomics, proper annotation is essential for data interpretation (Chaleckis et al., 2019). High-throughput metabolite profiling and annotation, as well as extremely difficult computational procedures used in metabolomic interpretation, are produced by specialized service centers (Rawlings et al., 2018).

The inconsistency in resolution and the difficulty of establishing universal benchmarks across different specialized laboratories and technology are the main sensitive problems. Despite the fact that the community has long established guidelines for metabolite identification, there hasn't been much progress in following these guidelines. The intricacy of metabolomic data from different combinations of chromatographic and mass spectrometric acquisition technologies has led to the establishment of diverse pipelines and workflows. Often, these processes entail manual curation that is not standard. Furthermore, the bioinformatics tools in the field still need to better address the enrichment analysis problem, accurately linking metabolomic data to the most reliable similar compounds and producing thorough pathway diagrams (Ambrosino et al., 2019).

Solaiman et al., (2014) used PCR, gene-walking, and cloning to express the sophorolipid-related glucosyltransferase gene (gtf-1) from *C. bombicola* in *Saccharomyces cerevisiae* (Thies et al., 2014; Jackson et al., 2015). The broad specificity of this lipid-glucosylation enzyme allows for the creation of novel glycolipids by using non-endogenous fatty acids or co-expressing cloned fatty acid genes. Using alkane or fatty acid-based substrates, the functionalities of three cytochrome P450 family monooxygenase genes (CYP52E3, CYP52M1, and CYP52N1) that were cloned into *S. cerevisiae* were examined. The glucosyltransferase UGTA1 was then used to analyze the biosynthetic products of all three enzymes. It was discovered that all fatty acids were glycosylated, although CYP52M1 was the most effective. According to Huang et al., (2014), CYP52M1 may be involved in the breakdown of alkanes, whereas CYP52E3 and CYP52N1 may be involved in the production of sophorolipids in *C. bombicola*.

3. The Potential of Bioinformatic Tools for the Application of Biosurfactants in Several Industries

The scientific community is beginning to recognize the importance of creating integrated marine data collections from multiple observations due to the wealth of information that the sea can provide. These collections should be established primarily to monitor and assess the health of marine ecosystems, but they should also serve as a multimodal means of elucidating the complexity of marine species of biological and biotechnological interest (Bourlat et al., 2013; Duffy et al., 2013). It is anticipated that these maritime observatories would facilitate the temporal and spatial gathering of a variety of data sets, from satellite imagery and multi-omics molecular data to traditional biogeochemical and oceanographic observations. Expanded mobile sensor networks and pervasive computing made affordable real-time data collecting possible. This has significantly improved metadata augmentation and enrichment. All of these elements, along with the expanding tendencies in artificial intelligence (mostly machine learning and deep learning) to generate knowledge from data, offer an exciting and difficult situation that can be used to unravel the innate complexity of large biological datasets. It is crucial to use this spatial-temporal data to create new in silico models and predictions in order to increase our understanding of marine species and their biotechnological significance. Adopting the concept of data fairness is another essential step in this approach, as it guarantees the accessibility and availability of the data produced by the scientific community (Wilkinson et al., 2016). Understanding biological systems requires the integration of omics data, which presents a barrier. For example, transcriptomic and proteomic data might enhance the precision of genome annotation (Prasad et al., 2017), and connecting metagenomics and meta-metabolomics will provide a connection between the manufacturers of bioactive compounds and their sources (Tuttle et al., 2019). The meta-analytical technique is another intriguing viewpoint (Gurevitch et al., 2018). Marine resources have yielded a wealth of data, particularly genomic and metagenomic databases. Typically, an overview of the primary detectable functionalities and a taxonomy profile are provided by the analysis of such data. But there are no systems that offer standardized data processing for marine biology, like the human metagenome's curated Metagenomic Data. In the rapidly developing field of marine biotechnology research, a critical comparison and suitable integration of findings from various projects will be valuable in identifying true patterns and sources of bias, advancing scientific discovery (Pasolli et al., 2017).

Silva et al., (2022) used short-read Illumina sequencing to gather six short-read metagenomic datasets from various fruit autochthonous fermentation batch reactors. Additionally, they employed the DIAMOND+MEGAN pipeline to identify and examine the genes implicated in the biosurfactant production pathways as well as to assess the microbial diversity found in the samples. Additionally, they used MEGAN to do a thorough taxonomy analysis based on DIAMOND alignments with the NCBI-nr protein reference database. Genes linked to the synthesis and degradation of biosurfactants were found using the domain-specific BiosurfDB database. The six fruit waste samples had a common core of Gamma-proteobacteria, according to the study, with the species *Klebsiella*, *Enterobacter*, *Stenotrophomonas*, *Escherichia*, and *Acinetobacter* predominating. The functional profiles and potential for biosurfactant production, particularly for lipopeptides, were comparable across all examined samples. Because of qualities including detergency and foam generation, biosurfactants are better suited for a variety of agricultural applications than chemical surfactants.

They can be used as biopesticides to manage pests, pathogens, phytopathogenic fungi and weeds. Moreover, plants can use biosurfactant molecules as substrates and microorganisms as biofertilizers (Silva et al., 2024).

Conclusion

The survival of many living things found in maritime settings is largely determined by the marine environment. It exhibits a broad variety of habitats where marine life has evolved with significant structural and functional diversity. Numerous bioactive substances found in marine habitat have potential uses in a broad range of biotechnological fields. Preserving the marine ecology is among the most crucial things. The risky event of a petroleum leak has a detrimental effect on it. To defend themselves against such events, a large number of marine microorganisms produce biosurfactant. By using hydrocarbons as substrates, forming stable emulsions, and lowering the interfacial tension between water and oil, marine biosurfactants aid in the bioremediation of hydrocarbon waste. The creation of sophisticated methods that attract molecular analyses at the genomic, metagenomic, transcriptomic, metatranscriptomic, proteomic, and metabolomic levels has been essential throughout the last 20 years. Studying the many biological systems, identifying the various kinds of biosurfactants that microorganisms create, and comprehending the crucial molecular mechanisms behind the synthesis and operation of the biosurfactants were all made possible by these methodologies. Because of the abundance of information that the sea can offer, the scientific community is starting to realize how important it is to create integrated marine data sets from multiple observations. Although the primary purpose of these collections should be to track and evaluate the condition of marine ecosystems, they should also be used as a multimodal tool to clarify the intricacy of marine species of biological and biotechnological significance.

The primary sensitive issues are inconsistent resolution and the challenge of creating universal standards across many specialized labs and technologies. Guidelines for metabolite identification have been created by the community for a long time, however adherence to these guidelines has not advanced considerably. Due to the complexity of metabolomic data from various chromatographic and mass spectrometric acquisition technology combinations, a variety of pipelines and workflows have been established. These procedures frequently include non-standard manual curation. Additionally, the field's bioinformatics tools still need to improve their ability to handle the enrichment analysis problem by precisely associating metabolomic data with the most trustworthy comparable substances and generating comprehensive pathway diagrams. Specialized service centers produce high-throughput metabolite profiling and annotation, as well as the computationally complex processes involved in metabolomic interpretation. In the near future, this could open a new horizon for study and depth of understanding of marine biosurfactants in terms of genomic and metabolomics approaches.

References

Ambrosino, L., Colantuono, C., Monticolo, F., Chiusano, M. L. (2017) Bioinformatics Resources for Plant Genomics: Opportunities and Bottlenecks in The -omics Era. *Curr. Issues Mol. Biol.* 2017:71–88.

Ambrosino, L., Tangherlini, M., Colantuono, C., Esposito, A., Sangiovanni, M., Miralto, M., Sansone, C., Chiusano, M. L. (2019) Bioinformatics for Marine Products: An Overview of Resources, Bottlenecks, and Perspectives. *Mar. Drugs.* 17:576.

Amos, G. C. A., Awakawa, T., Tuttle, R. N., Letzel, A.-C., Kim, M. C., Kudo, Y., Fenical, W., Moore, B. S., Jensen, P. R. (2017) Comparative transcriptomics as a guide to natural product discovery and biosynthetic gene cluster functionality. *Proc. Natl. Acad. Sci. USA.* 114: E11121.

Baltar, F., Bayer, B., Bednarsek, N., Deppeler, S., Escribano, R., Gonzalez, C.E., Hansman, R.L., Mishra, R.K., Moran, M. A., Repeta, D. J., et al., (2019) Towards Integrating Evolution, Metabolism, and Climate Change Studies of Marine Ecosystems. *Trends Ecol. Evol.* 34(11):1022-1033.

Banat, I. M., Makkar, R. S., Cameotra, S. S. (2000) Potential commercial applications of microbial surfactants. *Appl. Microbiol. Biotechnol.* 53:495–508.

Barzkar, N., Sukhikh, S., Babich, O. (2024) Study of marine microorganism metabolites: new resources for bioactive natural products. *Front Microbiol.* 14:1285902.

Béjà, O., Aravind, L., Koonin, E. V., Suzuki, M. T., Hadd, A., Nguyen, L. P., Jovanovich, S. B., Gates, C. M., Feldman, R. A., Spudich, J. L., et al., (2000) Bacterial Rhodopsin: Evidence for a New Type of Phototrophy in the Sea. *Science. 289*(5486): 1902–1906.

Béjà, O., Spudich, E. N., Spudich, J. L., Leclerc, M., DeLong, E. F. (2001) Proteorhodopsinphototrophy in the ocean. *Nature.* 411: 786–789.

Bhadoriya, S. S., Madoriya, N. (2013) Biosurfactants: A new pharmaceutical additive for solubility enhancement and pharmaceutical development. *Biochem. Pharmacol..* 2:113.

Blin, K., Shaw, S., Steinke, K., Villebro, R., Ziemert, N., Lee, S. Y., Medema, M. H., Weber, T. (2019) antiSMASH 5.0: Updates to the secondary metabolite genome mining pipeline. *Nucleic Acids Res.* 47(W1): W81–W87.

Bourlat, S. J., Borja, A., Gilbert, J., Taylor, M. I., Davies, N., Weisberg, S. B., Griffith, J. F., Lettieri, T., Field, D., Benzie, J., et al., (2013) Genomics in marine monitoring: New opportunities for assessing marine health status. *Mar. Pollut. Bull.* 74:19–31.

Bremges, A., McHardy, A. C. (2018) Critical Assessment of Metagenome Interpretation Enters the Second Round. *mSystems.* 3:4.

Brierley, A. S., Kingsford, M. J. (2009) Impacts of Climate Change on Marine Organisms and Ecosystems. *Curr. Biol.* 19:R602–R614.

Bundy, J. G., Willey, T. L., Castell, R. S., Ellar, D. J., Brindle, K. M. (2005) Discrimination of pathogenic clinical isolates and laboratory strains of *Bacillus cereus* by NMR-based metabolomic profiling. *FEMS Microbiol. Lett.* 242:127–136.

Chaleckis, R., Meister, I., Zhang, P., Wheelock, C. E. Challenges, progress and promises of metabolite annotation for LC–MS-based metabolomics. *Curr. Opin. Biotechnol.* 55:44–50.

Coutinho, F. H., Gregoracci, G. B., Walter, J. M., Thompson, C. C., Thompson, F. L. (2018) Metagenomics Sheds Light on the Ecology of Marine Microbes and Their Viruses. *Trends Microbiol.* 26:955–965.

Cruickshank, D. W. (1999) Remarks about protein structure precision. *Acta Crystallogr. Sect. DBiol. Crystallogr.* 55:583–601.

Dai, X., Shen, L. (2022) Advances and Trends in Omics Technology Development. *Front Med* (Lausanne). 9:911861.

Danovaro, R., Corinaldesi, C., Dell'Anno, A., Fuhrman, J. A., Middelburg, J. J., Noble, R. T., Suttle, C. A. (2011) Marine viruses and global climate change. *FEMS Microbiol. Rev.* 35: 993–1034.

De Santi, C., Tedesco, P., Ambrosino, L., Altermark, B., Willassen, N. P., de Pascale, D. (2014) A New Alkaliphilic Cold-Active Esterase from the Psychrophilic Marine Bacterium Rhodococcus sp.: Functional and Structural Studies and Biotechnological Potential. *Appl. Biochem. Biotechnol. 172*(6):3054–3068.

Dombrowski, N., Teske, A. P., Baker, B. J. (2018) Expansive microbial metabolic versatility and biodiversity in dynamic Guaymas Basin hydrothermal sediments. *Nat. Commun. 9*: 4999.

Doudna, J. A. and Charpentier, E. (2014) The new frontier of genome engineering with CRISPR-Cas9. *Science*. 346:1258096.

Duffy, J. E., Amaral-Zettler, L. A., Fautin, D. G., Paulay, G., Rynearson, T. A., Sosik, H. M., Stachowicz, J. J. (2013) Envisioning a Marine Biodiversity Observation Network. *BioScience*. 63:350–361.

El-Gebali, S., Mistry, J., Bateman, A., Eddy, S. R., Luciani, A., Potter, S. C., Qureshi, M., Richardson, L. J., Salazar, G. A., Smart, A., et al., (2019) The Pfam protein families database in 2019. *Nucleic Acids Res.* 47:D427–D432.

Esposito, A., Colantuono, C., Ruggieri, V., Chiusano, M. L. (2016) Bioinformatics for agriculture in the Next-Generation sequencing era. *Chem. Biol. Technol.* Agric. 3:1–12.

Feder, M.E., Walser, J.C. (2005) The biological limitations of transcriptomics in elucidating stress and stress responses. *J. Evol. Biol*. 18:901–910.

Finn, R. D., Attwood, T. K., Babbitt, P. C., Bateman, A., Bork, P., Bridge, A. J., Chang, H. Y., Dosztányi, Z., El-Gebali, S., Fraser, M., et al., (2017) InterPro in 2017—beyond protein family nd domain annotations. *Nucleic Acids Res*. 45: D190–D199.

Ghosh, S., Sarkar, T., Pati, S., Kari, Z. A., Edinur, H. A. and Chakraborty, R. (2022) Novel bioactive compounds from marine sources as a tool for functional food development. *Frontiers in Marine Science*. 9:832957.

Gurevitch, J., Koricheva, J., Nakagawa, S., Stewart, G. (2018) Meta-analysis and the science of research synthesis. *Nature*. 555:175.

Handelsman, J. (2004) Metagenomics: Application of genomics to uncultured microorganisms. *Microbiol.* Mol. Biol. Rev. Mmbr. 68: 669–685.

Hartmann, E. M., Durighello, E., Pible, O., Nogales, B., Beltrametti, F., Bosch, R., Christie-Oleza, J. A., Armengaud, J. (2014) Proteomics meets blue biotechnology: A wealth of novelties and opportunities. *Mar. Genom*. 17:35–42.

Hausmann, R., Syldatk, C. (2014) Types and classification of microbial surfactants. In: *Biosurfactants: Production and Utilization—Processes, Technologies, and Economics* (Kosaric, N., Varder-Sukan, F., eds.). CRC Press Taylor & Francis Group, Boca Raton, FL, USA. pp. 3–18.

Heigwer, F., Kerr, G., Boutros, M. (2014) E-CRISP: Fast CRISPR target site identification. *Nat. Methods*. 11:122.

Huang, F. C., Peter, A., Schwab, W. (2014) Expression and characterization of CYP52 genes involved in the biosynthesis of sophorolipid and alkane metabolism from *Starmerellabombicola*. *Appl Environ Microbiol*. 80(2):766-76.

Iskar, M., Zeller, G., Zhao, X. M., van Noort, V., Bork, P. (2012) Drug discovery in the age of systems biology: The rise of computational approaches for data integration. *Curr. Opin. Biotechnol.* 23:609–616.

Jackson, S., Borchert, E., O'Gara, F., Dobson, A. (2015) Metagenomics for the discovery of novel biosurfactants of environmental interest from marine ecosystems. *Curr Opin Biotechnol*. 33:176–182.

Johann, S., Seiler T.-B., Tiso T., Bluhm, K., Blank, L. M., Hollert, H. (2016) Mechanism-specific and whole-organism ecotoxicity of mono-rhamnolipids. *Sci. Total Environ*. 548:155–163.

Kubicki, S., Bollinger, A., Katzke, N., Jaeger, K. E., Loeschcke, A., Thies, S. (2019) Marine Biosurfactants: Biosynthesis, Structural Diversity and Biotechnological Applications. *Mar Drugs*. 17(7):408.

Lacerda, C. M. and Reardon, K. F. (2009) Environmental proteomics: Applications of proteome profiling in environmental microbiology and biotechnology. *Brief. Funct. Genom. Proteom*. 8: 75–87.

Lauritano, C. and Ianora, A. (2018) Grand Challenges in Marine Biotechnology: Overview of Recent EU-Funded Projects. In: *Grand Challenges in Marine Biotechnology*. Springer: New York, NY, USA, 2018, pp. 425–449.

Liang, X., Luo, D., Luesch, H. (2019) Advances in exploring the therapeutic potential of marine natural products. *Pharmacol. Res*. 147: 104373.

Lin, W. R., Tan, S. I., Hsiang, C. C., Sung, P. K., Ng, I. S. (2019) Challenges and opportunity of recent genome editing and multi-omics in cyanobacteria and microalgae for biorefinery. *Bioresour. Technol*. 291:121932.

Machado, H., Sonnenschein, E. C., Melchiorsen, J., Gram, L. (2015) Genome mining reveals unlocked bioactive potential of marine Gram-negative bacteria. *BMC Genom*. 16:158.

Montague, T. G., Cruz, J. M., Gagnon, J. A., Church, G. M., Valen, E. (2014) CHOPCHOP: A CRISPR/Cas9 and TALEN web tool for genome editing. *Nucleic Acids Res*. 42: W401–W407.

Nikolova, C., Gutierrez, T. (2021) Biosurfactants and Their Applications in the Oil and Gas Industry: Current State of Knowledge and Future Perspectives. *Front. Bioeng. Biotechnol*. 9:626639.

Ortega, S. S., Cara, L. C., Salvador, M. K. (2012) In silico pharmacology for a multidisciplinary drug discovery process. Drug Metab. *Drug Interact.* 27: 199–207.

Pasolli, E., Schiffer, L., Manghi, P., Renson, A., Obenchain, V., Truong, D. T., Beghini, F., Malik, F., Ramos, M., Dowd, J. B., et al., (2017) Accessible, curated metagenomic data through ExperimentHub. *Nat. Methods*. 14:1023.

Poomalai, P., Krishnan, J., Ravichandran, A. and Sureshkumar, R. (2024) Biosurfactants: Sustainable Alternative To Synthetic Surfactants And Their Applications. *International Journal of Applied Pharmaceutics*. 16(2): 34–43.

Poremba, K., Gunkel, W., Lang, S., Wagner, F. (1991) Toxicity testing of synthetic and biogenic surfactants on marine microorganisms. Environ. Toxicol. Water Qual. 6:157–163.

Prasad, T. S. K., Mohanty, A. K., Kumar, M., Sreenivasamurthy, S. K., Dey, G., Nirujogi, R. S., Pinto, S. M., Madugundu, A. K., Patil, A. H., Advani, J., et al., (2017) Integrating transcriptomic and proteomic data for accurate assembly and annotation of genomes. *Genome Res.* 27:133–144.

Rawlings, N. D., Barrett, A. J., Thomas, P. D., Huang, X., Bateman, A., Finn, R. D. (2018) The MEROPS database of proteolytic enzymes, their substrates and inhibitors in 2017 and a comparison with peptidases in the PANTHER database. *Nucleic Acids Res*. 46:D624–D632.

Rivera-de-Torre, E., Palacios-Ortega, J., Gavilanes, J. G., Martínez-del-Pozo, Á., García-Linares, S. (2019) Pore-Forming Proteins from Cnidarians and Arachnids as Potential Biotechnological Tools. *Toxins.* 11:370.

Ron, E. Z., Rosenberg, E. (2001) Natural roles of biosurfactants. *Environmental microbiology*. 3(4): 229–236.

Roumpeka, D. D., Wallace, R. J., Escalettes, F., Fotheringham, I., Watson, M. (2017) A Review of Bioinformatics Tools for Bio-Prospecting from Metagenomic Sequence Data. *Front Genet.* 8:23.

Russell, N. J. (2000) Toward a molecular understanding of cold activity of enzymes from psychrophiles. *Extrem. Life Under Extrem. Cond.* 4:83–90.

Seitz, K. W., Dombrowski, N., Eme, L., Spang, A., Lombard, J., Sieber, J. R., Teske, A. P., Ettema, T. J. G., Baker, B. J. (2019) Asgard archaea capable of anaerobic hydrocarbon cycling. *Nat. Commun.* 10:1822.

Silva, G. F. D., Gautam, A., Duarte, I. C. S., Delforno, T. P., Oliveira, V. M., Huson, D. H. (2022) Interactive analysis of biosurfactants in fruit-waste fermentation samples using BioSurfDB and MEGAN. *Sci Rep.* 12(1):7769.

Silva, M. d. G. C., Medeiros, A. O., Converti, A., Almeida, F. C. G., Sarubbo, L. A. (2024) Biosurfactants: Promising Biomolecules for Agricultural Applications. *Sustainability*. 16:449.

Solaiman, D. K., Liu, Y., Moreau, R. A., Zerkowski, J. A. (2014). Cloning, characterization, and heterologous expression of a novel glucosyltransferase gene from sophorolipid-producing Candida bombicola. *Gene.* 540(1):46–53.

Stemmer, M., Thumberger, T., del Sol Keyer, M., Wittbrodt, J., Mateo, J. L. (2015) CCTop: An Intuitive, Flexible and Reliable CRISPR/Cas9 Target Prediction Tool. *PLoS ONE*. 10: e0124633.

Teeling, H. and Glockner, F. O. (2012) Current opportunities and challenges in microbial metagenome analysis—A bioinformatic perspective. *Brief. Bioinform.* 13:728–742.

Thies, S., Santiago-Schübel, B., Kovačić, F., Rosenau, F., Hausmann, R., Jaeger, K. (2014) Heterologous production of the lipopeptide biosurfactant serrawettin W1 in *Escherichia coli*. *Journal of biotechnology*. 181:27–30.

Trindade, M., van Zyl, L. J., Navarro-Fernández, J., Abd Elrazak, A. (2015) Targeted metagenomics as a tool to tap into marine natural product diversity for the discovery and production of drug candidates. *Front. Microbiol*. 6:890.

Tully, B. J., Graham, E. D., Heidelberg, J. F. (2018) The reconstruction of 2631 draft metagenome-assembled genomes from the global oceans. *Sci. Data*. 5:170-203.

Tuttle, R. N., Demko, A. M., Patin, N. V., Kapono, C. A., Donia, M. S., Dorrestein, P., Jensen, P. R. (2019) Detection of Natural Products and Their Producers in Ocean Sediments. Appl. *Environ. Microbiol.* 85:e02830-e18.

Wang, M., Carver, J. J., Phelan, V. V., Sanchez, L. M., Garg, N., Peng, Y., Nguyen, D. D., Watrous, J., Kapono, C. A., Luzzatto-Knaan, T., et al., (2016) Sharing and community curation of mass spectrometry data with Global Natural Products Social Molecular Networking. *Nat. Biotechnol*. 34:828–837.

Whittaker, P. A. (2003) What is the relevance of bioinformatics to pharmacology? *Trends Pharmacol. Sci.* 24:434–439.

Wilkinson, M. D., Dumontier, M., Aalbersberg, I. J., Appleton, G., Axton, M., Baak, A., Blomberg, N., Boiten, J. W., da Silva Santos, L. B., Bourne, P. E., et al., (2016) The FAIR Guiding Principles for scientific data management and stewardship. *Sci. Data*. 3:160018.

Xie, B., Huang, Y., Baumann, K., Fry, B. G., Shi, Q. (2017) From Marine Venoms to Drugs: Efficiently Supported by a Combination of Transcriptomics and Proteomics. *Mar. Drugs*. 15: 103.

Chapter 11

Balancing Agriculture's Impact with Sustainable Practices to Protect Marine Ecosystems

Payal Kapoor*
Department of Medical Laboratory Sciences, School of Allied Medical Sciences, Lovely Professional University, Phagwara, India

Abstract

Agricultural activities such as nutrient runoff, sedimentation, and chemical pollution have a significant impact on marine ecosystems, threatening water quality, biodiversity, and ecological balance. Excess fertilizers contribute to nutrient overload in coastal waters, leading to eutrophication, harmful algal blooms and oxygen depletion. Sediment runoff from farms can blanket marine habitats, disrupting biodiversity and ecosystem functions. Pesticides and herbicides also enter marine environments, endangering aquatic life and disturbing food webs. Additionally, the expansion of agriculture often leads to habitat destruction, further weakening the resilience of marine ecosystems. To mitigate these effects, it is essential to adopt sustainable agricultural practices and implement effective management strategies aimed at protecting marine habitats and biodiversity. These strategies can reduce the environmental impact by lowering marine pollution levels and support the conservation and recovery of marine ecosystems. This chapter emphasizes the crucial need to balance agricultural practices with environmental sustainability to safeguard the long-term health and resilience of marine ecosystems.

Keywords: agriculture, marine ecosystems, nutrient runoff, sedimentation, chemical pollution, water quality

1. Impact of Agricultural Practices and Contaminants of Concern on Marine Ecosystem: An Introduction

When pollutants from human activities enter the marine environment, they bring about harmful effects on marine life and ecosystem, human health, fishing industries, maritime activities, and

---

* Corresponding Author's Email: payal.30893@lpu.co.in

In: Healing from the Depths Marine Metabolites in Disease Management
Editors: Pranav Kumar Prabhakar and Tarun Kumar Upadhyay
ISBN: 979-8-89530-832-5

the quality of seawater. These pollutants, originating from various sources, degrade the richness of marine ecosystems, lower water quality, and diminish the productivity of marine ecology [1]. As per the United Nations Convention on Law of the Sea (UNCLOS) 1982, marine pollution can be caused by land-based activities and coastal activities [2]. The land-based pollution mainly originates from municipal waste, industrial waste, ship-breaking activities and agriculture associated activities or agricultural waste. Through land runoff, rivers, pipelines, and other outfall structures, the land-based pollutants find way into the coastal and marine waters. Across the world, land-based sources of pollution account for around 80% of marine pollution of which ship associated activities contribute up to 33% of the marine pollution globally [3]. Marine environment is also significantly affected by human activity around the coasts. These coastal activities include the destruction of habitat caused by infrastructure projects and coastal development, overfishing, pollution from urban and industrial runoffs [4].

Pollution that enters into the water bodies can originate as a 'point' or as a 'non-point' source. A 'point source' pollution includes a single, identifiable source such as a drain or a pipe and hence easy to locate; whereas 'non-point' source pollution may come from many diffuse sources rather than a single point. Polluted runoff from agricultural regions that drain into rivers or the windborne trash that blows out to sea are examples of 'non-point sources' of water pollution. These pollutants have a negative impact on water bodies. Non-point source pollution can be challenging to manage since it originates from an extensive number of sources and no single solution will entirely address the issue. In this category, of particular concern is the contaminants that originate from agriculture source. Extensive contaminant discharge into marine ecosystems occur through agricultural practices such as excessive use of agrochemicals, improper handling of animal waste and post-harvest issues and usage of traditional irrigation techniques. Agricultural contaminants such as agri-nutrients, salts, pesticides, organic matter, sediments, heavy metals, and other newly discovered toxins when released into the environment can find their way to water bodies through various pathways, such as runoff, erosion, atmospheric deposition and leaching [5, 6]. Water contaminated with sediment along with other pollutants can deteriorate fish spawning gravels, decrease light penetration and biological production, lower recreational opportunities, raise the demand of water for residential and commercial use and may even erode pumping machinery [7, 8].

The past fifty years have seen a sharp increase in the demand for food due to the substantial growth in the world population. Food consumption is predicted to increase from 59% to 98% by 2050 [9]. The increase in agricultural activities to meet the increasing food demands is associated with increased usage of pesticides and fertilizers [10, 11]. Pesticides and fertilizers that get carried from the ground into the water can have detrimental effects on marine ecosystems leading to complex, frequently cascading eutrophication consequences [12]. To meet the rising food demand, the primary goal in the years to come will be to maximize agricultural efficiency and also to reduce the negative impact of agricultural activities on the environment [13].

Adopting 'Best Management Practices (BMP)' or a combination of practices can help reduce pollution on land, water and air that is generated through non-point sources [14]. In agriculture, practices that can be used to maintain sustainable agriculture include tillage, cropping techniques, converting cropland to less intense usage, applying pesticides and fertilizers more effectively. 'Structural BMPs' involve constructing or buildup a structure to prevent or reduce soil pollution. These includes contour farming, fencing, constructing sediment basins and other activities. In contrast, 'non-structural' BMPs do not require any specific build-up but aims to reduce pollution at the source by implementing practical measures such planting cover crops, controlling weeds, and water management [15]. Further, alteration or usage of farming methods that minimize the amount of nitrogen and phosphorus in agricultural regions can be beneficial. Leaching of applied fertilizers and pesticides can also be reduced with the application of micro-irrigation techniques like sprinklers and drip irrigation. In drip irrigation, water applied at the root zone of the crop is less subjected to loss which might otherwise carry away the applied chemicals [16].

It is also important to identify the effectiveness of the applied BMP. Some practices might be more effective than other on same type of non-point source of contaminant. Also, the economic viability to implement a particular BMP should also be assessed. Despite numerous benefits of BMPs, their adoption rate has been found to be slow, specifically in the developing countries. The reason for such lag include lack of technical knowledge and advice over the method and time of fertilizer application, lack of assistance to understand the application of required doses, high cost of fertilizers, and even the poor financial conditions of farmers [17]. A well-established organization that can assist farmers in determining the appropriate amount of fertilizer for a given crop is essential to implement the controlled use of fertilizers. Further, many best management practices can be expensive, therefore it is important to promote agricultural finance organizations and offer various incentives and/or subsidies to encourage the adoption of BMPs [18].

1.1. Submarine Groundwater Discharge (SGD): A Link between Land and Water Systems

Submarine groundwater discharge (SGD) serves as a vital connection between terrestrial and marine systems, playing a significant role in determining both water quantity and quality. It is responsible for discharging over 60% more freshwater into the oceans compared to riverine fluxes. In addition to freshwater, SGD carries a variety of substances into marine environments, including polluted ions, nutrients, trace elements, rare earth elements, dissolved ions, organic matter and biochemical particles. These inputs can significantly influence coastal water quality, marine ecosystems and biogeochemical cycles. The transport of such materials through SGD highlights its importance not only as a water source but also as a vector for pollution and essential nutrients, underscoring the need for better management of terrestrial activities, including agriculture and waste disposal to protect marine ecosystems [19].

2. Intensive Farming: Impact of Excessive Use of Fertilizers on Marine Ecosystem

Approximately 1/3rd of the land on earth is currently being used for agriculture purpose. Almost 20% of earth's forests and ~50% of its savannas, grasslands, and shrublands are modified for farming. By 2050, it may be necessary to double global food production due to the anticipated increase in the world's population and dietary changes brought about by rising income [20]. In traditional farming, a farmer considers all the different crop types to ensure diversity while in the intensive agricultural practices (also known as industrial agriculture), the understanding of a single crop or animal is typically required. The equipment for managing livestock and the methods employed to harvest crops is less as these practices involves a single type of crop or animal. There are multiple benefits of intensive farming because it is simple, efficient technique and produces more food variety. However, this technique involves higher use of inputs such as capital, labour, agrochemicals, water and also focuses on higher crop yields per unit land area [21].

Modern agricultural systems rely heavily on chemical inputs such as fertilizers, pesticides, and herbicides to boost productivity and achieve higher yields per unit area. Excessive or improper use of these chemicals can lead to numerous problems, including environmental pollution (affecting soil, water, and air quality), reduced efficiency of inputs and hampers food quality. Overuse of these chemicals can also cause resistance in weeds, emergence of diseases and pests, degrade soil health, and lead to micronutrient deficiencies. Furthermore, it harms beneficial organisms both above and below the soil surface, ultimately reducing farmer income due to unsustainable practices [22]. The availability of nutrients in ecosystems and biogeochemical cycles is profoundly affected by agriculture practices. Nitrogen (N) and phosphorus (P) are the two nutrients that are used extensively in agroecosystems. Nutrients enter coastal ecosystems through atmospheric deposition, surface water runoff, and groundwater. Fertilizer runoff from farming activities dumps tons of surplus nitrogen and other nutrients into rivers that eventually flow into lakes and coastal waterways [23, 24]. N-fertilizers contribute more than twenty percent of the used in the agricultural systems into the aquatic environments [23].

In Intensive farming, excessive or improper use of these chemical fertilizers can cause marine eutrophication and production of algal blooms [25] (See section 4). Sudden explosion in algal population can occur due to surplus nutrient availability. The intense algae population growth depletes the oxygen levels. Also, when these algae die, they are decomposed by bacteria that further lead to the formation of regions with low oxygen known as "dead-zones." Low oxygen availability results in hypoxia and fish mortality, elevated frequency and intensity of algal blooms, groundwater contamination, marine habitat compression and species diversity. These dead-zones cannot sustain fish and other marine populations, causing disruption in diversity and distribution of the aquatic and marine microbe and animals [26]. These pollutants also have numerous detrimental effects on the public health [27]. Even livestock manure, which is rich in nutrients such as nitrogen and phosphorous, if improperly applied may leach off into ground water or surface water leading to eutrophication. Livestock manure also contains a variety of microorganisms, such as bacteria (*E. coli.* and *Salmonella*), viruses, and parasites (*cryptosporidium*), that not only impacts humans but also aquatic ecosystems. These pathogens contaminate water bodies and cause the spread of water-borne diseases in humans [28].

Some helpful measures for the suppression of eutrophication and production of algal blooms include careful use of fertilizers, checking soil quality to determine the fertilizer quantity needed, avoiding fertilization before a rainstorm, using organic fertilizers that release nutrients more slowly, growing perennials like grasses or alfalfa instead of annuals like maize or soybeans. Perennials keep nitrogen fixed in the rooting zone, resulting in less N loss to groundwater and thus to the coastal and marine environments [29]. Soil erosion also contributes majorly to the loss of nutrients that get into surface water systems. Notably, N and P that are affixed to soil particles are immediately carried by stormwater runoff to surface waterways, including rivers, lakes, wetlands, streams, and oceans. Therefore, using cover crops helps shield soil that are susceptible to faster erosion loss. Furthermore, during seasons like winter and spring, maximum leaching from agricultural lands occurs in many regions owing to the antecedent moisture conditions when plant soil water and nutrient intake drastically reduces. Sowing winter cover crops help lower the amount of nutrients transfer into groundwater [30].

Practically in all ecosystems, the application of proper amount of fertilizer is essential to appropriate nutrient management. 'Buffers' such as grass hedges, stiff grass hedges, field buffers, riparian forests, forest buffers, and wetland forests can be frequently used to reduce the consequences of agricultural and urban land use practices. Forests increase infiltration rates by improving soil structure, enhance vegetative cover, produces a thick organic layer, increase surface roughness and evapotranspiration rates, thus are ideal for controlling nonpoint source pollution [31]. Rehabilitation and restoration of rivers and lakes also help increase their ability to hold onto nutrients and buffer the nutrient fluxes, thus lowering the concentration of N and P runoff in surface water thus preventing their leaching towards marine habitats [32].

3. The Impact of Marine Agriculture (Aquaculture) on Marine Ecosystem

The impact of marine agriculture on the marine ecosystem depends on the species being farmed, the farming method used, stocking density, feed type, local water conditions, and husbandry practices. Aquaculture operations utilize natural resources like land and water, which are a component of the ecosystem that all living things share. Across all types of aquaculture systems, a significant proportion of organic carbon and the nutrients introduced through feed are lost to the surrounding environment mainly due to feed wastage, fish excretion, faeces, and respiration [33]. This nutrient loss into aquatic systems can lead to pollution and raise environmental concerns, particularly in areas with limited water circulation and low carrying capacity. Thus, aquaculture can introduce and bring about the spread of aquatic diseases, water eutrophication, hamper water quality, and can also cause loss or modify natural environments. Fish farm's excrement and leftover food pellets disrupt the organic content in the sedimentary layer, thus altering the oxygen availability and consumption. This results in the local eutrophication [34]. High level of pollution is associated with fish farming. A study found that around 23% of carbon (C), 21% of nitrogen (N), and 53% of phosphorus (P) from feed input accumulate in the sediments beneath the aquatic farms, however, this environmental impact of accumulation is generally localized, typically within 1 to 1.5 kilometres of the farm. It is therefore unlikely that fish farming will cause widespread eutrophication across larger regions. The major effects that can be seen on the sea floor include increased sediment oxygen demand, development of anoxic conditions, release of toxic gases like methane and reduction in benthic (seafloor) biodiversity [35]. A Study by Leung et al., (1999), [36] estimated that about 53% of nitrogen introduced into grouper culture systems (via open sea cages) is released into the surrounding environment as ammonia, approximately 169.8 grams of ammoniacal nitrogen released into the environment per kilogram of fish produced. Ammonia in its un-ionized form is highly toxic to marine life, and its toxicity levels also vary based on the salinity, temperature and pH of water. Additionally, ammonia discharge may contribute to the growth of harmful algal blooms. In contrast, phosphorus is less influential in promoting algal growth in marine environments and is therefore unlikely to have a significant impact. Despite these impacts on marine ecosystem, there is no strong evidence that suggests that aquaculture associated practices significantly increase harmful algal blooms or pose significant environmental risks due to practices such as the use of therapeutants, vitamins, antibiotics, pigments and antifouling agents [37] but the introduction of genetically modified strains and other pathogens through aquaculture activities raises a concern about its impact on biodiversity and ecosystem health. These combined factors have led to the employment of increased scrutiny and drawn regulatory attention in many countries to manage the environmental impacts of marine fish farming effectively [38].

To reduce the impact of marine agriculture on marine ecosystem, it is important to monitor the feeding activity of the fish and adjust the amount of feed given according to its feeding behaviour. Feed wastage can be minimized by improving the stability of the feed and controlling its sinking rate, while also providing fish with appropriately sized feed at different growth stages. Ammonia excretion, which is linked to protein intake, can be reduced by optimizing the protein-to-energy ratio in the feed. By using highly digestible feed, the protein needs of each species at various developmental stages can be met efficiently. Additionally, fish can meet their energy requirements through carbohydrates and fats, allowing more protein to be directed toward tissue growth rather than being wasted as energy.

4. Eutrophication and Algal Blooms

Eutrophication occurs when limited growth resources such as sunlight, carbon dioxide, and nutritional fertilizers that are necessary for photosynthesis become more readily available causing plants and algae grow excessively [39]. A study has revealed a substantial correlation between algal bloom and the ratio of nitrate to phosphate nutrients availability. Furthermore, the nutrient ratio can be used to regulate the incidence of algal bloom in the aquatic environment [40]. In most aquatic ecosystems, nitrogen and/or phosphorus fertilizer accumulation causes eutrophication [41]. Eutrophication enriches food chains by increasing the primary production and delivering more energy up in the aquatic food chain. This results in an increase in the biomass and productive aquatic ecosystems [42]. Intensification in the biomass demands high oxygen that leads to the death of aerobic aquatic biota due to less oxygen availability. Further, eutrophication also causes the explosion of the species which are nutrient-tolerant compared to the nutrient-sensitive species. This creates a drastic reduction in bio-diversity and a shift in energy distribution in the aquatic food chain [43]. Certain aquatic creatures under eutrophication-stressed habitats may produce poisons to hamper the growth and survival/occurrence of other organisms while strengthening their own survival over competitors. Such toxins, like the cyanotoxins produced by cyanobacteria, have an influence on the food chain and a variety of the aquatic ecosystem as these get accumulated in the fish and shellfish [12]. Consumption of seafood contaminated with toxic algae might cause impairment of human health as it produces a neurological impact and can also cause death. Algal poisoning may cause paralysis, neurotoxicity, amnesia, and diarrhoea [44]. Skin irritation and respiratory tract irritation has also been reported when a human comes in contact with algal blooms [45]. Further, eutrophication leads to the appearance of cloudiness and a green shade of water resulting in reduced penetration of the sunlight that impacts the photosynthetic rates of the submerged plants [46] [47].

In general, the impact of eutrophication on both freshwater and marine ecosystems includes [48]:

- Accumulation of organic carbon produced from overgrowth of macroalgae and phytoplankton and growth of noxious cyanobacteria [49].
- Kelps and scleractinian corals get replaced with simple foliose or turf-alga [50].
- The imbalance in nutrient ratio becomes favourable for toxic algal blooms causing sea life mortality and shellfish poisoning.
- The biodiversity and biomass of the benthic (bottom-dwelling) population are altered, which ultimately results in a decrease in species variety and an increase in the dominance of gelatinous animals like jellyfish.
- Bacterial and algal blooms consume the majority of the oxygen in the water that results in 'dead-zone' at the seafloor. These regions can sustain only certain species that can survive in limited oxygen availability. The deficiency of oxygen in these regions strains the aquatic ecosystems, which can result in the death of marine life, emigration of marine populations, brings changes in the energy fluxes within the ecosystem, or even the complete collapse of the ecosystem [51].

It has been estimated that, nitrate and phosphate contribute nearly 20–50% and 35–55%, respectively, to the prevalence of water body eutrophication. Accordingly, the agricultural runoff of fertilizer applied to farmlands and wastewater from aquaculture and concentrated animal feed activities have a major influence on these percentage of inputs of nitrate and phosphate. Thus, in order to prevent eutrophication, it is crucial to remove phosphate and nitrate from wastewater [52, 53]. Dredging, a physical technique, eliminates the bottom debris, which are heavily enriched in phosphate and nitrates. Aeration, another physical technique establishes an anoxic environment that releases P from the soil to promote its internal cycling and reduces eutrophication [46]. Other physical techniques used to control eutrophication include infiltration-percolation [54], membrane separation technology [55]. Chemical treatments such as coagulation/flocculation [56] adsorption and ion exchange method can also be employed [57]. Along with this, control on the release of industrial wastage into oceans, limited application of fertilizers can help mitigate the development of harmful algal blooms.

5. The Impact of Metal Poisoning on Marine Ecosystems

Globally, contamination of soil, air, water bodies by heavy metals has been recognized as a significant environmental problem because these heavy metals can find their way to human food chains [58]. It has long been recognized that the marine ecosystem contains a significant concentration of heavy metals like copper, nickel, zinc, cobalt and cadmium. These metals are introduced into estuaries, coastal areas and marine habitats from medical facilities, industrial and agricultural waterways, home sewage systems, and air pollution. Factory waste contributes a variety of chemicals and toxins such as mercury to the oceans and coastal areas. During the production of fertilizers, variable amounts of different heavy metals get transferred to fertilizers. These contaminated fertilizers when applied to the soil eventually enters the food chain [59]. This leads to decreased species variety and abundance, metal buildup in living things and food chains, which can worsen the state of marine ecosystems [60, 61]. Further when introduced to the marine environment, heavy metal contaminants (either in soluble or particulate form) eventually end up on the bottom of the sea. Anions such sulfate, chloride, fluoride, and bicarbonates/carbonates precipitate the soluble metal cations. In the marine environment, these precipitates have very low solubilities at deep waters as well as surface waters [62]. Though the use of heavy metal in agriculture has been identified as a lesser cause of metal pollution, these can result from the use of fertilizers, insecticides, pesticides and other inputs. In agriculture practices, some fertilizers used may contain mercury as a contaminant if these are prepared from raw material exposed to mercury. Even certain pesticides and herbicides may have mercury or its compounds. Extensive use of such fertilizers/ pesticides/herbicides may find their way to waterways on leaching. These metals can accumulate in the aquatic environment, including water bodies, sediment and aquatic species like macroinvertebrates, even when their initial concentrations are low.

Studies have shown that long-term exposure to heavy metals results in increased young fish mortality and decreased adult fish reproductive capacity. Mercury, like many contaminants, tends to accumulate in the tissues of organisms. Predators like swordfish and tuna that consume smaller, mercury-contaminated fish, tend to accumulate extremely high mercury levels in them. Evidence suggest that other pollutants are also accumulating in marine organisms and cause negative impact on the upper trophic-level species, including humans [63].

Certain fertilizers and pesticides contain copper as a micronutrient that is applied to the crops for their growth. It can bind with soil particles, however leaching due to heavy rain/ runoff may cause its accumulation in aquatic ecosystems. Copper concentration in aquatic systems have also grown due to its usage and discharge by growing industrial activities. Excessive use of copper sulphate as a fungicide in agricultural techniques, and in the management of algae and infections in fish culture ponds cause severe fish poisoning [64]. Cu(II) harms marine dinoflagellates such *Amphidinium carterae* [65]. Copper ions can be removed by using remediation techniques such as membrane filtration, electrodialysis, ion exchange, physical adhesion, electrocoagulation, electroflotation, chemical precipitation, co-precipitation, nanofiltration and adsorption [66]. Diatoms are crucial for the elimination, breakdown, and detoxification of heavy metal ions from polluted areas. Therefore, diatom-based strategies are regarded as a practical and affordable way to remove heavy metal ions from tainted water [67]. Also, invertebrates, fishes and shellfish serve as effective indicators in determining the amounts of these metals and their impact on aquatic ecosystem as these creatures can accumulate metals in their body in a more concentrated manner than the actual concentration of these metals in sea water [68, 69].

Heavy metals provide a potential risk to water quality and, therefore, human health because of their toxicity and persistent nature [70]. The implementation of prevention and remediation strategies requires scientific investigation of heavy metal pathways, uptake mechanisms, speciation in seawater and sediments, and responses of marine organisms to heavy metal contamination. Strategies such as integrated crop management and conservation practices, policies and regulations to control metal pollution from agricultural activities, water treatment and sanitation systems, research and monitoring programs for metals in agricultural systems, can be adopted for managing and reducing metal pollution in agriculture [71, 72].

6. The Impact of Deforestation on Marine Ecosystem

Ensuring a sufficient food supply to meet the demands of a growing population and achieve "zero hunger" is a key objective within the Sustainable Development Goals (SDGs). This has ended up in the loss of forests for agricultural development and food production on large scale, rendering agriculture as one of the leading causes of deforestation [73]. Frequent destruction of forests produces a negative impact on the environment. These include increase in the risk of occurrence of natural disasters, reduced ability to absorb atmospheric carbon, hence causing damage to the biodiversity and also disturbs the natural water cycle. Loss of forest makes the soil more loose, unstable and exposed, rendering it more susceptible to natural calamities like floods and landslides. Further, large amounts of loose soil can easily clog waterways [74]. Sediment inflow induces the development of heat sensitive aquatic ecosystems and also decreases the resilience of corals towards climatic change. Significant loss of forest areas in developing nations, especially those in the tropical and subtropical latitudes, have significantly contributed to the development of greenhouse gas emissions [75]. The greenhouse gas emissions produce a profound effect on marine life and the marine environment. Greenhouse gases include carbon dioxide, methane, chlorofluorocarbons, ozone etc. The oceans absorb about 30% of the atmospheric carbon dioxide (CO_2), however, due to human activities such as deforestation, burning fossil fuels, the percentage of CO_2 absorption by water bodies increases. Increased uptake of the atmospheric carbon dioxide by the ocean causes 'ocean acidification.' Ocean acidification impacts marine life. It impacts the development of oysters and commercially important shellfish, threaten coral reefs and plankton by inhibiting the production of calcium carbonate exoskeletons essential for these species, although few species of coccolithophores have shown increased calcification under acidic conditions [76]. Altered fish physiology has also been observed in high CO_2 conditions that causes disruption in mechanism related to the Gamma-aminobutyric acid ($GABA_A$) neurotransmitter functioning [77]. Some zooplanktons taxa are also affected by ocean acidification [78]. Pelagic food webs are often altered by ocean acidification that causes either an increase in the number of species that produce toxic blooms or might intensify the toxicity of alga [79]. Reforestation is the key and the potential solution to slow down the impact of deforestation and the resulting climate change on marine ecosystems.

7. The Impact of Pesticides on Marine Ecosystem

The broad category of chemicals known as pesticides are designed to control or eradicate pests in agricultural practices. Based on the target and composition of these chemicals, they are divided into many groups such as insecticides, herbicides, fungicides, and nematodes, among others. Agriculture uses millions of tons of active pesticide ingredients annually [80]. Farmers are the primary pesticide users, using many pesticides to safeguard their crops and boost agricultural yields. Pesticide poisoning of water systems is a severe problem, particularly in areas with substantial agricultural practices. Wrong selection and misuse of pesticides can lead to the contamination of water sources. Heavy rains, irrigation or runoff might lead to the leakage of these dangerous chemicals into water systems causing detrimental impacts on aquatic life, animal health, and human health.

Ukhurebor et al., [81] found a clear link between pesticide usage and climate change. Crop development, insect pest migration and dispersion, pest abundance variations, pest outbreak frequency and vector dispersal, weed evolution, and disease stimulation are all impacted by climate change. Additionally, the properties of pesticides such as its volatilization, runoff, leaching, degradation (photolysis as well as chemical and microbiological breakdown) is also accelerated by climatic change [82]. All these factors lead to increased frequency and a wide range of pesticides usage by farmers. The impact of increased usage of pesticides to reduce pests in cultivation areas due to climate change causes shifts in temperature and precipitation in most regions of the world [83, 84].

Depending on their polarity and water solubility, these pesticides can enter aquatic systems in a various ways after being sprayed on crop areas [85]. These routes include agricultural runoff, spills, drifts, hazardous chemical pesticides, industrial effluents, equipment washing, aerial spraying, and transportation from pesticide-treated soil [86]. Also, pesticides carried by air get partially adsorbed to aerosols and find their way into the ocean by offshore winds. Through such operations, around 1-2 percent of the entire pesticides are dumped straight into the ocean [87]. Fish exposed to a high endosulfan pesticide concentration were found to have decreased red blood cell count. There has also been speculation that the decrease in the number of red blood cells might be the result of pollutants destroying red blood cells and inhibiting their synthesis [88]. Certain pesticides have a lengthy half-life in the environment; for instance, organochlorine insecticides were found in surface waters even after thirty years after its usage had stopped [89]. Long-term exposure of aquatic ecosystems to pesticides results in malignancies, tumors, and lesions, infertility, weakening of the eggshell, immune system suppression, damage to cells and DNA of organisms and other health issues for fish, including an abundance of slime on their scales and gills. Adverse impact on many endangered and rare fish like osprey, peregrine falcon, and the bald eagle by these pesticides have been seen [90]. Herbicides cause destruction of aquatic plants thereby resulting in reduced oxygen content and an impact on fish productivity.

Pesticides disrupt marine ecosystems by changing food chains and habitats. These chemicals cause the plants and insects that serve fish and other aquatic creatures as food and habitat to disappear from the aquatic ecosystem. Fish that feed on insects lose a lot of their diet due to pesticide use. Insect-eating fish become vulnerable to predators when these travel to far-away distances in search of food when insects abruptly disappear from the food chain. This causes a significant disruption to an ecosystem's food web. An increase in the number of predators or prey results from the disruptions of the food chain, which in turn disrupts the marine ecology. Insecticide discharge also kills useful, predatory arthropods such damselflies and dragonflies during their aquatic phases. Due to these agricultural activities, many aquatic species are affected because of the altered populations of tiny, crustacean-like animals as they represent an essential food source for the bottom end of the aquatic food chain. [91].

Due to the widespread usage of pesticides/ herbicides/ insecticides and a growing concern about their potentially harmful effects on non-targeting creatures, such as fish, birds and even humans, pesticide usage needs to be closely monitored and controlled in order to safeguard both the environment and public health due to their significant risks [92]. Frequent monitoring can help to identify the patterns and changes associated with the pesticide usage and provide information for control and management strategies. The implementation of regulations and monitoring systems, integrated pest management, pesticide reduction, and other best practices can all help to achieve efficient pesticide management and control. Licensing and pesticide registration are further crucial steps in guaranteeing the use of only safe and efficient pesticides.

Conclusion

Agriculture has a profound impact on marine ecosystems. Prevention and control of marine pollution will be beneficial in multiple ways as it boosts economy and tourism, renovates fisheries, and also positively impacts human well-being. Improving the health of marine life will be an advancement towards the achievement of Substantiable Development Goals. Agricultural runoff containing fertilizers, pesticides, and sediment, can lead to eutrophication, the emergence of harmful algal blooms and loss of biodiversity in marine environments. These pollutants can disrupt marine food chains, harm marine organisms, and degrade habitats like coral reefs and seagrass beds. To mitigate these impacts and promote sustainability, implementation of various agricultural practices is crucial. One key approach is the adoption of sustainable farming practices that minimizes chemical inputs and reduce soil erosion. Organic farming, integrated pest management, and agroforestry are such practices which can reduce the outflow of excess nutrients and pollutants from reaching marine environments. Furthermore, promoting riparian buffers and wetland restoration can help filter agricultural runoff before it enters waterways, thus reducing its negative impact on marine ecosystems. Investing in research and innovation to develop and promote eco-friendly agricultural technologies and practices is also essential for long-term sustainability and reduce negative environmental footprints on marine ecosystem. By promoting sustainable agricultural practices and implementing targeted conservation measures, we can work towards preserving marine biodiversity and ensuring the health of our oceans for future generations.

References

[1] Kumar PS, Prasannamedha GJMTSfMP, Elsevier: Amsterdam TN: Biological and chemical impacts on marine biology. *Modern Treatment Strategies for Marine Pollution; Elsevier: Amsterdam, The Netherlands* 2021:11-27.

[2] Tanaka YJTImoiml: Regulation of land-based marine pollution. *The IMLI manual on international maritime law* 2016, 3:139-168.

[3] Sands P, Peel J: Principles of international environmental law: Cambridge University Press; 2012.

[4] Kroon FJ, Thorburn P, Schaffelke B, Whitten SJGcb: Towards protecting the Great Barrier Reef from land-based pollution. *Global change biology* 2016, 22(6):1985-2002.

[5] Moss BJPTotRSBBS: Water pollution by agriculture. *Philosophical Transactions of the Royal Society B: Biological Sciences* 2008, 363(1491):659-666.

[6] Zia H, Harris NR, Merrett GV, Rivers M, Coles NJC, agriculture ei: The impact of agricultural activities on water quality: A case for collaborative catchment-scale management using integrated wireless sensor networks. *Computers and electronics in agriculture* 2013, 96:126-138.

[7] Zheng Y, Han F, Tian Y, Wu B, Lin Z: Addressing the uncertainty in modeling watershed nonpoint source pollution. In: *Developments in Environmental Modelling. Volume 26*, edn.: Elsevier; 2014: 113-159.

[8] Wiens JJCWRJ: Agricultural runoff and water pollution. *Canadian Water Resources Journal* 1980, 5(3):78-89.

[9] Elferink M, Schierhorn FJHbr: Global demand for food is rising. Can we meet it. *Harvard business review* 2016, 7(04):2016.

[10] Clark M, Tilman DJERL: Comparative analysis of environmental impacts of agricultural production systems, agricultural input efficiency, and food choice. *Environmental Research Letters* 2017, 12(6):064016.

[11] Folberth C, Yang H, Gaiser T, Liu J, Wang X, Williams J, Schulin RJERL: Effects of ecological and conventional agricultural intensification practices on maize yields in sub-Saharan Africa under potential climate change. *Environmental Research Letters* 2014, 9(4):044004.
[12] Papendick RI, Elliott LF, Dahlgren RBJAJoAA: Environmental consequences of modern production agriculture: How can alternative agriculture address these issues and concerns? *American Journal of Alternative Agriculture* 1986, 1(1):3-10.
[13] Harmel R, Kenimer A, Searcy S, Torbert HJPA: Runoff water quality impact of variable rate sidedress nitrogen application. *Precision Agriculture* 2004, 5:247-261.
[14] Loehr R: Pollution control for agriculture: *Elsevier*; 2012.
[15] Ritter WF, Shirmohammadi A: Agricultural nonpoint source pollution: watershed management and hydrology: CRC Press; 2000.
[16] Narayanamoorthy AJDPR: Impact assessment of drip irrigation in India: the case of sugarcane. *Development Policy Review* 2004, 22(4):443-462.
[17] Ramappa K, Kannan E, Lavanya BJIfS, Change E: Adoption of recommended doses of fertilisers on soil test basis by farmers in Karnataka. *Institute for Social Economic Change* 2015.
[18] Akkari C, Robin Bryant CJA: Toward improved adoption of Best Management Practices (BMPs) in the Lake Erie Basin: Perspectives from resilience and agricultural innovation literature. *Agriculture* 2017, 7(7):54.
[19] Prakash R, Loveson V, Kessarkar PM, Kumar A, Pesso C, Gomes CJEr: Fresh and recirculated submarine groundwater discharge zones along the central west coast of India. *Environmental research* 2024, 250:118406.
[20] Ramankutty N, Rhemtulla JJFiE, Environment t: Can intensive farming save nature? In., vol. 10: Ecological Society of America; 2012: 455-455.
[21] Shankar T, Praharaj S, Sahoo U, Maitra SJIB: Intensive farming: It's effect on the environment. *Int Bimon* 2021, 12:37480-37487.
[22] Kumar R, Kumar R, Prakash OJCE: Chapter-5 the impact of chemical fertilizers on our environment and ecosystem. 2019, 35(69):1173-1189.
[23] Galloway JN, Dentener FJ, Capone DG, Boyer EW, Howarth RW, Seitzinger SP, Asner GP, Cleveland CC, Green P, Holland EAJB: Nitrogen cycles: past, present, and future. *Biogeochemistry* 2004, 70:153-226.
[24] Vitousek PM, Aber JD, Howarth RW, Likens GE, Matson PA, Schindler DW, Schlesinger WH, Tilman DGJEa: Human alteration of the global nitrogen cycle: sources and consequences. *Ecological applications* 1997, 7(3):737-750.
[25] Boesch DF: Harmful algal blooms in coastal waters: Options for prevention, control and mitigation: *US Department of Commerce, National Oceanic and Atmospheric Administration …*; 1997.
[26] Michael Beman J, Arrigo KR, Matson PAJN: Agricultural runoff fuels large phytoplankton blooms in vulnerable areas of the ocean. *Nature* 2005, 434(7030):211-214.
[27] Bouwman AF, Beusen AH, Billen GJGbc: Human alteration of the global nitrogen and phosphorus soil balances for the period 1970–2050. *Global biogeochemical cycles* 2009, 23(4).
[28] Mateo-Sagasta J, Zadeh SM, Turral H, Burke J: Water pollution from agriculture: a global review. Executive summary. 2017.
[29] Randall GW, Mulla DJJJoeq: Nitrate nitrogen in surface waters as influenced by climatic conditions and agricultural practices. *Journal of environmental quality* 2001, 30(2):337-344.
[30] Howarth RWJSiCSCLS: The development of policy approaches for reducing nitrogen pollution to coastal waters of the USA. *Science in China Series C: Life Sciences* 2005, 48:791-806.
[31] Grace III JJTotA: Forest operations and water quality in the south. *Transactions of the ASAE* 2005, 48(2):871-880.
[32] Schoumans O, Chardon W, Bechmann M, Gascuel-Odoux C, Hofman G, Kronvang B, Rubæk GH, Ulén B, Dorioz J-MJSotte: Mitigation options to reduce phosphorus losses from the agricultural sector and improve surface water quality: a review. *Science of the total environment* 2014, 468:1255-1266.
[33] Ballester-Moltó M, Sanchez-Jerez P, Aguado-Giménez FJMer: Consumption of particulate wastes derived from cage fish farming by aggregated wild fish. An experimental approach. *Marine environmental research* 2017, 130:166-173.

[34] Mirto S, Bianchelli S, Gambi C, Krzelj M, Pusceddu A, Scopa M, Holmer M, Danovaro RJMER: Fish-farm impact on metazoan meiofauna in the Mediterranean Sea: analysis of regional vs. habitat effects. *Marine Environmental Research* 2010, 69(1):38-47.

[35] Holby O, Hall POJMEPS: Chemical fluxes and mass balances in a marine fish cage farm. II. Phosphorus. *Marine Ecology Progress Series* 1991:263-272.

[36] Leung K, Chu J, Wu RJMEPS: Nitrogen budgets for the areolated grouper Epinephelus areolatus cultured under laboratory conditions and in open-sea cages. *Marine Ecology Progress Series* 1999, 186:271-281.

[37] Wu R: **E**nvironmental impacts of marine fish farming and their mitigation. In: *Responsible Aquaculture Development in Southeast Asia Proceedings of the Seminar-Workshop on Aquaculture Development in Southeast Asia organized by the SEAFDEC Aquaculture Department, 12-14 October 1999, Iloilo City, Philippines: 2001*: Aquaculture Department, Southeast Asian Fisheries Development Center; 2001: 157-172.

[38] Weitzman JJES: Applying the ecosystem services concept to aquaculture: A review of approaches, definitions, and uses. *Ecosystem Services* 2019, 35:194-206.

[39] Sonarghare P, Masram S, Sonparote U, Khaparde K, Kharkate SJIJfER, Conservation X: Causes and effects of eutrophication on aquatic life (a review). *International Journal for Environmental Rehabilitation Conservation, XI* 2020:213-218.

[40] Sidabutar T, Srimariana E, Wouthuyzen S: The potential role of eutrophication, tidal and climatic on the rise of algal bloom phenomenon in Jakarta Bay. In: *IOP Conference Series: Earth and Environmental Science: 2020*: IOP Publishing; 2020: 012021.

[41] Howarth RW, Marino RJL, oceanography: Nitrogen as the limiting nutrient for eutrophication in coastal marine ecosystems: evolving views over three decades. *Limnology* 2006, 51(1part2):364-376.

[42] Camargo JA, Alonso ÁJEi: Ecological and toxicological effects of inorganic nitrogen pollution in aquatic ecosystems: a global assessment. *Environment international* 2006, 32(6):831-849.

[43] Koelmans A, Van der Heijde A, Knijff L, Aalderink RJWR: Integrated modelling of eutrophication and organic contaminant fate & effects in aquatic ecosystems. A review. *Water Research* 2001, 35(15):3517-3536.

[44] Tillmann U, Elbrächter M, Krock B, John U, Cembella AJEJoP: Azadinium spinosum gen. et sp. nov.(Dinophyceae) identified as a primary producer of azaspiracid toxins. *European Journal of Phycology* 2009, 44(1):63-79.

[45] Backer LC, Kirkpatrick B, Fleming LE, Cheng YS, Pierce R, Bean JA, Clark R, Johnson D, Wanner A, Tamer RJEhp: Occupational exposure to aerosolized brevetoxins during Florida red tide events: effects on a healthy worker population. *Environmental health perspectives* 2005, 113(5):644-649.

[46] Rathore S, Chandravanshi P, Chandravanshi A, Jaiswal KJIJoA, Science V: Eutrophication: Impacts of excess nutrient inputs on aquatic ecosystem. *IOSR Journal of Agriculture* 2016, 9(10):89-96.

[47] Sonarghare P, Masram S, Sonparote U, Khaparde K, Kharkate SJIJfER, Conservation X: Causes and effects of eutrophication on aquatic life (a review). *International Journal for Environmental Rehabilitation Conservation, XI* 2020:213-218.

[48] Selman M, Sugg Z, Greenhalgh S: Eutrophication and Hypoxia in Coastal Areas. 2008.

[49] Wurtsbaugh WA, Paerl HW, Dodds WKJWIRW: Nutrients, eutrophication and harmful algal blooms along the freshwater to marine continuum. *Wiley Interdisciplinary Reviews: Water* 2019, 6(5):e1373.

[50] Strain EM, Thomson RJ, Micheli F, Mancuso FP, Airoldi LJGcb: **I**dentifying the interacting roles of stressors in driving the global loss of canopy-forming to mat-forming algae in marine ecosystems. *Global change biology* 2014, 20(11):3300-3312.

[51] Altieri AH, Diaz RJ: Dead zones: oxygen depletion in coastal ecosystems. In: *World seas: An environmental evaluation.* edn.: Elsevier; 2019: 453-473.

[52] Zou H, Lu X, Abualhail SJJoC: Characterization of Denitrifying Phosphorus Removal Microorganisms in a Novel Two-Sludge Process by Combining Chemical with Microbial Analysis. *Journal of Chemistry* 2014, 2014(1):360503.

[53] Sui Q, Duan L, Zhang Y, Zhang X, Liu Q, Zhang HJW: Seasonal water quality changes and the eutrophication of lake Yilong in Southwest China. *Water* 2022, 14(21):3385.

[54] Bali M, Gueddari MJAWS: Removal of phosphorus from secondary effluents using infiltration–percolation process. *Applied Water Science* 2019, 9:1-8.

[55] Roy PK: Nanofiltration as a tertiary treatment for phosphate removal from wastewater. Master Thesis, Asian Institute of Technology, Bangkok, Thailand; 1995.

[56] Akinnawo SO, Ayadi PO, Oluwalope MTJOUAoC: Chemical coagulation and biological techniques for wastewater treatment. *Ovidius University Annals of Chemistry* 2023, 34(1):14-21.

[57] Ademiluyi F, Amadi S, Amakama NJJJoAS, Management E: Adsorption and Treatment of Organic Contaminants using Activated Carbon from Waste Nigerian Bamboo. *Journal of Applied Sciences Environmental Management* 2009, 13(3).

[58] Habib SS, Batool AI, Rehman MFU, Naz SJBTER: Evaluation and association of heavy metals in commonly used fish feed with metals concentration in some tissues of O. niloticus cultured in biofloc technology and earthen pond system. *Biological Trace Element Research* 2023, 201(6):3006-3016.

[59] Wei B, Yu J, Cao Z, Meng M, Yang L, Chen QJIJoER, Health P: The availability and accumulation of heavy metals in greenhouse soils associated with intensive fertilizer application. *International Journal of Environmental Research Public Health* 2020, 17(15):5359.

[60] Hosono T, Su C-C, Delinom R, Umezawa Y, Toyota T, Kaneko S, Taniguchi MJE, Coastal, Science S: Decline in heavy metal contamination in marine sediments in Jakarta Bay, Indonesia due to increasing environmental regulations. *Estuarine, Coastal Shelf Science* 2011, 92(2):297-306.

[61] Fu F, Wang QJJoem: Removal of heavy metal ions from wastewaters: a review. *Journal of environmental management* 2011, 92(3):407-418.

[62] Ansari T, Marr I, Tariq NJJoAS: Heavy metals in marine pollution perspective-a mini review. *Journal of Applied Sciences* 2004, 4(1):1-20.

[63] Olojo E, Olurin K, Mbaka G, Oluwemimo AJAJoB: Histopathology of the gill and liver tissues of the African catfish Clarias gariepinus exposed to lead. *African Journal of Biotechnology* 2005, 4(1):117-122.

[64] Hayes WJ: Pesticides studied in man. 1982.

[65] Lage O, Soares H, Vasconcelos M, Parente A, Salema RJEJoP: Toxicity effects of copper (II) on the marine dinoflagellate Amphidinium carterae: influence of metal speciation. *European Journal of Phycology* 1996, 31(4):341-348.

[66] Arbabi M, Golshani NJE, Journal HS: Removal of copper ions Cu (II) from industrial wastewater: A review of removal methods. *Epidemiology Health System Journal* 2016, 3(3):283-293.

[67] Marella TK, Saxena A, Tiwari AJBT: Diatom mediated heavy metal remediation: A review. *Bioresource Technology* 2020, 305:123068.

[68] Nammalwar PJSR: Heavy metals pollution in the marine environment. *Science Reporter* 1983, 20(3):158-160.

[69] Nammalwar PJSR: Heavy metals pollution in the marine environment. 1983, 20(3):158-160.

[70] Jastrzębska M, Kostrzewska MK, Saeid A, Jastrzębski WPJM: Do new-generation recycled phosphorus fertilizers increase the content of potentially toxic elements in soil and plants? *Minerals* 2021, 11(9):999.

[71] Ahmed RS, Abuarab ME, Ibrahim MM, Baioumy M, Mokhtar AJC: Assessment of environmental and toxicity impacts and potential health hazards of heavy metals pollution of agricultural drainage adjacent to industrial zones in Egypt. *Chemosphere* 2023, 318:137872.

[72] Jilani MI, Nadeem R, Hanif MA, Mahmood Ansari T, Majeed AJD, Treatment W: Utilization of immobilized distillation sludges for bioremoval of Pb (II) and Zn (II) from hazardous aqueous streams. *Desalination Water Treatment* 2015, 55(1):163-172.

[73] Gimah BG, Bodo TJIJoAR, Publications: Creation of awareness through environmental adult education as a solution to the problem of habitat loss in Ogoni, Rivers State, Nigeria. *International Journal of Advanced Research Publications* 2019, 3(1):22-28.

[74] Cook MJSScf-c-d-h: Four consequences of deforestation. *Sciencing* 2018.

[75] Achard F, Beuchle R, Mayaux P, Stibig HJ, Bodart C, Brink A, Carboni S, Desclée B, Donnay F, Eva HDJGcb: Determination of tropical deforestation rates and related carbon losses from 1990 to 2010. *Global change biology* 2014, 20(8):2540-2554.

[76] Fabry VJJS: Marine calcifiers in a high-CO2 ocean. *Science* 2008, 320(5879):1020-1022.

[77] Tresguerres M, Hamilton TJJJoEB: Acid–base physiology, neurobiology and behaviour in relation to CO2-induced ocean acidification. *Journal of Experimental Biology* 2017, 220(12):2136-2148.
[78] Bednaršek N, Feely RA, Reum JC, Peterson B, Menkel J, Alin SR, Hales BJPotRSBBS: Limacina helicina shell dissolution as an indicator of declining habitat suitability owing to ocean acidification in the California Current Ecosystem. *Proceedings of the Royal Society B: Biological Sciences* 2014, 281(1785):20140123.
[79] Riebesell U, Aberle-Malzahn N, Achterberg EP, Algueró-Muñiz M, Alvarez-Fernandez S, Arístegui J, Bach LT, Boersma M, Boxhammer T, Guan WJNCC: Toxic algal bloom induced by ocean acidification disrupts the pelagic food web. *Nature Climate Change* 2018, 8(12):1082-1086.
[80] Shah ZU, Parveen SJPR: A review on pesticides pollution in aquatic ecosystem and probable adverse effects on fish. *Pollut Res* 2020, 39(2):309-321.
[81] Ukhurebor KE, Aigbe UO, Onyancha RB, Adetunji COJMRoPEV: Climate change and pesticides: their consequence on microorganisms. *AU eJournal of Interdisciplinary Research* 2021:83-113.
[82] Strassemeyer J, Daehmlow D, Dominic A, Lorenz S, Golla BJCp: SYNOPS-WEB, an online tool for environmental risk assessment to evaluate pesticide strategies on field level. *Crop protection* 2017, 97:28-44.
[83] Delcour I, Spanoghe P, Uyttendaele MJFRI: Literature review: Impact of climate change on pesticide use. *Food Research International* 2015, 68:7-15.
[84] Ukhurebor K, Aigbe U, Olayinka A, Nwankwo W, Emegha JJAeoIR: Climatic change and pesticides usage: a brief review of their implicative relationship. *AU eJournal of Interdisciplinary Research* 2020, 5(1).
[85] Abrantes N, Pereira R, Gonçalves FJEm, assessment: First step for an ecological risk assessment to evaluate the impact of diffuse pollution in Lake Vela (Portugal). *Environmental monitoring assessment* 2006, 117:411-431.
[86] Faust SD, Aly OMJJAWWA: Water pollution by organic pesticides. *Journal-American Water Works Association* 1964, 56(3):267-279.
[87] Kookana RS, Baskaran S, Naidu RJSR: Pesticide fate and behaviour in Australian soils in relation to contamination and management of soil and water: a review. *Soil Research* 1998, 36(5):715-764.
[88] Anusha A, Cyril A, Kumar L, Selvanayagam M, Saroj MJIJoET: Effect of endosulfan on oxygen consumption, RBC count and food utilization of fresh water fish Clarius dussumieri. *Industrial Journal of Environmental Toxicology* 1994, 4:33-36.
[89] Chengala L, Singh NJIJLS: Botanical pesticides—A major alternative to chemical pesticides: A review. *Int J Life Sci* 2017, 5(4):722-729.
[90] Riaz U, Rafi F, Naveed M, Mehdi SM, Murtaza G, Niazi AG, Mehmood H: Pesticide Pollution in an Aquatic Environment. In: *Freshwater Pollution and Aquatic Ecosystems.* edn.: Apple Academic Press; 2021: 131-163.
[91] Pereira JL, Gonçalves FJSotte: Effects of food availability on the acute and chronic toxicity of the insecticide methomyl to Daphnia spp. *Science of the total environment* 2007, 386(1-3):9-20.
[92] Shah ZU, Parveen SJH: Pesticides pollution and risk assessment of river Ganga: A review. *Heliyon* 2021, 7(8).

Chapter 12

Future Directions in Marine Drug Development: Challenges, Opportunities and Sustainability

Dipjyoti Sharma[1]
Pollobi Porasar[1]
Abhilash Bharadwaj[1]
Nilutpal Sharma Bora[1]
Bhaswati Kashyap[2]
Bhargab Jyoti Sahariah[3]
Prakash Kishore Hazam[4]
and Koushik Nandan Dutta[1,*]

[1]Department of Pharmacognosy, NETES Institute of Pharmaceutical Science, NEMCARE Group of Institutions, Assam, India
[2]Department of Pharmacognosy, School of pharmacy, Pragjyotishpur University, Assam, India
[3]Department of Pharmaceutics, NETES Institute of Pharmaceutical Science, NEMCARE Group of Institutions, Assam, India
[4]Marine Research Station, Institute of Cellular and Organismic Biology, Academia Sinica, Jiaushi, Ilan, Taiwan

Abstract

Marine organisms have emerged as a promising source for novel therapeutic compounds, offering unique chemical structures and biological activities that differ significantly from terrestrial sources. This chapter provides a comprehensive overview of marine drug development, from historical perspectives to future directions. The ocean, covering more than 70% of Earth's surface, harbors an immense diversity of organisms that are known to contain complex biochemical compounds to survive extreme marine conditions. These compounds have demonstrated significant potential in treating various diseases, including cancer, infections, and neurological disorders. This chapter explores the journey of marine drug development, beginning with the discovery of arabino-nucleosides in the 1950s to modern FDA-approved drugs like cytarabine, ziconotide, and trabectedin. It addresses critical challenges in marine drug development, including supply chain issues, aquaculture

* Corresponding Author's Email: koushik5dutta@gmail.com

In: Healing from the Depths Marine Metabolites in Disease Management
Editors: Pranav Kumar Prabhakar and Tarun Kumar Upadhyay
ISBN: 979-8-89530-832-5

limitations, genetic engineering complexities and synthetic modification hurdles. This chapter also delves into various classes of marine metabolites—alkaloids, terpenoids, polyketides, peptides, carotenoids, fatty acids, and sterols—and explores examining their therapeutic potential and mechanisms of action. Special attention is given to sustainability aspects, exploring aquaculture, biotechnology, and innovative approaches for maintaining consistent supply while preserving marine biodiversity.
The chapter also outlines future directions in marine drug development, highlighting the potential of unexplored marine environments, emerging technologies, and the integration of sustainable practices. Emphasizing the vast untapped potential of marine natural products in drug discovery and the importance of balancing pharmaceutical development with environmental conservation, this chapter is aimed to assist the readers in understanding the impressive paradigm of marine drug development.

Keywords: marine organisms, therapeutic compounds, novel chemical structures

1. Introduction

1.1. Natural Products

Metabolites, or endogenous chemical ingredients, present in microorganisms, insects, animals, plant extracts, marine organisms are known as Natural Products [1]. For thousands of years, natural products have been a significant source of remedial agents. Over 85-90% of the world's population uses natural products as traditional medicine for their basic health care. Approximately 73% of the pharmaceutical drugs/products are made from natural products [2]. In ancient times, natural products were used for treating various diseases; however, the use was restricted to people like priests, shamans, magicians, witch doctors and medicine men. Various cultures from all over the world use individual herbs/animals as medication, but some traditions use powerful combinations with various ingredients such as poultices, tinctures, and mixtures [2]. Natural Products from plants have gained popularity in recent years for their availability, good biocompatibility and biodegradability [3]. Eighty percent of the diverse plant and animal species of the world are present in the ocean. Bioactive compound sources such as sponges, tunicates, fishes, soft corals, nudibranchs, sea hares, opisthobranch molluscs, echinoderms, bryozoans, prawns, shells, sea slugs, and marine microorganisms are marine organisms [4]. The bioactive agents of natural products have a significant role in the existence of the human population, and to this day they are used for treatment in both traditional and modern medications [2].

1.2. History of Marine Drugs

Two-thirds of the earth's surface and 95% of its biosphere [5], is covered with oceans, that contains a huge ecological, chemical and biological diversity [6]. The enormous ecological resources of the sea have been exploited since ancient times and included the use of marine animals like fish and algae preparation as the sources of medicine [4]. Human civilization has been consuming sea products such as sea salt as a food ingredient, which is obtained by direct evaporation of seawater, and other seafood throughout history [7]. More than 3,500 million years ago, the first living organism appeared in the ocean and evolutionary processes have equipped many marine organisms with the appropriate mechanisms to survive in extreme conditions in terms of extreme temperatures, changes in salinity and pressure, bacterial attack and viral pathogens [6]. In the 1950s, the interest and hope of various chemists and biomedical researchers peaked for the investigation of natural products of marine origin for therapeutic benefits with the discovery of a cytotoxic compound arabinosyl thymidine (spongothymidine). After this discovery, different marine compounds and marine compounds' synthetic analogues were discovered by various scientist around the world [7]. In the late 1950s, Bergmann formally reported the first biologically active marine natural product [4]. Till 1968, fewer than 50 examples of marine natural products were known to the world, which gradually increased to 5000 and are steadily increasing at present date [8]. The genetical and biochemical uniqueness of marine plants and animals was established in the late 1970s [4]. *Nereistoxin* was the first marine natural product that was used for industrial purposes as a prototype in the synthesis of insecticides that was exported to various countries for controlling insect plagues [9]. By the year 2015, 27,000 marine natural products had been reported to be isolated [5]. Even in the harsh conditions in the ocean, the marine organisms produce a great variety of novel molecules that are unique and differ from isolated compounds from terrestrial plants with respect to diversity, structure and functions [6].

1.3. Marine Natural Product Derived Drugs

Marine products are a rich source of nutraceuticals and a great candidate for curing various human illnesses. Fish oils are the most common example of marine-derived products that have been used since ages [4]. Marine-based natural products are found to be resourceful in different industrial sectors such as biotechnology, nutrition, cosmetics, and health care. From an economic point of view, marine natural products' market size reached up to USD 10 billion in 2021 and is expected to grow more than 8% by year 2026 [10]. Marine natural products are mostly secondary metabolites but are not produced by regular or biological pathways and do not have any primary function related to the development, growth or propagation of the species [4]. There are more than thousands of secondary metabolites that have been isolated from marine sources with the ability to treat various health complications such as bacterial, viral, inflammation, tumour, fungal, malaria, allergies [11]. Digitalis, morphine, and various other natural compounds obtained from terrestrial plants have been used as drug molecules since ancient times, but modern pharmaceutical approaches are directed towards developing new drug compounds obtained from marine sources because it is believed that they provide novel and potent drug molecules due to their ability to survive under extreme conditions like high temperature, oxidative stress and pressure of the seawater. However, producing drugs from marine natural products is a difficult process as it is hard to obtain and develop novel marine drug leads in bulk quantities [12].

In the early 1950s, Arabino-nucleosides spongothymidine and spongouridine were isolated from sponge *Tectitethya crypta* were the first marine natural products that had a huge impact on the development of drugs. This discovery started the synthesis and biological evaluation of various other nucleosides, which eventually led to the creation of *cytarabine*, an anti-leukemic drug and vidarabine, an anti-viral agent [5]. Marine natural compounds that are regarded as drugs originate from bacterial and cyanobacterial biosynthesis sources, including salinosporamide A and brentuximad vedotin; practically all of these products are presently under clinical trials [13]. In 1969, cytarabine was first approved by the FDA and is still in use for acute lymphocytic leukaemia, acute myeloid leukaemia, non-Hodgkins's lymphoma, and myelodysplastic syndrome treatment. In 1976, vidarabine was approved for Herpes infection treatment; however, it is no longer being used in the United States and Europe. In 2004, ziconotide was the first FDA-approved drug that was directly derived from marine source, which is used as pain medication [5]. Marine natural products have a huge potential in pharmaceutical field, but the major challenges in the drug discovery process are poor screening, isolation, characterization, and optimization techniques. However, these challenges have now been overcome by advanced techniques such as gene mining, genome engineering and microbial culturing. These techniques have restored the interest of various pharmaceutical industries in identifying and drug development of marine drugs, opening opportunities for several new treatments [12]. To fully discover the potential of the marine natural product's secondary metabolites for drug discovery and secure the supply chain, optimization of lead compounds along with broader assessment for various bioactivities and drug interaction will be required [5].

1.4. Collection, Extraction and Isolation of Marine Drugs

Marine organisms are collected from water bodies by snorkelling or wading in water for natural products in shallow locations, while for specimens in deep waters (2-30 m) scuba diving is required. To collect specimens from 80m depth, NITROX diving is used [14]. After collection, the crude drug extract is screened, followed by isolation and bioactive agent identification that is then fractionated to obtain the active moiety from the natural product. This process is done using advanced analytical techniques [12]. The process is depicted in Figure 1.

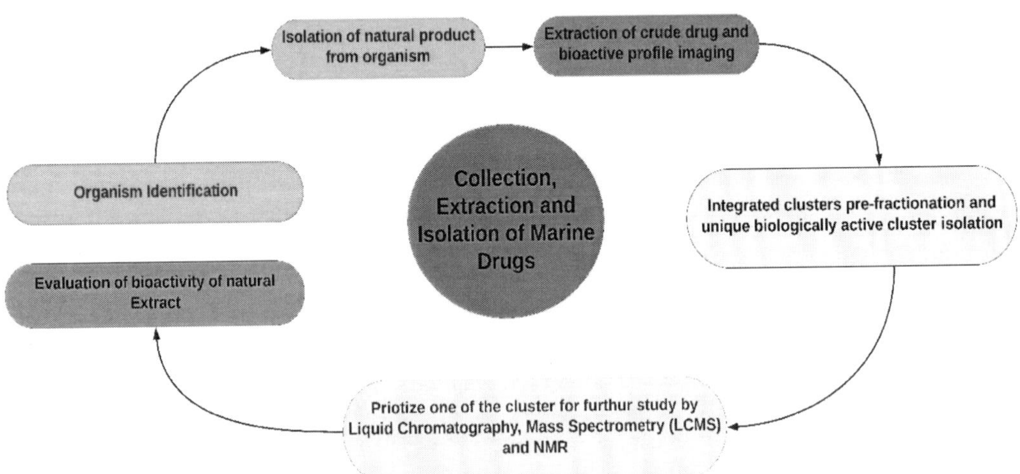

Figure 1. Collection, Extraction and Isolation of Marine Drugs.

Table 1. Classification of Marine drugs [12, 15, 16]

Classification of Marine drugs:

A. Anti-cancer Compounds:

S. No.	Name of Organism	Chemical Compound	Physiological Effect and Use
1.	Carribbean Sponge	Ara-C	It is used in the treatment of myelogenous leukemia and human acute leukemia
2.	*Pseudoplexaura porosa*	Crassin Acetate	Cytotoxic to human leukemia and hela cells
3.	*Aplidium* species	Geranyl hydroquinone	Cytotoxic to leukemia and mammary carcinoma
4.	*Aplysia angasi*	Aplysistain	Anti-leukemia
5.	*Halielona viridis*	Halitoxin	Anti-tumor
6.	*Lyngbya majuscula*	Majusulamine	Inhibitor of X-5563 myelome
7.	Trididemnum Species	Didemnins	Potent inhibitor of L-1210 leukemia cells in-vitro and P388 leukemia and B-16 melanoma in-vivo
8.	*Monanchora pulchra*	Monanchocidins A-E	Inhibits HL-60 human leukemia cells at very low concentration.
9.	*Stylostella* Species	Stylopeptide 2	Displays mild activity against breast cancer cell lines BT-549 and HS 578T
10.	*Phakellia fusca*	Phakellistatins 15 and16	Displays mild activity against P388 cancer cell line
11.	*Dolabella auriculria*	Dolastatin	Anti-proliferative activity against tumor cells.

B. Neurodegeneration Compounds:

S. No.	Name of Organism	Chemical Compound	Physiological Effect and Use
12.	Brown algae	Fucoidan	Preserves the morphology of the neuronal cells and increases the mitochondrial activity
13.	Brown seaweed	Phlorotannins	Prevents dopaminergic neurotoxicity and increases cell viability.
14.	Sea Cucumber	Cerebrosides	Improves cognitive functioning and synaptic plasticity and regulates neuropathic pathway.

C. Anti-Viral Compounds:

S. No.	Name of Organism	Chemical Compound	Physiological Effect and Use
15.	Ascidian *Lissocilium*	Patellazole-B	Shows very high in-vitro activity against herpes simplex viruses.
16.	*Eudistoma olivaceum*	Eudistomin A	It has the ability to cross the blood-brain-barrier and has high therapeutic value. It is potentially used in AIDS
17.	*Disidea avara* (Sponge)	Evarol and Evarone	
18.	Laminaria (brown algae)	Sulfated polysaccharide fucoidan	It inhibits both the DNA and RNA envelop viruses including herpes simplex and HIV.
19.	*Tethya crypta*	Ara- A	Anti-viral activity
20.	*Trididemnum* Species	Diademnins	In-vivo anti-viral and anti-tumor activity

D. Cardiovascular Activity Substances:

S. No.	Name of Organism	Chemical Compound	Physiological Effect and Use
21.	*Anthopleura xanthorgrannica*	Anthopleurin A and B	Without acting on Na^+, ATPase, heart rate and BP, it shows strong positive ionotropic effect.
	Anthopleura elegantissima	Athopleurin C	It is 35 times potent than digitoxin and very less toxic. It produces cardiotonic effect in conscious dog.

Table 1. (Continued)

S. No.	Name of Organism	Chemical Compound	Physiological Effect and Use
22.	*Eptatretus stoutii* (hogfish)	Eptatretin	It directly stimulates the mammalian myocardium. It is comparable to the adrenaline.
23.	*Laminaria angustata*	Laminine	Hypotensive effect
24.	Microalgae	Astaxanthin	Has strong antioxidant activity and quench oxygen in singlet states and release free radicals.
25.	Octopus Macropus *O. vulgaris* *Eledone moschata*	D(-)octopamine	Adrenergic cardiovascular response
26.	*Saxidomus gigantus*, California mussel, *Mytilus californianus*	Saxitoxin	Hypotensive effect
27.	*Verongia fistularis*	Autonomium	Produces α and β adrenergic effects and also exhibits cholinergic activity. CNS stimulant effect shown in mice exhibited by high increase in spontaneous motor activity. This compound proved useful in regulating the behaviour of the heart due to both adrenergic and cholinergic activity.
28.	Sea cucumbers	Holothurins	Cardiotonic, ichthyotoxic and also shows haemolytic activity. Holotoxin A, B, C shows anti-fungal activity.
29.	*Crypotethia crypta*	Spongosine	The rate and force of contraction of the heart is reduced
30	Anthopleurins	ATX-II	Potent and dose dependent cardiotonic effect on mamaliation heart, due to the indirect action on the pump activity induced by increased NA^+ transport as a result of a delayed activation of fast NA^+ current.
31.	*Eledone moschata*	Eledosin	Powerful hypotensive compound.

E. Anti-parasitic Compounds:

S. No.	Name of Organism	Chemical Compound	Physiological Effect and Use
32.	*Digenia simplex*	A-kainic acid	Broad spectrum anthelmintic and it is effective against parasitic round worm, whip worm and tape worm
33.	*Laminaria angustata*	Laminine	Anthelmintic as a smooth muscle relaxant and also has hypotensive effect
34.	*Chondria armata*	Domoic acid	Anthelmintic activity
35.	Nudibranch and a zoanthid	Bengamide	Anthelmintic activity
36.	Sea cucumber	Cucumechinoside F	Antiprotozoal activity

F. Anti-microbial Agents:

S. no	Name of Organism	Chemical Compound	Physiological Effect and Use
37.	Cephalosporium	Cephalosporin	Antibiotic
38.	*Streptomyces tenjimariensis*	Isotamycins	Shows activity against Gram-positive and gram-negative bacteria, including those resistant to aminoglycoside antibiotics.
39.	*Ptilonia austrulasica*	Bromopyrones	These compounds contain halogens, in particular bromine but halogens are not very potent or are too toxic as antimicrobial agents
40.	*Delisea fimbriata*	Fimbrolides	
41.	Laurenica	laurene	
42.	Sea cucumber	Holotoxin A, B and C	Anti-fungal activity
43.	*Dictyopteris zonaroides*	Zonarol and isozonarol	Anti-microbial activity
44.	Red algae	Tetrabromoheptanone	Anti-microbial activity

S. no	Name of Organism	Chemical Compound	Physiological Effect and Use
45.	*Asparogopsis taxiformis*	Four isomers of polyhalo 3-butene-2-one and 7 isomers of polythaloacetones	Anti-microbial activity
46.	*Verongia aerpholoa*	Aeropolysnin-1 (+) Aeropolysnin-1 (-)	Anti-microbial agent
47.	*Laurencia pacifia, L. filformis*	Prepacifenol	Anti-microbial agent
48.	*Thelepsus setosul*	Thelpin	Anti-microbial agent
49.	*Aplysia california*	Debromolaurentenol	Anti-microbial agent
50.	*Eunicia mammosa*	Eunicin	Anti-microbial agent
51.	*Acanthella acuta*	Acanthelin	Shows activity against mycobacterium

G. Anti-inflammatory and Antispasmodic Agents:

S. No	Name of Organism	Chemical Compound	Physiological Effect and Use
52.	*Luffariella variabilis*	Manoalide	Inactivates phospholipidase A_2, which is present in some neurotoxins and also has a role in synthesis of PGs in humans. It also has analgesic and selective anti-inflammatory activity.
53.	*Phyllospongia dendyl*	Dendalone 3-hydroxy butyrate	Shows anti-inflammatory activity
54.	*Sinularia flexibilis*	Flexibilide	Shows anti-inflammatory activity
55.	*Caulosystis cephalor-nithos*	6-n-tridecyl salicylic acid	Shows anti-inflammatory activity
56.	*Flustra foliaceal*	Flustramine A and B	Muscle relaxant
57.	Puffer fishes	Tetradotoxin	Strong antispasmodic
58.	*Rivularia firma*	Series of novel biindoles	Oedema and also acts on CNS

H. Antibiotic Compounds:

S. No	Name of Organism	Chemical Compound	Physiological Effect and Use
59.	*Pseudomonas bromutilis*	2,4-Dibromo-6-(3,4,5-tribromopyrrole-2-yl) phenol	Antibiotic
60.	*Chondria oppositiclada*	Cycloeudesmol	Antibiotic
61.	Sponge	Variabilin Irinin-I	Antibiotic
62.	*Caulosystis cephalor-nithos*	6-n-tridecyl salicylic acid	Antibiotic

I. Marine Toxins:

S. No	Name of Organism	Chemical Compound	Physiological Effect and Use
63.	*Amphidium carteri*	Unknown	Acetylcholine like activity neurotoxin, fast death factor
64.	*Anacystis cyanea*	Unknown	
65.	*Babylonia japonica*	Neosurugatoxin	It is a specific blocker of nicotinic ach receptor and a potent blocker of sympathetic receptor
66.	*Caulerpa prolifera*	Ichthyotoxic	Inhibits feeding by fish
67.	*Dinophysis fortii*	Dinophysistoxin	Diarrhetic shellfish toxins
68.	*Gambierdisics toxicus*	Maintotoxin, Ciguatoxin	Inotropic and depolarizes excitable membrane, respiratory depression and bradycardia.
69.	*Gymnodinium breve*	Unknown	Depolarises membrane, RBC haemolysis and neurotoxic
70.	*Gymnopodium veneficum*	Unknown	Depolarises membrane permeability alteration to Na^+
71.	*Gonyaulax catenella* *Alexandrium tamarense*	Sexitoxin gonyautoxin	Blocks Na^+ conductance
72.	*Haliclona rubens*	Halitoxin	Potent neuromuscular blocker
73.	*Halimeda opuntia*	Halimedatrial	Anti-microbial and cytotoxic
74.	*Lophogorgia rigida*	Lophotoxin	Neuro muscular blocker

Table 1. (Continued)

S. No	Name of Organism	Chemical Compound	Physiological Effect and Use
75.	Lophorgia sp.	Lophotoxin	Irreversible postsynaptic blockage at neuromuscular junction
76.	*Lyngbya a majuscula*	Lyngbyatoxin	'Swimmer itch' contact dermatitis majusculamide
77.	*Psuedonitzschia pungens* *P. australis*	Domoic acid	Poisoning of GIT, nausea, vomiting, anorexia, gastric bleeding, diarrhoea, and abdominal cramps followed by neurological disorder such as confusion disorientation, loss of short-term memory, coma. Maitotoxin is a powerful Ca^{+} channel
78.	*Prymnesium parvum*	Unknown	Neurotoxin, blocks end plate transmission, haemolytic agent
79.	*Polythoa species*	Palytoxin	Most potent coronary vasoconstrictor
80.	*Ptychodiscus brevis*	Brevetoxin	Positive inotropic and arrhythmogenic

2. Challenges for Marine Drugs

2.1. Supply Chain

A fundamental aspect of drug development from marine creatures is the consistent availability of enough quantities of organisms without jeopardising the aquatic ecosystem. Supply must be handled in a financially and environmentally sustainable manner in order to maintain a steady opportunity in the marketplace. Chemical synthesis, semi-synthesis, modification or marine biotechnology processes which include aquaculture, mariculture, fermenter cultivation, genetic engineering, enzymatic synthesis or modification can address the supply issue if natural environmental collection cannot be carried out sustainably [17]. Natural products are often extracted from higher marine species, such as marine invertebrates, because of their unique properties. These products are the result of a symbiotic relationship and are difficult to produce independently from cultures because their development is dependent, either directly or indirectly, on the host. Therefore, if these products are attempted to be developed in vitro, it is likely that a significant number of the genes that are essential will stay significant [18]. A critical factor is the accessibility of sufficient marine environmental resources for a certain chemical. The pharmaceutical industry will not attempt to fund clinical studies including marine-derived lead compounds unless they are certain or they can effectively solve the availability problem associated with such compounds. Considering the astronomical expenses of the clinical studies, the ecological feasibility is a legitimate worry. Inducing structural alterations to improve the drug's qualities is necessary, but the cost-effectiveness of this process is a major consideration [19].

2.2. Aquaculture/Cultivation

In the beginning, aquaculture was predominantly a freshwater food production system in Asia. However, it has since extended to other continents, including all aquatic settings and using a wide variety of aquatic species. Despite the fact that production is still primarily based on small-scale operations and is still primarily based in Asia, there is a widespread consensus that aquaculture has the potential to meet the growing demand for nutritious food fish around the world, as well as to contribute to the growth of national economies and to support the sustainable livelihoods of a great number of communities [20]. Aquaculture increases food availability by supplying domestically grown aquatic goods and purchasing food via foreign exchanges. Aquaculture produces nutritious, energetic aquatic food items that are rich in protein, vital fatty acids, vitamins, and minerals by strengthening food quality. Fish products are very beneficial for heart disease prevention and for susceptible populations, including pregnant and breastfeeding mothers, newborns, and preschoolers. Thus, aquaculture has a role in promoting dietary health [21]. A government's commitment to giving proper assistance to the aquaculture industry is one of the requirements that must be met before aquaculture can make a contribution to sustainable development. This commitment is made visible by the clear articulation of policies, plans, and strategies, as well as the availability of sufficient money and capacity development for the purpose of putting these into action. One of the most significant challenges is to prevent a pattern that is seen across the agricultural sector, which is that such commitment is losing its strength as a result of changing priorities in government development [22]. The following are some of the trends that can be identified in the development of the aquaculture sector: (i) the ongoing intensification of aquaculture production; (ii) the ongoing diversification of the species that are used; (iii) the ongoing diversification of production systems and practices; (iv) the growing influence of markets, trade, and consumers; (v) the enhancement of regulation and improvement of governance of the sector; and (vi) the increasing attention on better management of the sector [23]. The examination of particular aquaculture sectors in respect to these aspects is a fascinating and educational endeavour. The key determinants are the demand in the market (and the competition in that market), the availability of environmental resources, the creation or transfer of relevant technology, and a favourable business climate that enables entrepreneurs to benefit from their investments in the sector.

On the other hand, there are many instances of development failing, particularly in Africa and some regions of Latin America. This is mostly owing to the absence of well-developed markets or the inability to access them due to infrastructural problems, which include the absence of acceptable quality standards for export. One of the most significant limitations in many nations has been the absence of investment, in addition to the existence of weak institutional frameworks.

In many cases, the solid wastes that are generated by such systems may be used as fertilisers for other types of crops. When appropriate water replacement regimens are implemented, dissolved nutrients may often be lost, which can occasionally pose difficulties in regions that have a significant amount of aquaculture output or that are characterised as being oligotrophic or mesotrophic communities. If freshwater production systems were better optimised with regard to water and feed management, it might be possible to triplicate productivity without increasing the amount of freshwater currently being used [24].

2.3. Genetic Engineering

Over the course of the last decade, there have been substantial advancements achieved in the area of genetic modification carried out on microalgae. As part of this, the effective production of transgenes via the utilisation of the unique mechanism of riboswitches for the control of genes is included [25]. Deng et al., have studied the CrPEPC1 [C-reactive protein and enhancer of polycomb homolog 1] gene, which is a component of the p109 protein, was cloned by two classes of the PEPC [phosphoenolpyruvate carboxylase] complex: PEPC1 and PEPC2. According to the findings, the levels and activities of PEPC1 and PEPC2 were altered, and the lipid accumulation in the green alga *Chlamydomonas* was indirectly controlled. This was discovered by the use of CrPEPC1 knockdown or overexpression. The knockdown of CrPEPC1 resulted in the artificial obstruction of the TCA [tricarboxylic acid] pathway, which in turn led the carbon flow to enter the oil accumulation in *C. reinhardtii.* The results of this study revealed that the expression of CrPEPC1 in microalgae may be suppressed in order to boost oil production [26]. Moreover, another study done by Derelle et al., has shown that the genomic structure of *Ostreococcus tauri*, the smallest free-living eukaryote, often matches expectations of compaction and streamlining. These predictions may be influenced by the organism's particular lifestyle and ecology. However, the heterogeneity that has been discovered in this instance with regard to two chromosomes presents a difficulty in terms of clarifying its origin. This origin might either be a remnant of the ancient character of this alga or, on the other hand, more recent adaptations to the environmental niche that it occupies. In addition, it raises the issue of whether this particular sort of heterogeneity is in fact not exclusive to *Ostreococcus tauri*, but rather a characteristic that is shared by several eukaryotic species. This is because the existing knowledge of eukaryotic genomes is dependent on a genome database that has been dominated by "higher organisms" up until this point. Developing new ideas and conducting further experiments are necessary in order to get an understanding of the characteristics of the evolutionary processes that occur within the green lineage. This full genome sequence serves as a great resource for both of these endeavours [27]. Single-cell green algae (microalgae) are becoming a viable fuel source in the present day. The most efficient way to fuel microalgae is solar-driven H2 generation from H_2O. The H2-producing hydrogenase (HYDA) is directly connected to the photosynthetic electron transport chain. Oey et al., have reported on an RNAi [RNA interference] triple knock-down method to knock down three light-harvesting complex proteins (LHCMB1, 2, and 3) in the high H2-producing a *Chlamydomonas reinhardtii* mutant Stm6Glc4. The Stm6Glc4L01 mutant had a bright green phenotype, reduced LHCBM1 (20.6% 60.27%), LHCBM2 (81.2% 60.037%), and LHCBM3 (41.4%) expression relative to 100% control levels, and enhanced light to H2 (180%) and biomass (165%) conversion. Since light is the limiting element for microalgae production, greater solar flux densities (450 instead of 100 mE m22 s 21) and high cell densities enhanced H2 production efficiency. Thus, the Stm6Glc4L01 phenotype holds critical lessons for high-efficiency photobiological H2 production systems [28]. Although there are numerous projects that basically aims to produce genome, transcriptome, and proteome data for different marine species, it is very crucial to also understand what the actual gene annotation and metabolic pathway networks are to maximise their potential because that might create a big challenge while producing the species.

2.4. Challenges in Synthesis and Semi Synthesis Modification

Natural products are of tremendous relevance in the process of prescription drug development. Natural products played a significant factor in the development of more than fifty percent of the small-molecule medicines that were created worldwide between the years 1981 and 2019 [29]. Marine microorganisms provide natural products with unique qualities and significant value, differing from terrestrial natural goods found on land. Marine natural products have significant variety, considerable therapeutic usefulness, and many biological activities [30].

Mayer et al., have reported that the synthesis of vidarabine (arabinofuranosyladenine, adenin arabinoside, Ara-A, originated from the first research on sponge nucleosides as potential pharmaceuticals in the 1950s. Vidarabine is swiftly transformed into adenine arabinoside triphosphate inside cells. This active form inhibits DNA polymerase and the DNA synthesis of herpes, vaccinia, and varicella zoster viruses. In the EU, it is administered as eye drops for the treatment of acute keratoconjunctivitis and recurrent epithelial keratitis induced by herpes simplex types 1 and 2. In the USA, its clearance was revoked by June 2001, perhaps owing to its narrow therapeutic window [31]. The use of opioid analgesics, on the other hand, is associated with a number of drawbacks, including the development of dependence and tolerance, the potential for misuse, and severe adverse effects such as respiratory suppression and constipation. Because of this, it is necessary to have alternate or extra powerful analgesics. The synthetic peptide ziconotide, which is derived from a toxin of the potentially lethal marine cone snail Conus magus, has the potential to be such an application. the alternative. The snails' envenomation machinery is responsible for the production of a venom that is composed of several peptide toxins, which are referred to as conotoxins (or conopeptides). This venom is used for defence and for capturing prey. When the venom is administered to fish in their natural environment, it renders them paralysed in a matter of seconds. It is also toxic to humans [32]. The limited supply, worries about the environment, and the complexity of chemical synthesis are all challenges that must be overcome. In order to transform molecules originating from marine sources into medications that are commercially viable, rigorous testing, preclinical investigations, and clinical trials are required. Looking to the future, developments in science and technology, like metagenomics and synthetic biology, provide exciting paths for the discovery and development of marine-derived pharmaceuticals in a more expedient and effective manner. Continued investigation of marine natural products offers the promise of substantial advancements in drug development while taking into consideration sustainable techniques and environmental issues. The potential for discovering new therapeutic agents is still considerable, and the potential for discovering new therapeutic agents remains large. Also, according to Lyu et al., the lack of approved drugs derived from marine sources is most definitely not because of the small chemical diversity of MNPs (marine natural products). On the contrary, problems with material supply continue to be the biggest hurdle in the way of marine drug discovery. Since there is no universal standard for total synthesis or mariculture, it is especially crucial for researchers working on MNPs to share information [33, 34].

Although there are big challenges in exploring the potential of marine drug development, in between the process of drug development, marine species are also considered to be a major source of inspiration, following in the footsteps of terrestrial plants and microbes. One half of the MNPs that have been found exhibit a variety of biological actions. Ziconotide, Trabectedin, and Eribulin are just a few examples of the marine-derived medications that have been approved by the FDA, and further candidates are currently in clinical testing [35].

3. Future Directions in Marine Drug Development

Algae and sea creatures have attracted much attention in the process of searching for new drugs, the cause being the unusual chemical constituents and bioactive activities of marine organisms

[36]. Oceans cover more than 70% of the earth's surface, and it is not surprising that most of the marine organisms have developed highly complex biochemical compounds as a form of resistance to environmental, predatory, and pathogenic infections [37]. It has come to be that these compounds have active roles in drug formulation, making marine organisms a great source of new drugs [38]. In this context, the future of marine drug development presents vast opportunities, particularly in exploring metabolites, recent advancements, and bioactivity for therapeutic purposes [36]. This review highlights the future directions in marine drug development, focusing on metabolites, recent technological advances, and bioactivity.

3.1. Marine Metabolites

Marine metabolites are chemicals synthesized by oceanic entities like bacteria, fungi, algae, and marine animals. Over the ages, these metabolites have evolved specific roles and have developed many properties, including diversification of their pharmacological effects [39]. Marine metabolites often have unique chemical structures and mechanisms of action that are distinct from terrestrial metabolites, making them valuable for drug discovery [40]. As our understanding of marine organisms expands and research techniques improve, the identification and utilization of marine metabolites for therapeutic purposes are expected to increase [41].

Marine metabolites refer to a broad group of chemical compounds obtained from different marine organisms such as algae, sponges, fungi, bacteria, and corals. Most of these metabolites have been grouped into alkaloids, terpenoids, polyketides, peptides, carotenoids, fatty Acids and sterols.

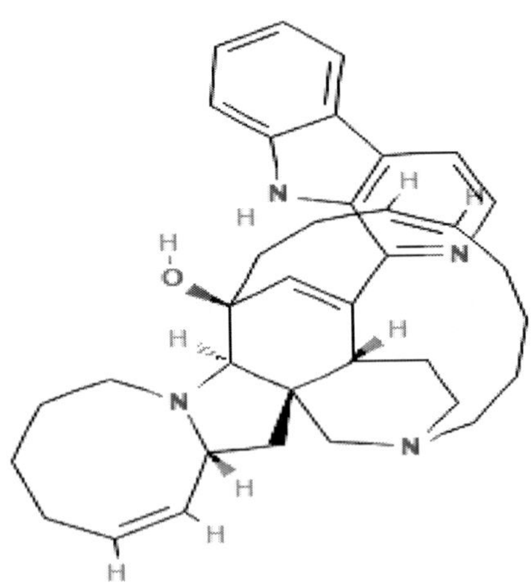

Figure 2. Manzamine A.

Alkaloids derived from marine organisms have shown significant pharmacological potential One such compound is 'Manzamine A' which is obtained from marine sponges [42]. This particular alkaloid possesses potent antimicrobial, antiviral and anti-malarial activity, making it an interesting compound that warrants further studies with regard to treatment of infectious diseases [43].

Marine polyketides are classified as secondary metabolites formed by marine-associated syncytial bacteria, fungi, and sponges [44]. One of the notable examples is 'Trabectedin' (ET-743) which binds to DNA and brings cell cycle arrest, exhibiting strong anticancer activity.

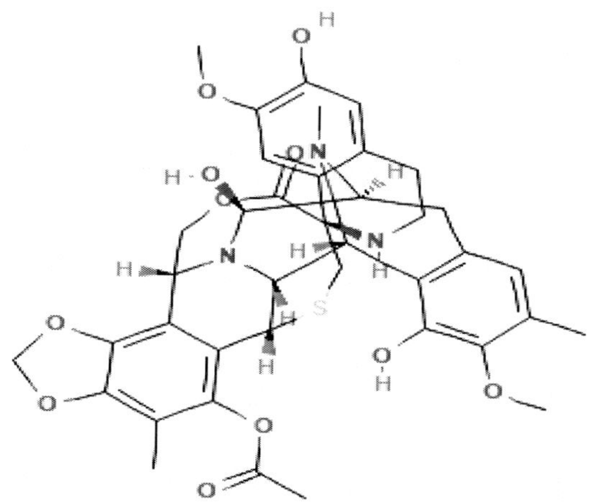

Figure 3. Trabectedin.

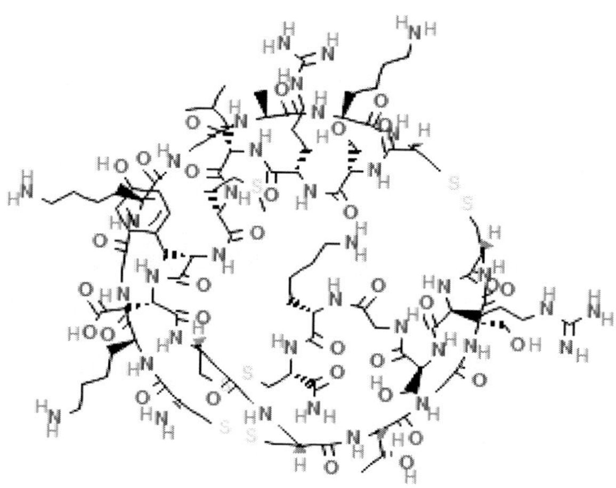

Figure 4. Ziconotide.

The development of polyketides for anticancer, antimicrobial and anti-inflammatory purposes seems plausible [45]. Peptides sourced from marine life have emerged as crucial candidates for medication development intended for the treatment of pain, infections, and disorders of the nervous system. For instance, Ziconotide, which is isolated from the toxic secretions of the marine cone shell *Conus magus* is an effective peptide that inhibits calcium channels. It is used in chronic pain management. These peptides have now high specificity and potency in action and therefore are sufficiently good for targeted precision medicine [46].

Another group of bioactive metabolites present in marine life, especially in microalgae, is carotenoids. 'Astaxanthin,' a carotenoid obtained from *Haematococcus pluvialis*, is a strong antioxidant with promising uses in the treatment of skin aging, oxidative stress, and diseases such as cardiovascular diseases [47].

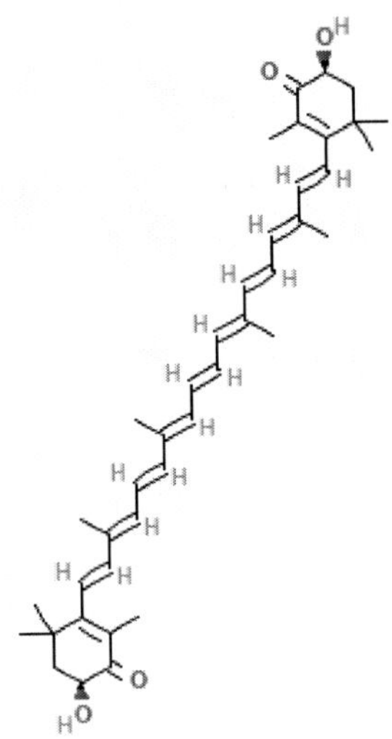

Figure 5. Astaxanthin.

Marine organisms, in particular algae, synthesize unique fatty acids and sterols with anti-inflammatory, antioxidative, and anticancer properties. These substances have assimilated research geared toward eradicating non-communicable diseases such as obesity, diabetes, and heart-related problems [48].

Table 2. Marine metabolites and their potential applications

Metabolite	Source	Therapeutic Potential	Mechanism of Action
Trabectedin	*Ecteinascidia turbinata*	Anticancer	DNA binding, cell cycle arrest
Manzamine A	*Haliclona* sp.	Antibacterial, Antiviral, Antimalarial	Microbial inhibition
Ziconotide	*Conus magus*	Neuroprotective, Pain relief	Voltage-gated calcium channel blocker
Fucoxanthin	*Undaria pinnatifida*	Anti-obesity, Anti-diabetic	Modulation of glucose metabolism
Astaxanthin	*Haematococcus pluvialis*	Antioxidant, Anti-aging, Skin health	Reduces oxidative stress, protects skin cells

3.2. Recent Advances in Marine Drug Development

Marine drug discovery has been looking up in recent years, primarily due much to new screening technologies, genomic exploration, and synthetic biology. Such technologies help find new metabolites with beneficial activities and produce such compounds effectively. The combination of bioinformatics, high-throughput screening, and metagenomics has boosted the rate of the discovery of products of the sea for pharmaceutical use [49].

High-throughput screening is a crucial approach that have propelled the new era of drug discovery by allowing the assessment of the biological activity of a large number of compounds. This technology is mostly advantageous for the development of marine drugs in that it helps to screen marine metabolites rapidly against various disease targets such as cancers, infectious diseases, and neurological disorders. HTS also facilitates the identification of lead compounds that can undergo further optimization [50]. The fields of genomics and metagenomics have advanced the development of marine pharmaceuticals by offering a better understanding of the genetic characteristics of marine organisms, in particular marine microbes [51]. An application of metagenomics is the analysis of environmental samples to sequence their DNA and characterize their potential for producing novel bioactive compounds. This method has been useful in the research of previously uncharacterized substances with possible pharmacological effects [52]. Synthetic biology means adapting the biochemical pathways of simple-structured organisms to replicate the complex metabolites exclusive to certain marine species. Hence, microbial production systems for significant marine-derived bioactive substances have been created for laboratory use [53]. By transferring marine bioengineering genes into such model organisms as yeast or bacteria, researchers can increase the quantities of metabolites that can be otherwise sustainably or economically impossible to extract from marine habitats [54]. AI and machine learning have found their applications in marine drug design by working on the assessment of the biological activity of new chemicals. Such technologies are capable of working with advanced genomic, metabolomic and chemical data and even making assumptions on the end variance, which will be useful for medication [55].

Table 3. Recent advances in Marine drug discovery technologies

Technology	Description	Impact on Marine Drug Discovery
High-throughput Screening	Automated testing of large compound libraries for bioactivity	Enables rapid identification of lead compounds for further development
Genomics and Metagenomics	Sequencing the genomes of marine organisms and microbial communities	Identifies novel biosynthetic pathways for producing bioactive metabolites
Synthetic Biology	Engineering microorganisms to produce marine-derived metabolites	Sustainable production of bioactive compounds in large quantities
Artificial Intelligence	Machine learning models to predict bioactivity and optimize compound design	Accelerates the drug discovery process and improves the accuracy of predictions

3.3. Bioactivity of Marine Metabolites

Marine metabolites present a remarkable range of bioactivity, and their use is being expanded to include the treatment of more diseases. The research on marine metabolites and their chemical constituents has identified a range of bioactivities, most of which include anticancer, antimicrobial, anti-inflammatory, neuroprotective, and antioxidant. These bioactive molecules frequently exhibit novel mechanisms of action, which may be beneficial compared to the conventional chemical drugs [56].

Marine compounds have shown strong anticancer activity by disrupting various important processes of a cancer cell, such as its DNA, apoptosis and its cycle control. Trabectedin (ET-743), a marine-derived compound from the tunicate *Ecteinascidia turbinata* for instance, is commonly used because this compound is able to inhibit DNA replication and causes cell cycle arrest, thereby being active against certain types of cancer. Other compounds obtained from the sea like bryostatins target and activate protein kinase C which is responsible for the proliferation and survival of cancer cells [57]. A recent trend is the study of marine-derived compounds as possible agents of infections due to drug-resistant pathogens. For instance, manzamine A has both antibacterial and antiviral activities, whereas marine peptides called piscidins are highly effective antimicrobials. Therefore, these substances are important allies in the conquest against the emergence of new antibiotic and antiviral resistant strains [58]. Chronic inflammation is central to the etiopathogenesis of numerous diseases, such as arthritis, cardiovascular illnesses, and neurodegenerative complications. For instance, some compounds derived from marine sources, such as Scytonemin derived from cyanobacteria, has been reported to inhibit pro-inflammatory cytokines, thus aiding in the alleviation of inflammation [59]. Furthermore, compounds such as Ziconotide which are marine-derived peptides have protective effects on the nervous system by inhibiting calcium channels, thus proving useful in alleviating some types of pain, including chronic pain and other degenerative disorders [46]. One such example is the antioxidant compound astaxanthin from microalgae, which prevents over oxidative stress, which contributes towards aging and degenerative diseases as it is a primal cause. Astaxanthin has been found to affect cells, decreasing oxidative stress, shield the skin from the sun's rays, and alleviate swelling; thus, it presents a perfect candidate for the skin health and anti-aging regimen [59].

4. Sustainability of Marine Drugs

Natural processes have produced secondary metabolites with a wide variety of chemical structures and biological functions, which have aided in the whole process of medication creation and discovery. There is a wealth of potentially dangerous chemicals found in marine creatures, making it a huge untapped potential for marine drug discovery because of the recently established specialized procedures and equipment. Despite covering 70% of the earth's surface, only 5% of deep-sea creatures have been examined. Marine creatures that fall within the order Alcyonacea include soft corals that have produced thousands of secondary metabolites that exhibit a broad variety of biological functions, including defense against bacteria, tumours, inflammation, infections, viruses, allergies, and malaria [18, 60]. Marine creatures are attracting the attention of bioprospectors looking for western medicinals, and nearly 14,000 novel chemical entities have been discovered in marine environments and marine natural products have been the subject of over 300 patents [61]. Consistent supply of these compounds produced from the sea may be achieved by whole synthesis or semi-synthesis but the need for precise identification of the chemical structures of the marine drugs may provide a number of challenges. Misassignments of the newly found compounds still occur even with improved structural elucidation methods, which may include the formula, composition, configuration of double bonds, absolute configuration and even one or more stereocenters [62]. Industry and research-associated entities should self-regulate to reduce the potential ecological effect of bioprospecting and the following steps could be taken by industries to ensure they are following environmental best practices: i) Creating standards or procedures for collecting; ii) Conducting precautionary and timely evaluations of economic and ecological feasibility; and iii) Coming up with plans for alternative sources of supply. Additionally, environmental requirements may be attached to research funding and publications that publish related research may require authors to follow an environmental code of conduct [63]. Synthesis and culture both alleviate issues with small populations and inconsistent bioactivity levels while decreasing reliance on natural sources. As a result of developments in chemistry, such as the proliferation of combinatorial approaches, natural products cease to be used as a raw material for a final commercial product but rather as a framework for further drug research [64]. In combinatorial chemistry, a natural product's scaffold or pharmacophoric unit is used as a starting point to construct a diversified chemical library and their goal is to find analogues and derivatives of the product that are more active and less hazardous [64, 65]. Aquaculture has proven to be a remarkable solution for fulfilling the increasing needs for reliable and sustainable sources of marine-derived natural products. The coral culture began in the late 1950s, primarily aimed at commercial and conservation efforts across various regions, including Japan, the Philippines, the Central Pacific, the South Pacific, North America, Sri Lanka, Australia, Taiwan and Thailand [66]. A variety of methods have been utilized to enhance the mass production of soft corals and to increase the diversity of cultured corals through direct transplantation, which includes coral gardening, micro-fragmentation and larval enhancement with certain *ex situ* and *in situ* techniques [67]. The main component of the culture system is the aquaria or cultivation tanks, which are typically furnished with coral sand, artificial rocks, artificial coral reefs, live sea sand, live sea rocks, marine creatures, some soft corals, snails, sea urchins, sea cucumbers and fish. The soft coral's primary medium for its cultural environment is seawater and the salinity of the ocean is additionally maintained at a level of around 36 psu where the temperature is kept between 24° and 28°C using a chiller with an LED light source, which is used to simulate natural sunlight while a foam fractionator is employed to filter out any debris that may have accumulated in the system [68–70]. Achieving a higher quality of life without exceeding sustainably high levels of resource use has emerged as the primary objective of sustainability. A clear association exists between the number of species being cultured and the search for viable bioactive chemicals, where the marine creatures may be sustainably used for industrial purposes via managed maintenance. When the resources are present, most organisms develop in a heterotrophic fashion, which means they don't need light to thrive and in order to determine the product profiles under each scenario, the biomass of the organisms of interest will be generated autotrophically, heterotrophically or mixotrophically. The method of nutrients also influences the bioactive component profile, while the benefits of heterotrophic cultures over photosynthetic ones include the production of high-value bioactive chemicals and hence enable easy scaling up and larger potential densities. Searching for sustainable fuel has become an emergency from an economic and environmental benefit standpoint in order to guarantee access to biofuel from a stable, economical and sustainable bioresource. While microalgae and macroalgae are dormant, they could be utilised to produce bioenergy by using products made from bio-refined algae, which would have a dual benefit of improving life and increasing their market value. This could pave the way for future sustainable green fuels [71]. Marine species have recently been considered as potential collagen sources because of their lack of religious restrictions and the absence of reports of any potentially contagious illnesses. Biomass from fisheries and fish processing units, including waste products like fish and sea urchins, undersized fish and by-catch animals including jellyfish, sharks, starfish and sponges, has the potential to become a significant yet underutilized collagen source and hence building a sustainable pipeline to produce collagen with much less of an effect on the environment is possible with the help of wasted and underutilized biomass [72–77]. Marine collagen and collagen-derived peptides, in addition to gelatin obtained from sustainable sources, have been the subject of a number of research studies undertaken with the purpose of determining their possible uses in the fields of biomedicine and pharmaceuticals. Collagen obtained from fish waste and bioactive peptides sourced from collagen-rich materials exhibit various bioactivities, hence presenting potential applications in cosmeceuticals and pharmaceuticals, along with serving as biomaterials for tissue engineering [75, 78]. A more current strategy involves looking at how space biotechnology and marine resources may work together for a more sustainable future. As mankind pushes beyond Earth's boundaries with the vast field of space exploration driven by the pursuit of sustainable solutions. An exciting new frontier that opens up new possibilities for sustainable space habitation under this framework is the merging of biotechnology with marine resources with a wealth of answers to the problems of space colonisation that may be found in the remarkable adaptability and resilience of marine creatures which has been shaped by the vast evolutionary history of Earth's seas, with a showcase of remarkable skills in nutrient recycling and wastewater treatment offering practical answers to the challenge of managing resources in constrained spaces [79].

Conclusion

The upside to the development of drugs from therapeutic agents of marine origin is immaculate. This is due to the unexplored depths in the marine sector, which have organisms that present a new category of bioactive metabolites. New technologies in genomics, bioinformatics, and synthetic biology are speeding up the exploration and utilization of these compounds. As time goes by, more attention will be placed on the development and utilization of marine metabolites to provide solutions to medical challenges prevailing today that do not have available treatment options, such as cancer, antibiotic resistance, neurological disorders, and chronic illnesses. Supported by enhancing technological capabilities and knowledge concerning marine environments, there is indeed a bright future for the creation of drugs from marine sources.

The potential for marine natural products in drug discovery, biomedicine, and sustainability is vast and still largely untapped. The oceans, which encompass 70% of the Earth's surface, serve as a rich source of secondary metabolites exhibiting various biological functions. Marine organisms such as soft corals generate compounds that hold promise for addressing a variety of health issues, including infections, cancer and inflammation. While there has been notable advancement in marine bioprospecting, a large portion of the deep-sea ecosystem is still uncharted and only a limited number of marine species have been examined for their potential medicinal uses. Marine bioprospecting nonetheless brings forth significant ecological issues that can be addressed through self-regulation within the industry and the adoption of sustainable practices and this includes implementing careful collection standards and initiatives aimed at cultivating marine organisms such as soft corals in controlled environments like aquaculture. Recent developments in combinatorial chemistry and culture methods enable the large-scale production of bioactive compounds, minimizing dependence on natural resources and promoting a more sustainable supply. Additionally, advancements in biotechnology, including the utilization of marine organisms for collagen generation and biofuel production, signify encouraging prospects for sustainable industrial uses. While the shift towards sustainable practices in marine bioprospecting is vital for protecting marine ecosystems and is key to tackling global issues such as responsible resource utilization, bioenergy generation and advancements in space exploration. Integrating marine biotechnology with advanced practices allows us to address the growing need for bioactive compounds while also safeguarding the environment and promoting the responsible use of marine resources for societal benefit and hence the potential of sustainable marine bioprospecting is bright, offering significant benefits for the planet's health and enhancing human well-being.

References

[1] Xing P, Zhong Y, Cui X, Liu Z, Wu X. Natural products in digestive tract tumors metabolism: Functional and application prospects. *Pharmacol Res* [Internet]. 2023;191(March):106766. Available from: https://doi.org/10.1016/j.phrs.2023.106766.

[2] Wangchuk P. Therapeutic Applications of Natural Products in Herbal Medicines, Biodiscovery Programs, and Biomedicine. *J Biol Act Prod from Nat*. 2018;8(1):1–20.

[3] de Almeida Bertassoni B, de Abreu Garófalo D, de Souza Bustamante Monteiro MS, Santos-Oliveira R, Gomes ACC, Toledo ALMM, et al., In vivo wound healing activity of electrospun nanofibers embedding natural products. *Arab J Chem*. 2024;17(12).

[4] Malve H. Exploring the ocean for new drug developments: Marine pharmacology. *J Pharm Bioallied* Sci. 2016;8(2):83–91.

[5] Altmann KH. Drugs from the oceans: Marine natural products as leads for drug discovery. *Chimia* (Aarau). 2017;71(10):646–51.

[6] Romano G, Costantini M, Sansone C, Lauritano C, Ruocco N, Ianora A. Marine microorganisms as a promising and sustainable source of bioactive molecules. *Mar Environ Res* [Internet]. 2017;128:58–69. Available from: http://dx.doi.org/10.1016/j.marenvres.2016.05.002.

[7] Haque N, Parveen S, Tang T, Wei J, Huang Z. Marine Natural Products in Clinical Use. *Mar Drugs*. 2022;20(8).

[8] Gribble GW. Biological activity of recently discovered halogenated marine natural products. *Mar Drugs*. 2015;13(7):4044–136.

[9] Rangel M, Falkenberg M. An overview of the marine natural products in clinical trials and on the market. *J Coast Life Med*. 2015;3(6):421–8.

[10] Papon N, Copp BR, Courdavault V. Marine drugs: Biology, pipelines, current and future prospects for production. *Biotechnol Adv*. 2022;54.

[11] Nguyen NBA, Chen LY, El-Shazly M, Peng BR, Su JH, Wu HC, et al., Towards Sustainable Medicinal Resources through Marine Soft Coral Aquaculture: Insights into the Chemical Diversity and the Biological Potential. *Mar Drugs*. 2022;20(10).

[12] Bhatia S, Makkar R, Behl T, Sehgal A, Singh S, Rachamalla M, et al., Biotechnological Innovations from Ocean: Transpiring Role of Marine Drugs in Management of Chronic Disorders. *Molecules*. 2022;27(5):1–23.

[13] Pereira F. Have marine natural product drug discovery efforts been productive and how can we improve their efficiency? *Expert Opin Drug Discov* [Internet]. 2019;14(8):717–22. Available from: https://doi.org/10.1080/17460441.2019.1604675.

[14] Facey P. Marine metabolites: oceans of opportunity. *Pharmacogn Fundam Appl Strateg Second Ed*. 2023;(January):411–39.

[15] Vinothkumar S, Parameswaran PS. Recent advances in marine drug research. *Biotechnol Adv*. 2013;31(8):1826–45.

[16] Ansari SH. Essentials of pharmacognosy. *Birla Publ Pvt Ltd New Delhi*. 2006;1:367–9.

[17] Lindequist U. Marine-derived pharmaceuticals - challenges and opportunities. *Biomol Ther*. 2016;24(6):561–71.

[18] Martins A, Vieira H, Gaspar H, Santos S. Marketed marine natural products in the pharmaceutical and cosmeceutical industries: Tips for success. *Mar Drugs*. 2014;12(2):1066–101.

[19] Kanase HR, Singh KNM. Marine pharmacology: Potential, challenges, and future in India. *J Med Sci*. 2018;38(2):49–53.

[20] Asia-Pacific N of AC in. *Regional review on aquaculture development 3. asia and the pacific – 2005.* 2005;(1017):110.

[21] Subasinghe R, Soto D, Jia J. Global aquaculture and its role in sustainable development. *Rev Aquac.* 2009;1(1):2–9.

[22] Bartley DM, Brugere C, Soto D, Gerber P, Harvey B. Comparative assessment of the environmental costs of aquaculture and other food production sectors. Methods for meaningful comparisons. FAO/WFT Expert Workshop, Vancouver, Canada, 24-28 April, 2006. 2007;

[23] Fisheries FAO, Paper T. 1484808844_Study-and-Analysis-of-Feeds-and-Fertilisers-for-Aquaculture.

[24] Verdegem MCJ, Bosma RH. Water withdrawal for brackish and inland aquaculture, and options to produce more fish in ponds with present water use. Water Policy. 2009;11(SUPPL. 1):52–68.

[25] Croft MT, Moulin M, Webb ME, Smith AG. Thiamine biosynthesis in algae is regulated by riboswitches. *Proc Natl Acad Sci USA*. 2007;104(52):20770–5.

[26] Deng X, Cai J, Li Y, Fei X. Expression and knockdown of the PEPC1 gene affect carbon flux in the biosynthesis of triacylglycerols by the green alga chlamydomonas reinhardtii. *Biotechnol Lett.* 2014;36(11):2199–208.

[27] McAdams HH. Bacterial stalks are nutrient-scavenging antennas. *Proc Natl Acad Sci USA.* 2006;103(31):11435–6.

[28] Oey M, Ross IL, Stephens E, Steinbeck J, Wolf J, Radzun KA, et al., RNAi Knock-Down of LHCBM1, 2 and 3 Increases Photosynthetic H2 Production Efficiency of the Green Alga Chlamydomonas reinhardtii. *PLoS One*. 2013;8(4).

[29] Newman DJ, Cragg GM. Natural Products as Sources of New Drugs over the Nearly Four Decades from 01/1981 to 09/2019. *J Nat Prod*. 2020;83(3):770–803.

[30] Yin JN, Wang CF, Zhang XL, Cheng YJ, Wu YW, Zhang Q, et al., Semisynthesis, Structure Elucidation and Anti-Mycobacterium marinum Activity of a Series of Marine-Derived 14-Membered Resorcylic Acid Lactones with Interesting Ketal Groups. *Mar Drugs*. 2024;22(10).

[31] Mayer AMS, Glaser KB, Cuevas C, Jacobs RS, Kem W, Little RD, et al., The odyssey of marine pharmaceuticals: a current pipeline perspective. *Trends Pharmacol* Sci. 2010;31(6):255–65.

[32] Olivera BM, Gray WR, Zeikus R, McIntosh JM, Varga J, Rivier J, et al., Peptide neurotoxins from fish-hunting cone snails. *Science* (80-). 1985;230(4732):1338–43.

[33] Lyu C, Chen T, Qiang B, Liu N, Wang H, Zhang L, et al., CMNPD: A comprehensive marine natural products database towards facilitating drug discovery from the ocean. *Nucleic Acids Res.* 2021;49(D1):D509–15.

[34] Montuori E, Hyde CAC, Crea F, Golding J. Marine Natural Products with Activities against Prostate Cancer : *Recent Discoveries*. 2023;

[35] Jiménez C. Marine Natural Products in Medicinal Chemistry. *ACS Med Chem Lett*. 2018;9(10):959–61.

[36] Banday AH, Azha N ul, Farooq R, Sheikh SA, Ganie MA, Parray MN, et al., Exploring the potential of marine natural products in drug development: A comprehensive review. *Phytochem Lett* [Internet]. 2024;59(November 2023):124–35. Available from: https://doi.org/10.1016/j.phytol.2024.01.001

[37] Karthikeyan A, Joseph A, Nair BG. Promising bioactive compounds from the marine environment and their potential effects on various diseases. *J Genet Eng Biotechnol* [Internet]. 2022;20(1):14. Available from: https://doi.org/10.1186/s43141-021-00290-4.

[38] Atanasov AG, Zotchev SB, Dirsch VM, Supuran CT. Natural products in drug discovery: advances and opportunities. *Nat Rev Drug Discov*. 2021;20(3):200–16.

[39] Srinivasan R, Kannappan A, Shi C, Lin X. Marine bacterial secondary metabolites: A treasure house for structurally unique and effective antimicrobial compounds. *Mar Drugs*. 2021;19(10):1–36.

[40] Giordano D, Coppola D, Russo R, Denaro R, Giuliano L, Lauro FM, et al., Marine Microbial Secondary Metabolites: Pathways, Evolution and Physiological Roles. Vol. 66, *Advances in Microbial Physiology*. 2015. 357–428 p.

[41] Bayona LM, de Voogd NJ, Choi YH. Metabolomics on the study of marine organisms. *Metabolomics* [Internet]. 2022;18(3):1–24. Available from: https://doi.org/10.1007/s11306-022-01874-y.

[42] Munekata PES, Pateiro M, Conte-Junior CA, Domínguez R, Nawaz A, Walayat N, et al., Marine alkaloids: Compounds with in vivo activity and chemical synthesis. *Mar Drugs*. 2021;19(7):1–24.

[43] Letchuman S, Madhuranga HDT, Kaushalya MBLN, Premarathna AD, Saravanan M. Alkaloids Unveiled: A Comprehensive Analysis of Novel Therapeutic Properties, Mechanisms, and Plant-Based Innovations. *Intell Pharm* [Internet]. 2024 Sep; Available from: https://www.sciencedirect.com/science/article/pii/S2949866X24001047.

[44] Kamat S, Kumar S, Philip S, Kumari M. Secondary metabolites from marine fungi: current status and application. In: *Microbial biomolecules*. Elsevier; 2023. p. 181–209.

[45] Gupta AP, Pandotra P, Sharma R, Kushwaha M, Gupta S. Marine resource: A promising future for anticancer drugs [Internet]. 1st ed. Vol. 40, *Studies in Natural Products Chemistry*. Copyright © 2013 Elsevier B.V. All rights reserved.; 2013. 229–325 p. Available from: http://dx.doi.org/10.1016/B978-0-444-59603-1.00008-4

[46] Cheung RCF, Ng TB, Wong JH. *Marine peptides: Bioactivities and applications*. Vol. 13, Marine Drugs. 2015. 4006–4043 p.

[47] Galasso C, Corinaldesi C, Sansone C. Carotenoids from marine organisms: Biological functions and industrial applications. *Antioxidants*. 2017;6(4).

[48] Ghallab DS, Ibrahim RS, Mohyeldin MM, Shawky E. Marine algae: A treasure trove of bioactive anti-inflammatory compounds. *Mar Pollut Bull*. 2024;199:116023.

[49] Tang A, Woldemariam S, Roger J, Sirota M. Translational Bioinformatics to Enable Precision Medicine for All: Elevating Equity across Molecular, Clinical, and Digital Realms. *Yearb Med Inform*. 2022;31(1):106–15.

[50] Blay V, Tolani B, Ho SP, Arkin MR. High-Throughput Screening: today's biochemical and cell-based approaches. *Drug Discov Today* [Internet]. 2020;25(10):1807–21. Available from: https://doi.org/10.1016/j.drudis.2020.07.024.

[51] Chen J, Jia Y, Sun Y, Liu K, Zhou C, Liu C, et al., Global marine microbial diversity and its potential in bioprospecting. *Nature* [Internet]. 2024;633(8029):371–9. Available from: http://dx.doi.org/10.1038/s41586-024-07891-2.

[52] Zhang L, Chen FX, Zeng Z, Xu M, Sun F, Yang L, et al., Advances in Metagenomics and Its Application in Environmental Microorganisms. *Front Microbiol*. 2021;12(December):1–15.

[53] Cravens A, Payne J, Smolke CD. Synthetic biology strategies for microbial biosynthesis of plant natural products. *Nat Commun* [Internet]. 2019;10(1):1–12. Available from: http://dx.doi.org/10.1038/s41467-019-09848-w

[54] Hosseini H, Al-Jabri HM, Moheimani NR, Siddiqui SA, Saadaoui I. Marine microbial bioprospecting: Exploitation of marine biodiversity towards biotechnological applications—a review. *J Basic Microbiol*. 2022;62(9):1030–43.

[55] Singh S, Kumar R, Payra S, Singh SK. *Artificial Intelligence and Machine Learning in Pharmacological Research: Bridging the Gap Between Data and Drug Discovery*. 2023;15(8).

[56] Ameen F, AlNadhari S, Al-Homaidan AA. Marine microorganisms as an untapped source of bioactive compounds. *Saudi J Biol Sci* [Internet]. 2021;28(1):224–31. Available from: https://doi.org/10.1016/j.sjbs.2020.09.052

[57] Dalisay DS, Tenebro CP, Sabido EM, Suarez AFL, Paderog MJ V., Reyes-Salarda R, et al., Marine-Derived Anticancer Agents Targeting Apoptotic Pathways: Exploring the Depths for Novel Cancer Therapies. *Mar Drugs*. 2024;22(3).

[58] Yasuhara-Bell J, Lu Y. Marine compounds and their antiviral activities. *Antiviral Res*. 2010;86(3):231–40.

[59] Kumar S, Kumar R, Kumari A, Panwar A. Astaxanthin: A super antioxidant from microalgae and its therapeutic potential. *J Basic Microbiol*. 2022;62(9):1064–82.

[60] Rodrigues IG, Miguel MG, Mnif W. A brief review on new naturally occurring cembranoid diterpene derivatives from the soft corals of the genera Sarcophyton, Sinularia, and Lobophytum since 2016. *Molecules*. 2019;24(4).

[61] Kerr RG, Kerr SS. Marine natural products as therapeutic agents. *Expert Opin Ther Pat*. 1999;9(9):1207–22.

[62] Maier ME. Structural revisions of natural products by total synthesis. *Nat Prod Rep*. 2009;26(9):1105–24.

[63] Marsh H, Kenchington R. The role of ethics in experimental marine biology and ecology. *J Exp Mar Bio Ecol*. 2004;300(1–2):5–14.

[64] Ganesan A. Integrating natural product synthesis and combinatorial chemistry. *Pure Appl Chem.* 2001;73(7):1033–9.

[65] Cragg GM, Newman DJ. Natural product drug discovery in the next millennium. *Pharm Biol.* 2001;39(SUPPL.):8–17.

[66] Delbeek, J C. Coral farming: past, present and future trends. *Aquarium Sci Conserv*. 2001;3:171–81.

[67] Boström-Einarsson L, Babcock RC, Bayraktarov E, Ceccarelli D, Cook N, Ferse SCA, et al., Coral restoration – A systematic review of current methods, successes, failures and future directions. Bianchi CN, editor. *PLoS One* [Internet]. 2020 Jan 30;15(1):e0226631. Available from: https://dx.plos.org/10.1371/journal.pone.0226631

[68] Huang TY, Huang CY, Chao CH, Lin CC, Dai CF, Su JH, et al., Aquaculture soft coral lobophytum crassum as a producer of anti-proliferative cembranoids. *Mar Drugs* [Internet]. 2020 Aug 29;18(9):452. Available from: https://www.mdpi.com/1660-3397/18/9/452

[69] Huang TY, Huang CY, Chao CH, Lin CC, Dai CF, Su JH, et al., New Hydroquinone Monoterpenoid and Cembranoid-Related Metabolites from the Soft Coral Sarcophyton tenuispiculatum. *Mar Drugs* [Internet]. 2020 Aug 29;18(9):452. Available from: https://www.mdpi.com/1660-3397/18/9/452

[70] Huang TY, Huang CY, Chao CH, Lin CC, Dai CF, Su JH, et al., New biscembranoids sardigitolides A–D and known cembranoid-related compounds from Sarcophyton digitatum: Isolation, structure elucidation, and bioactivities. *Mar Drugs*. 2020;18(9):1–13.

[71] Thakur M, Belwal T. Bioactive Components: A Sustainable System for Good Health and Well-Being. *Bioactive Components: A Sustainable System for Good Health and Well-Being*. 2022. 1–609 p.

[72] Avila Rodríguez MI, Rodríguez Barroso LG, Sánchez ML. Collagen: A review on its sources and potential cosmetic applications. *J Cosmet Dermatol*. 2018;17(1):20–6.

[73] Felician FF, Xia C, Qi W, Xu H. Collagen from Marine Biological Sources and Medical Applications. *Chem Biodivers*. 2018;15(5).

[74] Langasco R, Cadeddu B, Formato M, Lepedda AJ, Cossu M, Giunchedi P, et al., Natural collagenic skeleton of marine sponges in pharmaceutics: Innovative biomaterial for topical drug delivery. *Mater Sci Eng C* [Internet]. 2017;70:710–20. Available from: http://dx.doi.org/10.1016/j.msec.2016.09.041

[75] Lim YS, Ok YJ, Hwang SY, Kwak JY, Yoon S. Marine collagen as a promising biomaterial for biomedical applications. *Mar Drugs*. 2019;17(8).

[76] Parisi JR, Fernandes KR, Avanzi IR, Dorileo BP, Santana AF, Andrade AL, et al., Incorporation of Collagen from Marine Sponges (Spongin) into Hydroxyapatite Samples: Characterization and In Vitro Biological Evaluation. *Mar Biotechnol*. 2019;21(1):30–7.

[77] Azizur Rahman M. Collagen of extracellular matrix from marine invertebrates and its medical applications. *Mar Drugs*. 2019;17(2):1–12.

[78] Nasri M. Bioactive peptides from fish collagen byproducts: A review. Byprod from Agric Fish Adding Value Food, *Feed Pharma Fuels*. 2019;309–33.

[79] Lemos MFL. *Beyond Earth : Harnessing Marine Resources for Sustainable Space Colonization*. 2024;

About the Editors

Dr. Pranav Kumar Prabhakar

Dr. Pranav Kumar Prabhakar is currently working as a Professor in the Department of Biotechnology, School of Engineering and Technology, Nagaland University, Meriema, Kohima, Nagaland, India. He is amongst the World's Top 2% Scientists (list published by Stanford University, USA, 2021, 2022, 2023, and 2024). He completed his PhD in Biotechnology from IIT Madras. His main area of research focuses on elucidating the molecular mechanisms and strategies for oral insulin delivery and mimicking signaling pathways in metabolic disorders, such as diabetes, using natural products. He is member of the Royal Society of Chemistry, and the Asia-Pacific Chemical, Biological & Environmental Engineering Society. He also serves as an editorial board member and reviewer for many reputed national and international journals. Dr. Pranav's honors include receiving a travel grant to attend ATTD 2009 in Greece from the Indian Institute of Technology Madras and the Council for Scientific and Industrial Research (CSIR), which was approved by the Department of Science and Technology (DST). He has published over 125 research articles in journals and has authored or edited over 20 books and 45 book chapters. He has also delivered nine oral and poster presentations at scientific meetings.

Dr. Tarun Kumar Upadhyay

Dr. Tarun Kumar Upadhyay has an MSc and a PhD in biotechnology, as well as a diploma and a postgraduate diploma in bioinformatics. He is an expert in polymers-based drug delivery systems, host-pathogen interaction, immunology, and cancer drug discovery. He is currently working as an associate professor of biotechnology in the Department of Life Sciences and Research and Development Cell at Parul University (NAAC A++), in Vadodara, Gujarat, India. He has over eight years of teaching and research experience at the university level, excluding four and a half years as a full-time Ph.D. researcher and six months of industrial experience. Dr. Upadhyay has mentored two Ph.D. students and over forty undergraduate and graduate students who have received degrees under his supervision in the last six years. He is currently mentoring five Ph.D. students and one postdoctoral fellow who are targeting different cancers and neurodegenerative diseases using cell lines and zebrafish as models. Dr. Upadhyay has published over 112 Scopus-indexed international research and review articles in journals ranking Q1, Q2, Q3, and Q4. He is the author of three textbooks and sixteen international book chapters, and he serves as an editor for two international books. He has a total of 1,700 Google Scholar citations, an h-index of 22, and an i10 of 50. Dr. Upadhyay is an editorial board member of Scientific Reports, BMC Infectious Diseases, Discover Molecules, and Frontiers in Immunology, and is associated with many other national and international journals as a reviewer. He is a member of the Royal Society of Biology, the Indian Society of Cell Biology (ISCB), the International Society for Infectious Diseases, the International Natural Product Sciences Taskforce (INPST)-Poland, and the Society for Biotechnologists (India) and the Society for Immunology & Immunopathology. He is an honorary faculty member at the Novel Global Community Educational Foundation of Australia's Department of Science and Engineering and a Bentham Science ambassador from India (2022–2024). He is the Member Secretary of the Institutional Biosafety Committee (Letter No. BT/IBKP/629/2022) at Parul University, which is approved by the Department of Biotechnology (DBT) of the Government of India. He is also a Contributing Member of the Global Outreach of the American Society for Microbiology. He is the recipient of the 2020 Best Biotechnologist Award, the 2020 Young Scientist Award in Biotechnology, the 2021 Associate Scientist Award in Biotechnology, the 2021 Best Faculty Teaching and Research Excellence Award, the 2022 Overall Best Research Performer Award, the 2022 Highest Impact Factor Publication Award, the 2023 Overall Best Research Performer Award, and the 2023 Highest Cumulative Impact Factor Publication Award.

Index

D

E

F

G

H

I

O

P

R

S

T

W